Proceedings of the International Symposium on

ADVANCED STRUCTURAL MATERIALS

Pergamon Titles of Related Interest

Ashby ENGINEERING MATERIALS 1
Ashby ENGINEERING MATERIALS 2
Bee MATERIALS ENGINEERING
Ryan WHITEWARES: PRODUCTION, TESTING & QUALITY CONTROL

Other CIM Proceedings Published by Pergamon

Bickert REDUCTION AND CASTING OF ALUMINUM
Jonas DIRECT ROLLING AND HOT CHARGING OF STRAND CAST BILLETS
Kachaniwsky IMPACT OF OXYGEN ON THE PRODUCTIVITY OF NON-FERROUS METALLURGICAL PROCESSES
Macmillan QUALITY AND PROCESS CONTROL IN REDUCTION AND CASTING OF ALUMINUM AND OTHER LIGHT METALS
Plumpton PRODUCTION AND PROCESSING OF FINE PARTICLES
Rigaud ADVANCES IN REFRACTORIES FOR THE METALLURGICAL INDUSTRIES
Ruddle ACCELERATED COOLING OF ROLLED STEEL
Salter GOLD METALLURGY
Thompson COMPUTER APPLICATION IN CHEMICAL AND EXTRACTIVE METALLURGY
Tyson FRACTURE MECHANICS

Related Journals
(Free sample copies available upon request)

MATERIALS RESEARCH BULLETIN
PROGRESS IN MATERIALS SCIENCE
CANADIAN METALLURGICAL QUARTERLY
SCRIPTA METALLURGICA
ACTA METALLURGICA

Proceedings of the International Symposium on

ADVANCED STRUCTURAL MATERIALS

Montréal, Canada
August 28-31, 1988

Vol. 9 Proceedings of the Metallurgical Society
of the Canadian Institute of Mining and Metallurgy

Edited by

D.S. WILKINSON

McMaster University, Hamilton, Ontario, CANADA

PERGAMON PRESS

New York Oxford Beijing Frankfurt São Paulo Sydney Tokyo Toronto

Pergamon Press Offices:

U.S.A.	Pergamon Press, Inc., Maxwell House, Fairview Park, Elmsford, New York 10523, U.S.A.
U.K.	Pergamon Press plc, Headington Hill Hall, Oxford OX3 0BW, England
PEOPLE'S REPUBLIC OF CHINA	Pergamon Press, Qianmen Hotel, Beijing, People's Republic of China
FEDERAL REPUBLIC OF GERMANY	Pergamon Press GmbH, Hammerweg 6, D-6242 Kronberg, Federal Republic of Germany
BRAZIL	Pergamon Editora Ltda., Rua Eça de Queiros, 346, CEP 04011, São Paulo, Brazil
AUSTRALIA	Pergamon Press (Aust.) Pty Ltd., P.O. Box 544, Potts Point, NSW 2011, Australia
JAPAN	Pergamon Press, 8th Floor, Matsuoka Central Building, 1-7-1 Nishishinjuku, Shinjuku-ku, Tokyo 160, Japan
CANADA	Pergamon Press Canada Ltd., Suite 271, 253 College Street, Toronto, Ontario M5T 1R5, Canada

---

First printing 1989

Library of Congress Cataloging in Publication Data

International Symposium on Advanced Structural Materials (1988 : Montréal, Québec)
Proceedings of the International Symposium on Advanced Structural Materials, Montréal, Canada, August 28-31, 1988 / edited by D.S. Wilkinson.
p. cm. -- (Proceedings of the Metallurgical Society of the Canadian Institute of Mining and Metallurgy ; vol. 9)
Includes index.
ISBN 0-08-036090-4
1. Materials--Congresses. I. Wilkinson, D.S. II. Title.
III. Series.
TA401.3.I587 1988
620.1′1--dc19 88-38801
CIP

In order to make this volume available as economically and as rapidly as possible, the authors' typescripts have been reproduced in their original forms. This method unfortunately has its typographical limitations but it is hoped that they in no way distract the reader.

Printed in the United States of America

The paper used in this publication meets the minumum requirements of American National Standard for Information Sciences -- Permanence of Paper for Printed Library Materials, ANSI Z39.48-1984

INTRODUCTION

This volume contains the Proceedings of an International Symposium on Advanced Structural Materials, held in conjunction with the 27th Annual Conference of Metallurgists in Montreal, 28-31 August, 1988. The symposium was sponsored by the Materials Engineering Section and the Basic Science Section of the Metallurgical Society of CIM, with the co-sponsorship of the Canadian Ceramic Society.

This was the first symposium at the Conference of Metallurgists to specifically address the field of advanced materials for structural applications. In light of this the symposium was opened by a tutorial session in which three renowned experts gave overviews on topics related to the focus of the symposium - namely, polymeric composites, metal matrix composites and structural ceramics.

The sessions were organized by an able committee consisting of:

Dr. R.A.L. Drew, McGill University
Dr. U. Erb, Queen's University
Dr. J.D. Lloyd, Alcan International
Dr. M. Munro, University of Ottawa

The success of the symposium is due to their efforts and to the high calibre of the presentations themselves.

D.S. Wilkinson
McMaster University
Symposium Chairman

TABLE OF CONTENTS

Interfaces

Polymeric Composite Materials - Advanced Processing Methods and Applications

Powder Metallurgical Materials

METAL MATRIX COMPOSITES - AN OVERVIEW

Dr. D.J. Lloyd

Alcan International Limited
Kingston Laboratories
P.O. Box 8400, Kingston, Ontario
K7L 4Z4

ABSTRACT

An overview of the present status of metal matrix composites is given. The different types of reinforcements used in metal matrix composites are considered, as well as the factors involved in choosing which reinforcement to use. The different fabrication methods for composites and some of their limitations and advantages are discussed. The property improvements that can be achieved are demonstrated by considering the reinforcing of the aluminium alloy AA6061 with SiC.

KEYWORDS

reinforcement, composite, matrix, metal, fibre, whisker, particulate, strength, modulus, thermal expansion, aluminum, SiC, C.

INTRODUCTION

Metal Matrix composites (MMCs) were initially of interest as a means of obtaining high strength and modulus, which were of particular interest to the aerospace industry. Since maximum strength is achieved with continuous reinforcement, the major early effort in MMCs was to develop methods of fabricating continuous fibre reinforced metals. Unfortunately, high strength fibres are very expensive, as are the fabrication methods necessary , and this seriously restricts the possible commercial applications of continuously reinforced MMCs. In addition, continuously reinforced composites have no secondary forming capability, and the"as fabricated" shape must be the shape in which they are used. As a result, use cannot be made of the usual metal working fabrication processes, such as extrusion, rolling and forging, which are efficient methods of obtaining final shape parts. As a result of these limitations, considerable effort has been expended in the last ten years to develop discontinuous reinforcements and fabrication methods for composites utilising these reinforcements. Discontinuously reinforced composites do not have the theoretical strength levels achievable with continuous fibres, but still give useful enhancement of properties at lower cost, and are more amenable to conventional metal processing methods. There has also been a growing appreciation that properties other than strength can be enhanced in composites, and improvements in properties, such as wear resistance and thermal stability could provide important market applications for MMCs.

In this paper the present status of MMCs is reviewed, with particular attention being given to the more recent developments.

REINFORCEMENT

A wide range of reinforcements are now available and they can be categorized into four groups:

(1) continuous fibres, either as tows or monofilaments
(2) discontinuous short fibres
(3) whiskers
(4) particulates.

In terms of cost, continuous fibres and whiskers are the most expensive and particulates are the cheapest.

Continuous Fibres

Some of the presently available continuous fibres, with their typical properties are given in Table 1.

TABLE 1 Typical Properties of Continuous Fibres

Fibre Type	Diameter μm	Density gm/cc	Specific Modulus	Specific Strength
B(SiC) fil.	240	2.4	162	1660
SiC(C) fil.	140	2.0	140	1330
Nicalon tow.	15	2.6	71	1040
αAl_2O_3 tow.	20	3.9	97	>360
C, HM tow	8.4	1.96	265	950
C, HT tow	7.0	1.75	130	2030

Continuous fibres are either in the form of monofilaments, or multifilament yarns or tows. B(SiC), which is a B filament on a W core with an outer coating of SiC, was one of the first fibres to be produced for metal matrix composites, and was produced by Avco corporation as BORSIC. While this fibre is still available, Avco have now developed a monofilament SiC(C) with a higher C content on the outside, which is more stable in contact with molten metal. Continuous fibres are available as individual reels of monofilament and tows, or as weaves.

Discontinuous Fibres

Discontinuous fibres or staple are available in a range of lengths, but have diameters of 3 to 5 μms, and those of interest for MMCs are mixed with oxides with the properties shown in Table 2.

TABLE 2 Typical Properties of Discontinuous Fibres

Fibre Type	Diameter μm	Density gm/cc	Specific Modulus	Specific Strength
$Al_2O_3/5\%/SiO_2$ (Saffil RF)	3.0	3.3	90	600
$50\%SiO_2/Al_2O_3$ (Fiberfrax)	2-3	2.73	38	360
Mullite (Fibermax)	2-3	3.0	50	280

The properties are greatly inferior to the continuous fibres considered in Table 1, but the cost is much less. The short staple are produced by a solution spinning process, and an important feature of the processing is to limit the non-fibrous content, or "shot", which is deleterious to the mechanical properties of the composite. Shot levels of less than 1% can now be achieved by control of the spinning process and washing techniques.

Whiskers

Whiskers have diameters less than 1μm and can have lengths up to several hundred microns, so they are a discontinuous reinforcement with high aspect ratio. The main commercially available whiskers are SiC or Si_3N_4 with the properties in Table 3.

TABLE 3 Typical Properties of Whiskers

Whisker Type	Diameter μm	Density gm/cc	Specific Modulus	Specific Strength
SiC (Silar)	0.6	3.2	220	2200
SiC (Tokamax)	0.1-0.5	3.19	170	4000
Si_3N_4	0.5-2.0	3.18	120	4000

Because of their small diameters, whiskers have very few defects to initiate fracture, and as a result have much higher strengths than discontinuous fibres.

Particulate

The cheapest form of reinforcement is particulate. Many different ceramics are readily available in a wide range of sizes, but those of interest for MMCs are SiC and Al_2O_3m which are produced for grinding media. A tight particle size and control of purity are the main additional requirements for composites. SiC particles in the size range 10 to 15 μms which are used to reinforce Al are shown in Fig.1.

A more detailed consideration of reinforcements for composites has recently been published by Stacey (1988).

Fig. 1. SiC particulate used in reinforcement.

CHOICE OF REINFORCEMENT AND MATRIX

Obviously the particular application will influence the matrix used, but in general most property gains are achieved in the low density matrix alloys, Mg, Al and Ti, because with the exception of the carbon fibres, the other reinforcements have densities above 3 gm/cc and there will be a severe weight penalty if they are incorporated into the higher density metals. Most of the composite work to date has involved Al and Ti, but there is increasing interest in reinforced Mg (Nunes and co-workers, 1985) and also the aluminides (Bose and German, 1988).

The choice of reinforcement is influenced by:

(1) the application, which dictates the reinforcement form and cost level;
(2) the compatibility between the reinforcement and the matrix;
(3) the interfacial strength between the fibre and the matrix.

If the application requires the highest possible strength, continuous fibres, or possibly whisker are necessary, and this will involve considerable cost. If the part has to be secondary formed to shape, by processes such as extrusion, rolling, forging etc., whiskers and particulates are preferred because these undergo the least damage during any working process. If the application is aimed for a low cost, mass market, particulates are the cheapest reinforcement.

Compatibility between the matrix and the reinforcement is the most difficult condition to achieve. During the initial fabrication of the composite compatibility is required so that the matrix can spread easily over the surface of the fibres, so that the fibres are well "wet". This will insure a void free composite. In general, ceramics are not easily wet by metals, and the wetting process is complicated (see (Delannay, Froyen and

Deruyltere, 1987) for a review). To achieve wetting, fibres are often surface treated prior to incorporation into the matrix. In the case of Al reinforced with carbon fibres, the fibres are first coated with titanium diboride, using a chemical vapour deposition process. This coating, which is typically about 200 Å thick, activates the carbon fibre surface so that it is more easily wet by Al. The titanium diboride layer also protects the fibre from interaction with Al, which is the other aspect of compatibility. Depending on the fabrication process and use of the composite, the fibres are subjected to high temperatures, and interaction can occur between the matrix and the fibre, resulting in intermetallic compound formation and degradation of the fibres. As shown in Fig. 2, from Khan (1976), carbon fibres will react with Al at temperatures considerably below the melting point to form aluminium carbide, Al_4C_3. Baker and co-workers, 1972), showed that this reaction resulted in a considerable decrease in strength. Similarly, boron fibres will react with a Ti matrix to form TiB_2, and this reaction can be inhibited by coating the boron with SiC which is the approach taken by AVCO with their "BORSIC" fibre.

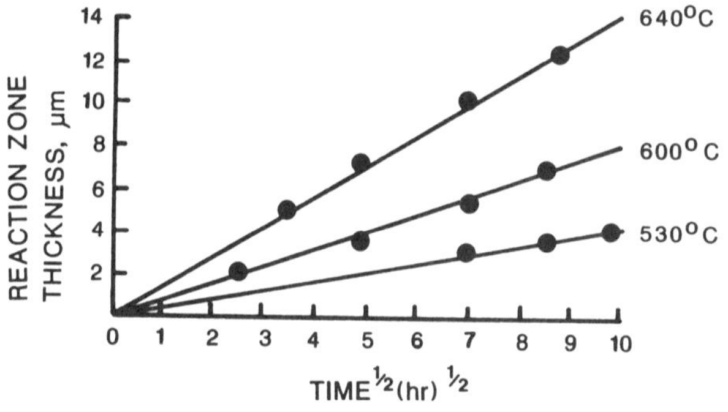

Fig. 2. Reaction of C fibres with Al. From Khan, 1976.

It is particularly difficult to achieve good compatibility with an Al matrix, because Al readily interacts to form intermetallics with most ceramics, and it is also very difficult to achieve good wetting. SiC is essentially stable below the melting point of Al, but reacts with most alloys at temperatures above the liquidus. The reactions is (Warren and Andersson, 1984).

$$4Al + 3SiC \rightarrow Al_4C_3 + 3Si \qquad (1)$$

The reaction produces aluminium carbide, which degrades the reinforcement and Si, which changes the composition of the matrix. However, if the Si content of the alloy is sufficiently high, the reaction will tend to go to the left in equation (1), and the reinforcement will be stable (Lloyd, 1988). Figure 3 shows the plates of Si and Al_4C_3 produced by reacting between SiC particles and molten Al at 800 °C. This means that if the Al-SiC composite is going to be involved with extensive liquid phase processing, the SiC will have to be protected or a high Si alloy must be used for the matrix.

The final factor that is important is the achievement of high interfacial strength, since the strength of the composite is obtained by transferring matrix loading to fibre loading through the interface. To obtain maximum loading of the fibre, according to the shear lag theory the fibre has to be longer than a critical length, l_c, which is dependent on the interfacial strength, γ (Piggott, 1980)

$$l_c = \frac{\sigma_f d}{2\tau} \qquad (2)$$

where σ_f is the fibre strength, and d is the fibre diameter. For particle reinforcement the critical length is double that given by equation (2).

Obviously, continuous fibres will satisfy equation (2), but a high interfacial strength is still important to give good transverse properties. With discontinuous composites, the higher the interfacial strength the lower will be the necessary fibre length to achieve maximum strengthening. This is important because fibres and whiskers are often broken up to some extent during fabrication. Unfortunately, it is very difficult to obtain a direct measurement of the interfacial strengths in the MMCs of interest, although some recent progress has been made using an indentation technique, where the stress required to push a fibre into the matrix is measured. (In many theoretical calculations the interfacial strength is assumed to be equal to the shear strength of the matrix)

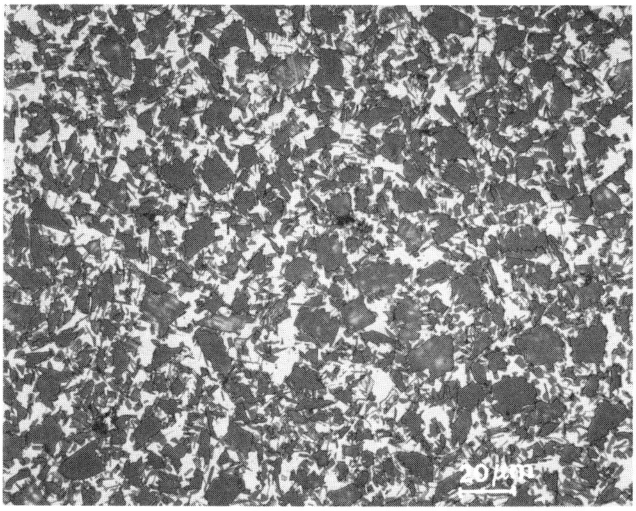

Fig. 3. Microstructure of Al-SiC composite showing Si needles and Al_4C_3 crystals produced by reaction between SiC and molten Al.

COMPOSITE FABRICATION

Many different methods have been used to fabricate composites, and process details are often proprietary, but the different methods can be considered in terms of the two basic types of reinforcement: continuous and discontinuous.

Continuous Reinforcement

Methods for fabricating continuously reinforced composites must satisfy the following:

(1) achieve fibre alignment and uniform distribution. Once the MMC is made the arrangement of reinforcement cannot be modified.

(2) avoid fibre damage. High strength fibres can be very sensitive to the kind of mechanical damage introduced by mutual contact between fibres, or fibres rubbing over tooling. Obviously, extreme damage of fibre breaking must be avoided.

(3) interfacial reactions between the fibre and the matrix must be controlled to achieve high interfacial strength.

As suggested by Harris, (1988), the fabrication routes for continuous fibre composites depends on whether the reinforcement is in the form of monofilaments or fibre tows, Fig. 4. The large diameter monofilaments can be handled as individual fibres and filament wound as alternate layers between sheets of the matrix alloy or each fibre layer can be covered by plasma spraying the matrix alloy over them. This later method was developed initially for boron fibres (Lynch and Kershaw, 1972). These methods then require final consolidation by solid state diffusion bonding or liquid phase sintering. Fibre tows are more difficult to infiltrate and the individual filaments have to be dispersed before infiltration. Often the dispersion is retained by use of a fugitive binder, such as polyvinyl acetate or silicates, to produce a preform which is manageable, and can then be infiltrated to make an infiltrated preform for further consolidation, or can be infiltrated directly to give the final shape. In the latter route the matrix must have high wettability with the fibre, and this has been obtained with αAl_2O_3 fibres in Al-Li alloys (Champion and co-workers, 1978).

Al-C composites are fabricated by consolidating wire preforms of individually infiltrated tows. Fig. 5 shows a cross section of the wire preform, and Fig. 6 shows a cross section of the composite containing many wire preforms, after liquid phase sintering in vacuum. The fibre distribution and composite soundness shown in these micrographs is typical for this fabrication method.

Discontinuous Reinforcement

Fabrication routes for discontinuously reinforced composites also have to avoid damage of the reinforcement, which in this case is concerned with fracture of the reinforcement, and reaction with the matrix, but alignment of the reinforcement is not generally a factor because this can be achieved by secondary processing. Four methods have been used to make discontinuously reinforced MMCs:

(1) Powder Metallurgy
(2) Liquid Metal Infiltration of Preforms - Squeeze Casting
(3) Molten Metal Mixing
(4) Spray Deposition

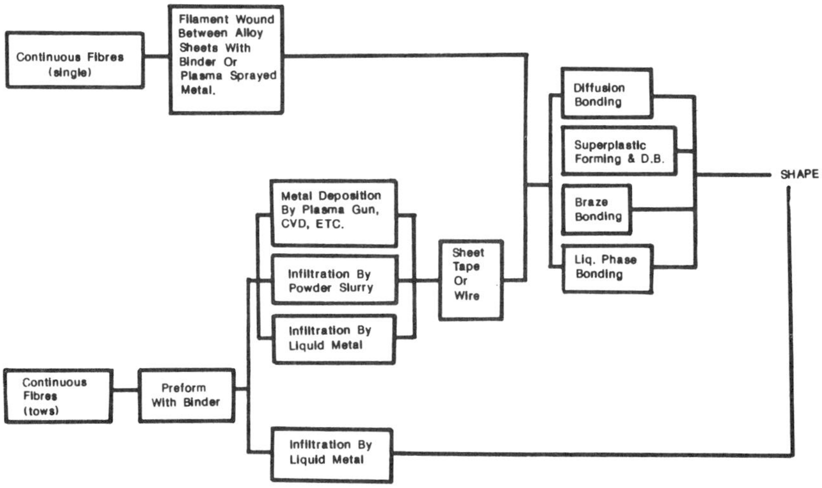

Fig. 4. Fabrication routes for continuous fibre composites. From Harris, 1988.

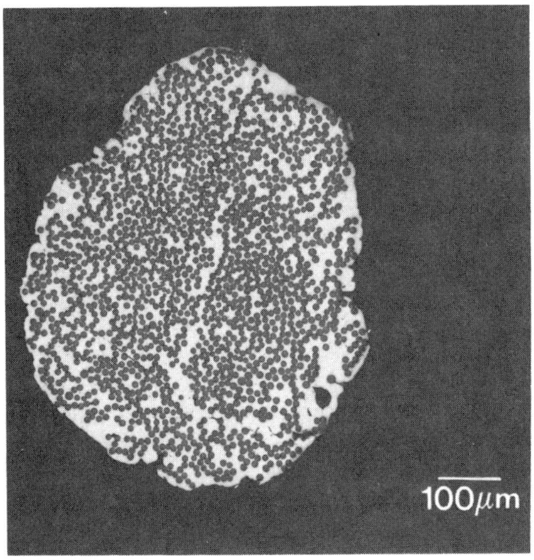

Fig. 5. Microstructure of Al-C fibre tow.

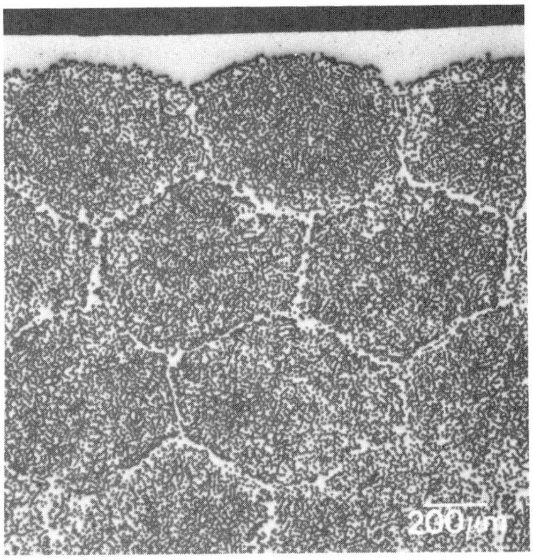

Fig. 6. Microstructure of consolidated preform. Al-C fibre composite.

Powder Metallurgy

This fabrication method has been used extensively for SiC whisker and SiC particulate reinforced MMCs, and is shown in Fig. 7. It is important that the metal powder size and the ceramic particle size are arranged so that as uniform a mix, without extensive reinforcement clustering is obtained. Any clustering will be carried over to some extent to the final product. Hunt, Richmond and Young, 1987, found that composites with an Al to SiC size ratio of 0.7:1 gave improved toughness properties over composites with a size ratio of 0.24:1. With whisker reinforcement very fine metal powders are required to penetrate between the whisker clumps and produce a uniform dispersion, and often liquid slurry mixing is used. Liquid slurry mixing also reduces the explosive hazard with fine Al powders. In spite of all the precautions that are taken extensive whisker break up occurs during the blending and consolidation, and the initial whisker aspect ratio of 50 to 100 is reduced to about 5 in the final extrusion. A detailed review of whisker reinforced Al alloys has been given by Nair, Tien and Bates, 1985.

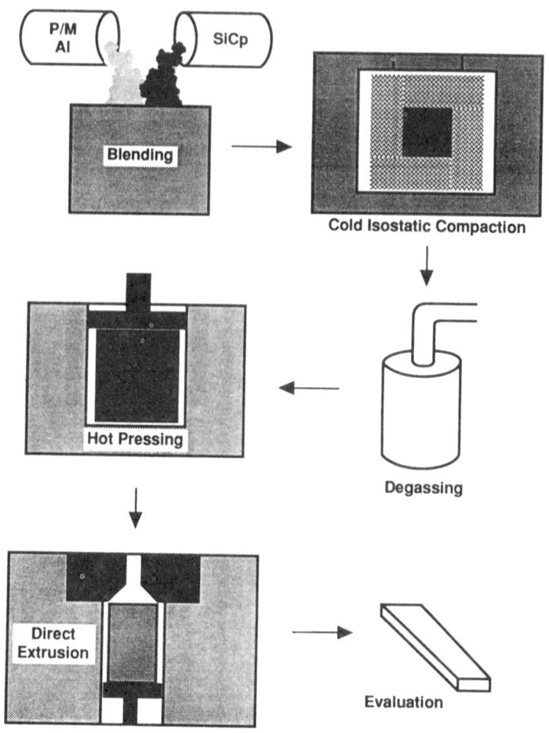

Fig. 7. Powder-metallurgy fabrication route for discontinuously reinforced composites.

Liquid Metal Infiltration

This method has been used for SiC whiskers (Fujita, Fukumoto and Kurita, 1987) and, most extensively for discontinuous fibres (Dinwoodie and co-workers, 1986, Chi and co-workers, 1987). A preform is first made to the desired shape, usually using a binder, and this is then placed in a die to which molten metal is added and the preform is infiltrated. Infiltration is either achieved by evacuating the die, or more, usually, by applying a pressure as in squeeze casting, Fig. 8. To obtain a sound product careful control of preform temperature, pressure cycle, melt superheat and solidification conditions is needed. Preform temperature influences the pressure required for the liquid to penetrate through the preform, as does the melt temperature. However, the applied pressure also produces an initial compression of the preform, and if it is too high preform and fibre fracture can occur. The infiltration process has been extensively modelled by Clyne, 1987. Solidification conditions need to be controlled to prevent solidification fronts meeting within the preform to give porosity.

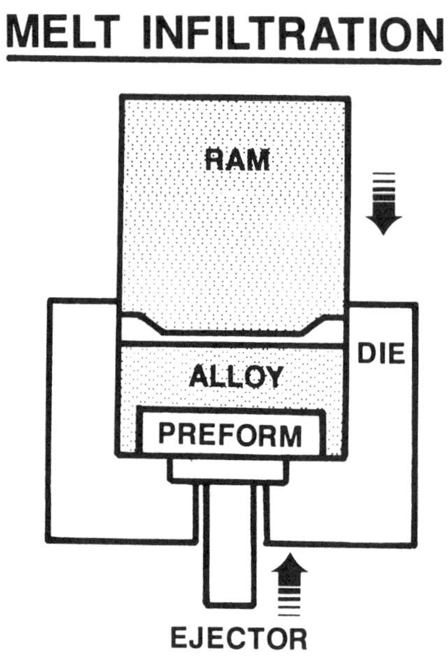

Fig. 8. Squeeze casting method for discontinuously reinforced composites.

The typical microstructure of a composite produced by squeeze casting is shown in Fig. 9, and the fibres are orientated randomly in the plane of the preform. For "Saffil" alumina fibres the binder is normally silica containing and this can present problems in some Al alloys because SiO_2 is readily reduced by Al, increasing the Si content of the melt. Reaction with the binder can also interfere with the modification of Al-Si casting alloys (Gallerneault, Huggins and Lloyd, 1988). As seen from Fig. 10 a normal strontium addition to the alloy does not produce modification in the reinforced region, and this has been attributed to binder reaction. Going to much strontium binder levels, or using a non-silica containing binder avoids the problem.

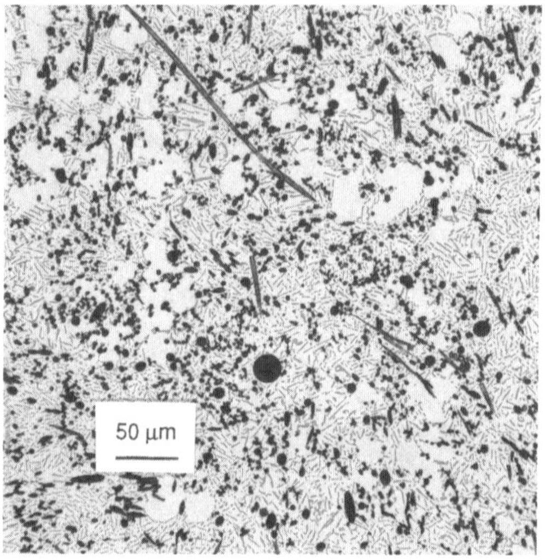

Fig. 9. Microstructure of Al_2O_3 short fibre reinforced Al-Si

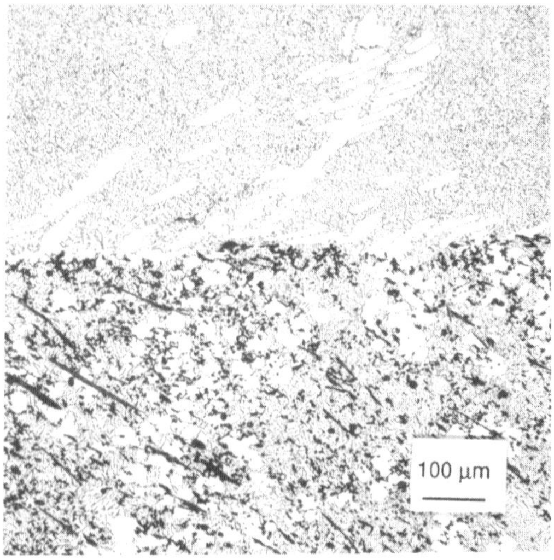

Fig. 10. Microstructure of Al_2O_3-AlSi composite showing modification outside reinforced region but lack of modification within the composite.

Preform infiltration by squeeze casting can occur during the squeeze casting of the final part, and this method has been used to produce such parts as pistons with a local region of reinforcement. Alternatively, they can be further fabricated to wrought shape with considerable break up of the reinforcement. Another alternative is that the composite can be used as an insert in a casting produced by a normal casting process, such as gravity die casting. To do this a technology has to be developed which will allow wetting of the insert in the casting process, and this type of technology looks quite possible.

Molten Metal Mixing

Molten metal mixing methods of producing composites are, in principle, the most direct and cheapest method of producing MMCs, but it is only recently that viable technologies to do this with some reinforcements have been developed by companies such as Dural Aluminum Composite Corporation (DACC, a division of Alcan). Early work on molten metal mixing added reinforcement to a semi-solid melt under high shear

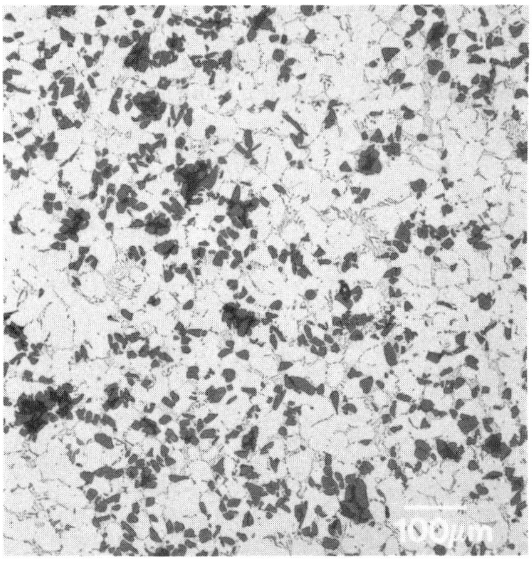

Fig. 11. In-complete wetting with resultant microporosity in a SiC particulate composite produced by molten metal mixing.

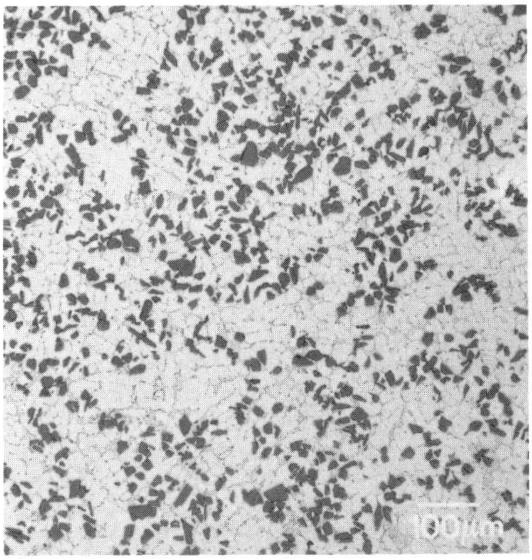

Fig. 12. Complete wetting and absence of microporosity in a SiC particulate composite produced by molten metal mixing.

(Flemings and Mehrabian, 1974). When a melt is held at a temperature between the liquidus and solidus, and subjected to high shear, the primary solid particles are spheroidal rather than dendritic, and ceramic particles added to the melt are trapped in the liquid regions between the primaries. The process is suited to alloys with a long freezing range, but in large melts it is difficult to achieve the temperature control required to maintain an appropriate liquid fraction. Surappa and Rohatgi 1981, have used a vortex method to incorporate low volume fractions of several different types of coarse ceramic particles. With both of these methods it is difficult to obtain volume fractions above 10%, and they tend to incorporate high gas levels in the composite. The basic problem with the melt mixing methods is overcoming the poor wettability between the melt and the reinforcement. DACC have developed a proprietary process which involves treatment of the reinforcing particles and a mixing method which can produce composites with up to 20 vol% particulate. Composites of Al alloys with SiC,B_4C and Al_2O_3 are routinely being produced with particles sizes in the 10 to 15 micron range.

The microstructural aspects of particulate composites produced by the molten metal route have been considered in some detail in recent publications (Lloyd, 1988, Lagacé and Lloyd, 1988). Because the contact time between the reinforcing particulate and the melt can be quite long, interaction between the two is of particular concern, and reinforcements have to be chosen that are stable in contact with the melt, or the ceramic particles have to be treated to provide a protective coating. The particle distribution is dependent on a variety of factors, including the mixing process and the solidification rate. Because the particles do not nucleate solidification, and are rejected at the solidification meniscus, they are pushed to the interdendritic regions, and this leads to a solidification clustering which will depend on the solidification cell size, and hence the solidification rate. The mixing process has to obtain complete wetting of the particles. Figs. 11 & 12 show two composites of SiC in an Al-Si alloy; in one complete wetting has not been achieved and there is microporosity in the melt, whereas the other is essentially porosity free. The microporosity present in Fig. 11 can be removed by secondary fabrication, such as extrusion, forging or hot isostatic pressing.

The commercial attraction of the metal mixing method is that conventional metal casting and fabrication processes can be used in conjunction with the process, and large scale billets can be produced and processed thorough conventional plants. With the correct choice of reinforcement and alloy, billet can be remelted and cast to shape using traditional shape casting methods. However, it must be appreciated that when material is remelted, the reinforcing particles will tend to sink to the bottom of the melt, since they are typically more dense than liquid metal. The settling tendency of SiC in an Al-Si alloy is shown in Fig. 13, where it is seen that the settling effect tends to saturate in the first hour, due to the particles interfering with one another. This data was obtained from a large melt, but settling will occur in a shape casting if the solidification rate is sufficiently slow, and in principle, it is a means of producing different volume fractions if different parts of a shape casting by selectively using different solidification rates.

Spray Deposition

Spray deformation was initially developed by Singer[27] and was brought to commercialization in the "Osprey" process (Brooks and co-workers, 1983) which was initially developed for steels. Alcan has taken this technology and modified it to produce MMCs (Willis and co-workers, 1987). Much of this technology is proprietary, but essentially the process consists of inert gas atomising a melt and subsequently solidifying it against a cooled substrate. The reinforcing particles are introduced into the atomising stream to be co-deposited and incorporated into the solidified alloy. Depending on the form of the substrate collector system, a variety of shapes, including sheet, can be produced. Metal flow rates are typically about 5kg/min. and the ingot is 95% dense or higher. However, a secondary processing step is required to achieve full density. The contact time between the reinforcement and the molten metal droplets is very brief, so reactivity is not a problem. Since the solidification rate is quite rapid, the microstructure is refined and this helps to homogenise the distribution of the reinforcement, Fig. 14. The as-deposited ingot can be rolled, extruded and forged to final shape, and the process can handle a range of alloys, including the high strength aerospace alloys.

PROPERTIES

The properties that are of particular interest in MMCs are:

Modulus
Strength
High Temperature Strength
Coefficient of Thermal Expansion
Wear Resistance

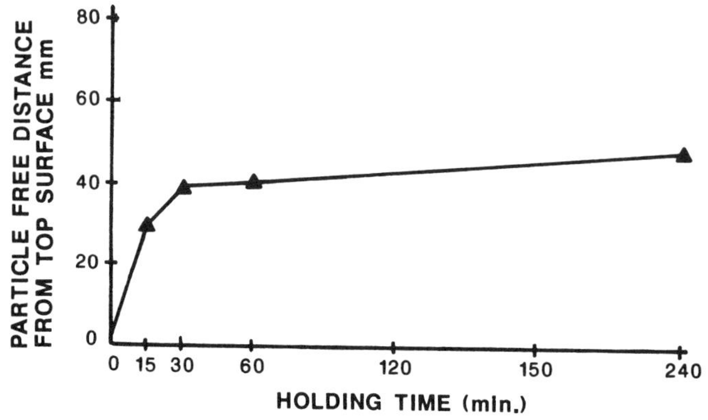

Fig. 13 **Extent of reinforcement settling in a SiC particulate composite.**

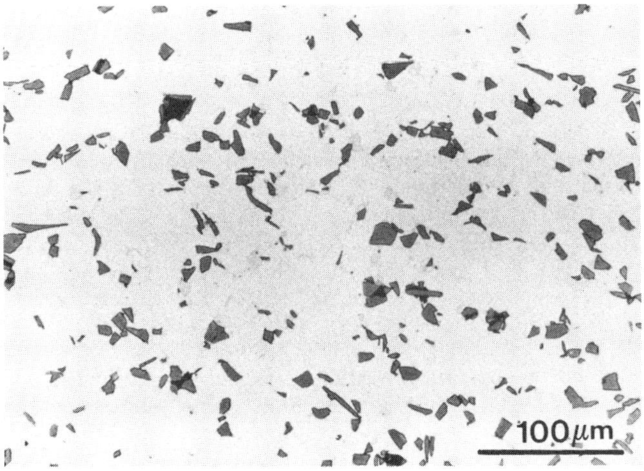

Fig. 14. **Microstructure of spray deposited Al-SiC composite.**

Properties which tend to be adversely affected by reinforcement are:

Elongation
Toughness
Electrical and Thermal Conductivity (although a decrease in these properties can be used to advantage)

Continuously reinforced composites are expected to obey the Rule of Mixtures for both strength and modulus:

$$E_C = E_f V_f + E_m V_m \qquad (3)$$

and

$$E_C = \sigma_f V_f + \sigma_m V_m \qquad (4)$$

where E is the modulus, σ is the strength, and V is the volume fraction, and the subscripts c, f and m refer to the composite, fibre and matrix respectively. The properties will increase with volume fraction as expected from the Rule of Mixtures, but usually, even in continuously reinforced composites they fall below the theoretically expected values. This is because of misalignment of the fibres and inhomogeneity in the fibre distribution. In Fig. 15 the influence of loading the composite at an angle to the fibre direction on the strength and Young's modulus of the composite is shown for an Al-B continuously reinforced composite. The properties fall off drastically with loading direction, and loading at 90° to the fibre axis produces no strengthening. This directionality is reflected in the nature of the fracture. The fracture surface of an Al-C fibre composite which has been loaded parallel to the fibre axis shows fibre fracture and fibre pull-out, Fig. 16. When the same composite is loaded perpendicular to the fibres the fracture surface shows almost complete interfacial failure, leaving the fibres intact, Fig. 17. It is, therefore, essential to insure that a composite is going to be loaded primarily in the fibre direction in unidirectional fibre composites, or use composites with different fibre ply orientations for multi-directional loading.

In short fibre reinforced composites the properties are less than the Rule of mixtures; the actual value depending on whether the fibre length is greater or smaller than the critical length, equation (2). When the fibre length is less than the critical length it cannot be fully loaded to the breaking strength of the fibre. As pointed out by Friend (1987), this effect, together with fibre orientation effects, can lead to essential no strengthening in some short fibre reinforced composites.

The strengthening mechanisms in whisker and particulate composites are still not fully understood, and often these materials give higher strengths than expected applying the Rule of Mixtures to the aspect ratios involved. These higher strengths have been attributed to edge loading effects (Nardone and Prewo, 1986), enhanced dislocation densities and internal misfit strain effects resulting from the different moduli of the matrix and the particle and differences in coefficients of expansion between the reinforcement and the matrix (Arsenault and Taya, 1987). Space does not permit any detailed consideration of these theories, but in general, whisker reinforced composites give strengths higher than particulate reinforced material and lower than continuous fibre composites.

To demonstrate the general mechanical property trends with the different reinforcements, the aluminium alloy AA6061, in the T6 temper, reinforced with SiC may be considered. Fig. 18 shows the influence of the different reinforcements on the tensile modulus, and Fig. 19 considers the influence of reinforcement on tensile strength. As expected, the continuously reinforced material gives the highest properties, followed by whisker reinforcement and then particulate. All of the composites with volume fractions higher than 10 vol% have a modulus above the highest modulus, unreinforced commercial Al alloys, which are the Al-Li based alloys. Fig. 18 also shows that the matrix does not significantly influence the modulus. The microstructures, and to some extent, the particle shape and size are different in the material processed by powder metallurgy (PM) and that produced by molten metal mixing (MMM). Within the particulate size range of 10 to 20 μm, the tensile modulus is relatively insensitive to particulate scale, distribution and microstructure, as can be seen from the comparison of the composites produced by powder metallurgy and molten metal mixing in Fig. 18. The PM material has a more uniform particle distribution, but there are also particle size and microstructural differences, and these effects influence strength, as seen in Fig. 19.

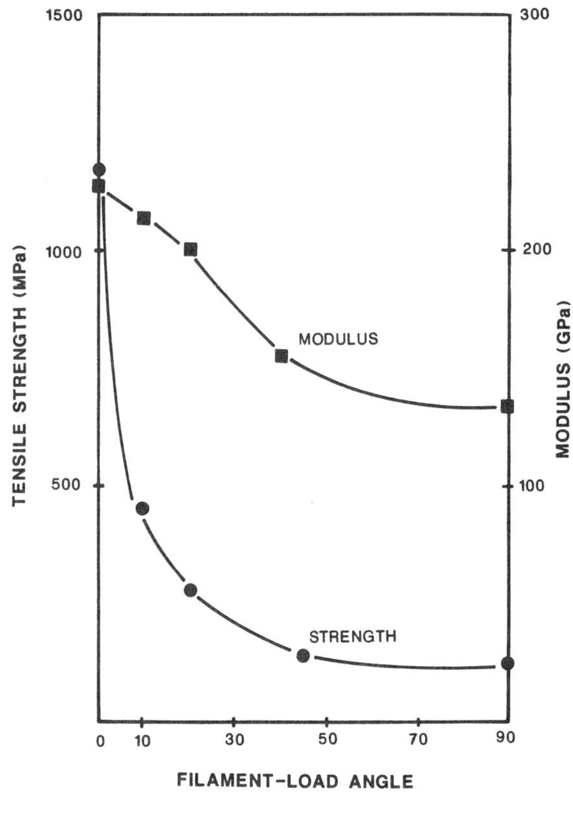

Fig. 15. Influence of loading angle on the mechanical property of an Al-B composite. Data replotted from Lynch and Kershaw, 1972.

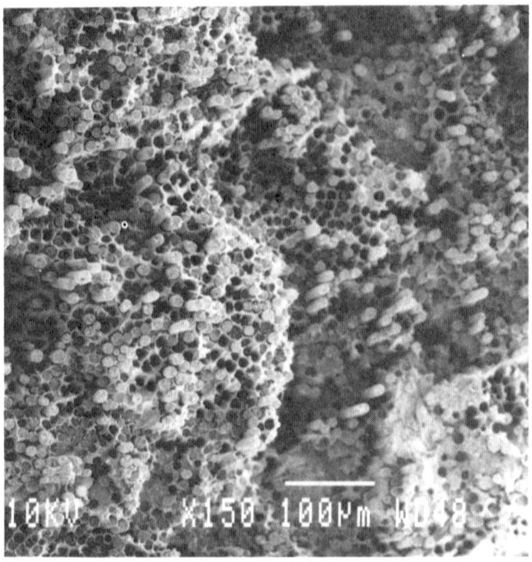

Fig. 16. Fracture surface of Al-C composite loaded in the fibre direction.

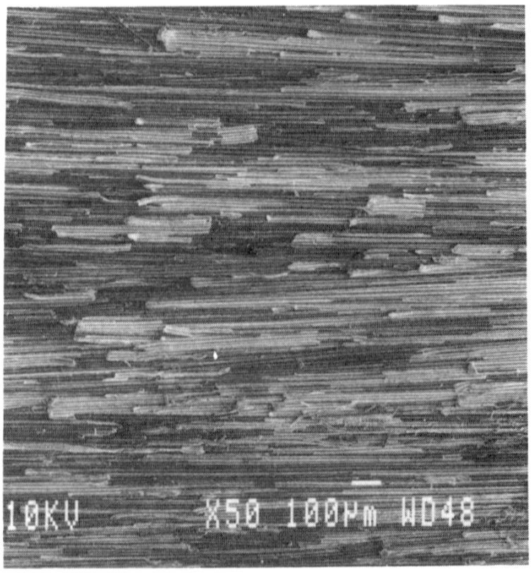

Fig. 17. Fracture surface of Al-C loaded perpendicular to the fibre direction.

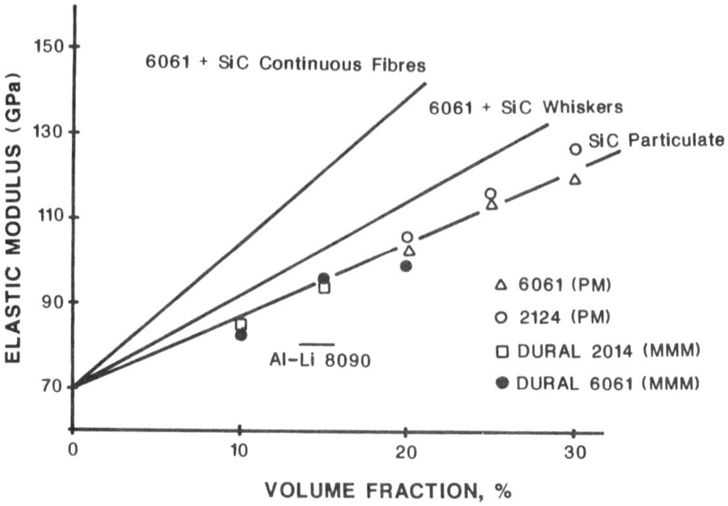

Fig. 18. Elastic properties of Al-SiC composites.

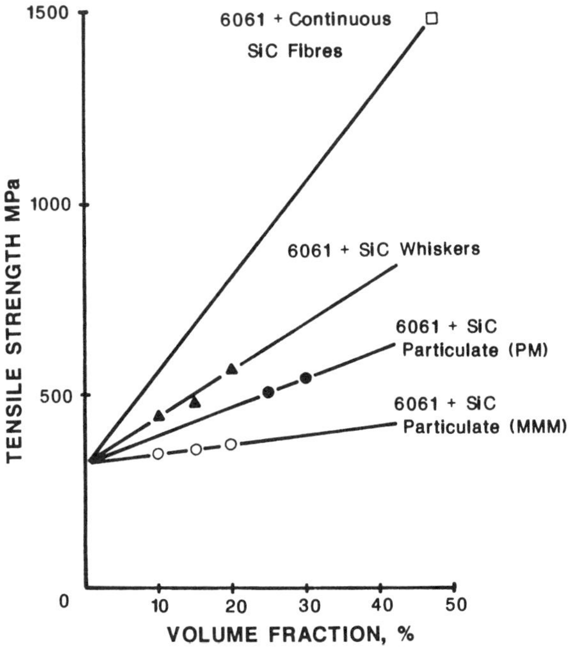

Fig. 19. Tensile properties of Al-SiC composites.

The retention of strength at high temperatures is demonstrated in Fig. 21, which compares the properties of AA6061-T6 with 6061 containing 15 vol % SiC whiskers. This composite was produced by pressure infiltration of a whisker preform, which limits any whisker breakage during fabrication, and was tested in the as-fabricated condition, so the high inherent aspect ratio of the whiskers was retained in this composite. (Properties of whisker reinforced material produced by powder metallurgy and subsequently extruded will have lower aspect ratio whiskers, and hence somewhat lower strengthening as a result, but this is offset to some extent by improved fibre alignment). Composite strengthening is essentially athermal over the temperature range considered, because the strength of the high melting point SiC does not decrease significantly in this temperature regime. The decrease in the composite strength is due to a decrease in the matrix strength provided by the alloy.

Another property of potential interest is the coefficient of thermal expansion of the composite. Fig. 21 shows how the CTE of 6061 is reduced by SiC particulate. By going to a sufficiently high volume fraction of reinforcement the CTE of 6061 can be reduced to that of steel. The lower the CTE of the reinforcement the greater is the reduction that can be achieved. With C fibres, which have a negative CTE, close to a zero thermal expansion can be achieved in Al alloys.

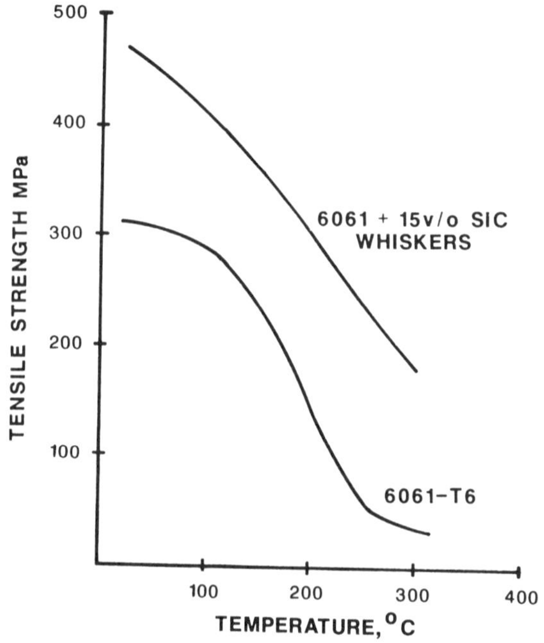

Fig. 20. Influence of temperature on the strength of 6061-15 v/o SiC whisker reinforced composite and AA6061-T6 unreinforced alloy.

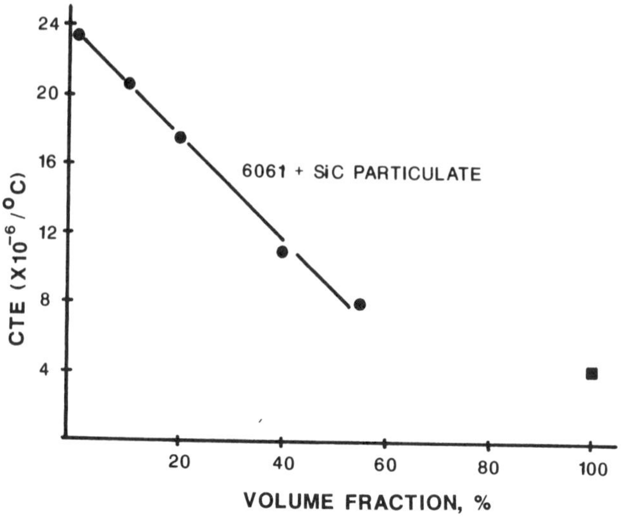

Fig. 21 Influence of particulate SiC on the coefficient of Thermal Expansion (CTE) of 6061 alloy.

SUMMARY

The present status of metal matrix composites has been reviewed, together with the different types of reinforcements. High cost continuous fibre reinforcement gives the highest mechanical properties, but these are difficult to fabricate and have no secondary forming capability. Discontinuous fibres and whiskers are somewhat cheaper, and can be more easily fabricated using squeeze casting, and powder metallurgical methods for whiskers. Squeeze casting enables near net shape parts to be produced, and some large commercial applications, such as automotive parts are being considered. Composites reinforced with particulates are the lowest cost composites, but also have the lowest properties. Fabrication processes such as molten metal mixing and spray deposition are now at the pilot plant scale and offer considerable potential as a source for low cost composites. While the properties achieved in particulate reinforced composites are less than for continuous fibre reinforcement, the properties are considerably above those of the unreinforced matrix. They can also be cast to shape or secondary fabricated with the correct choice of matrix and reinforcement, which are important factors if mass market applications of composites are to be achieved.

ACKNOWLEDGMENTS

The author is grateful to Alcan International for permission to publish this work, and to his colleagues in the metal matrix group at Kingston and Banbury for their contributions to various aspects of this work.

REFERENCES

Arsenault, R. J. and R. M. Fisher (1983). *Scripta Met., 17,* 67-71.

Arsenault, R. J. and M. Taya (1987). *Acta Met., 35,* 651-659.

Baker, A. A., P. A. Cripwell, P.W. Jackson and C. Simpson (1972). *Fibre Science Techn. 5,* 285.

Bose, A., and R. M. German (1988). *Adv. Mat. and Manuf. Proc. 3,* 37-56.

Brooks, R. G., A. G. Leatham, G. R. Dunstan and C. Moore (1983). *Metal Powders Rep., Vol. 38, No. 10.*

Champion, A. R., W. H. Krueger, H. S. Hartmann and A. K. Dhingra (1978) *In ICCM-II, The Metallurgical Soc., Inc.,* 882-904.

Chi, F. K., R. D. Maier, T. W. Krucek and P. M. Boymel (1987). *In ICCM-VI and ECCM-2, Vol. 2, Elsevier Applied Science,* 449-458.

Clyne, T. W. (1987). *In ICCM-VI and ECCM-2, Vol. 2, Elsevier Applied Science,* 275-286.

Delannay, F., L. Froyen, and A. Deruyttere (1987). *J. of Mater. Sci., 22,* 1-16.

Dinwoodie, J., E. Moore, C. A. J. Langman and W.R. Symes. (1986). *In ICCM-V, The Metallurgical Society, Inc.,* 671-685.

Flemings, M. C. and R. Mehrabian (1974). *Met. Trans. 5,* 1899.

Friend, C. M. (1987). *J. Mat. Sci., 22,* 3005-3010.

Fujita, Y., H. Fukumoto and Y. Kurita (1987). *In ICCM-Vi and ECCM-2, Vol. 2, Elsevier Applied Science,* 340-349.

Galleneault, M., U. Huggins and D. J. Lloyd, (1988). *Microstructural Science, 16,* 75-83.

Harris, S. J. (1988) *Mat. Sci. and Techn. 4,* 231-239.

Hunt Jr., W. H., O. Richmond, and R. D. Young (1987). *In ICCM-VI and ECCM-2, Vol. 2, Elsevier Applied Science,* 209-223.

Khan, I. H. (1976). *Met. Trans 7A,* 1281.
Lagace, H. and D. J. Lloyd (1988). *Canadian Met. Quart.,* in press.
Lloyd, D. J. (1988). *Composite Science and Technology,* in press.
Lynch, C. T. and J. P. Kershaw (1972). *In Metal Matrix Composites,* CRC Press.
Nair, S. V., J. K. Tien, and R. C. Bates (1985). *Int. Met. Rev., 30,* 275-290.
Nardone, V. C. and K. M. Prewo (1986). *Scripta Met. 20,* 43-48.
Nunes, J., E.S.C. Chin, J. M. Slepetz, and N. Tsangarakis (1985). *In ICCM-V, The Metallurgical Society, Inc.* 723-745.
Piggott, M.R. (1980). In "Load Bearing Fibre Composites", Pergamon Press.
Singer, A. R. E. (1968). *British Pat., No. 1262471.*
Stacey, M. H. (1988). *Mat. Sc. and Techn., 4,* 227-230.
Surappa, M. K. and P. K. Rohatgi (1981). *J. Mat. Sci., 16,* 983-993.
Warren, R., and C-H. Anderson (1984). *Composites, 15,* 101-111.
Willis, T. C., J. White, R. M. Jordan and I. R. Hughes (1987). "3rd International Solidification Processing Conference", Sheffield, in press.

MICROSTRUCTURAL EFFECTS ON FRACTURE MICROMECHANISMS IN LIGHTWEIGHT METAL MATRIX COMPOSITES

C. Liu and J.J. Lewandowski
Department of Materials Science and Engineering
Case Western Reserve University
Cleveland, Ohio 44106

ABSTRACT

Microstructural effects on the deformation and fracture behavior of an aluminum alloy metal matrix composite has been studied. The matrix microstructure was systematically varied via both heat treatment and changes in the SiC particle size. Testing included smooth tensile and smooth bend specimens, in addition to notch toughness specimens. It is shown that both the matrix microstructure and the SiC particle size affect the failure micromechanisms. SiC particle fracture predominates in underaged (UA) specimens containing large (i.e. average size 13 μm) SiC particles while matrix and "interface" failure predominate in UA specimens containing smaller (i.e. average size 5 μm) SiC. Overaged (OA) microstructures exhibited a preference for fracture in the matrix and near the SiC/matrix interface, regardless of the SiC particle size. Analyses were performed using a combination of scanning and transmission electron microscopy. The fracture results are discussed in light of recent work on fracture of composite materials.

INTRODUCTION

Considerable work has recently been devoted to the area of metal matrix composites (MMC) due to the need for materials with high specific strength and stiffness. Although work in the 1960's had indicated that the potential for MMC's was great, problems with both the reproducibility of properties as well as high costs of production reduced the attractiveness of these materials. Recent advances in processing techniques and inexpensive sources of both particulate and whisker reinforcement have renewed interest in these class of materials, with particular interest in Aluminum and Titanium matrices (1-3).

The desired form of the reinforcement (i.e. filament, fiber, particulate, whisker) depends strongly on the intended use and requirements, with continuous reinforcement offering the highest potential strengths and stiffnesses but with marked anisotropy. Although the strength and stiffness of discontinuous reinforced composites are somewhat less than continuous reinforced materials containing equivalent volume fraction of reinforcement, the discontinuous reinforced MMC's hold the additional advantages of possessing more isotropic properties (4), and may be fabricated via powder metallurgy (P/M) or casting techniques (5), followed by conventional metalworking processes such as extrusion, rolling, etc.

Although it is well accepted that changes in microstructure may control the deformation and fracture processes occurring for both ferrous (6,7) and non-ferrous (8,9) unreinforced materials, fewer studies have focussed on the role(s) of microstructure in the deformation and fracture of MMC's. The microstructure of a composite material, however, consists not only of the size, size distribution and volume fraction of reinforcement but additionally includes the details (i.e. aging condition, microstructure) of the matrix. For example, recent work

(4,10) has revealed significant effects of changes in matrix material (11) as well as matrix temper (10), volume percent, and size distribution of reinforcement on subsequent properties in monotonic loading.

Additional researchers (4,10,12) have begun to study the effects of particle size at constant volume fraction on deformation and fracture under monotonic loading conditions. The intent of the present work was to perform systematic investigations into the effects of microstructure (i.e. matrix temper) and reinforcement particle size on the micromechanisms of fracture in SiC_p reinforced aluminum alloys, and represents a continuation of previous works reported elsewhere (4,10,13).

EXPERIMENTAL PROCEDURE

Materials and Heat Treatment

The powder metallurgy matrix alloy composition used in this work, designated Alcoa MB78, contains 7%Zn, 2%Cu, and 0.14%Zr, balance Al, and was reinforced with 20% by volume of either F-600 grade (average size prior to blending = 16 μm) or F-1000 grade (average size = 5 μm) SiC_p produced by Norton Co. Additional details of the processing of this material via powder metallurgy techniques can be found elsewhere (4). The as-extruded materials were subsequently solution heat treated, cold water quenched, and aged to systematically vary the matrix microstructure. In particular, the aging practices were selected to match the strengths of the materials in the UA and OA tempers. In the present work, this was accomplished by solution heat treating for 4 hours at 500°C followed by cold water quenching. Aging to the underaged (UA) temper was accomplished by aging at 120°C/20 minutes, while the overaged (OA) temper was produced by a double aging treatment: 120°C/24 hours, followed by 170°C/36 hours.

Microstructural Evaluation

Quantitative metallographic techniques were utilized to characterize the average particle size and size distribution of SiC in the composite. Micrographs obtained on a JEOL-35SM Scanning Electron Microscope (SEM) were analyzed with a Zeiss Videoplan 2 Image Analyzer, while foils for TEM examination were prepared using an ion miller equipped with a cold stage.

Mechanical Testing

Tensile testing was performed in the longitudinal orientation on cylindrical tensile specimens with diameter of 5mm and gage length of 25.4mm at a strain rate of 1×10^{-3}/sec on an INSTRON 1125 Universal Testing Machine equipped with special fixturing designed to reduce bending moments. Strain was monitored via an extensometer affixed to the specimen surface.

Bend testing was conducted in four point bending on both smooth and notched bend specimens of nominal dimensions 12.7mm x 12.7mm x 75mm . The tensile surfaces on the smooth specimens were polished prior to loading so that the surface deformation (i.e. SiC fracture, matrix failure, interface failure) could be monitored after increasing increments of strain, including the failure strain. The details of the notched bend experiments conducted on both monolithic materials as well as on composites have been described elsewhere (4,10, 13-16) where they have been utilized to examine the fracture initiation details ahead of a blunt stress concentrator. Notch toughness testing was additionally conducted on bend specimens containing notches of root radius smaller than 60μm. The notches were introducted via a high speed wire saw which utilized an 80 μm diameter wire. The bend specimens were notched to half their depth and were subsequently tested in accordance with E-399. A clip gage was attached to the specimen surface to enable continuous monitoring of load vs crack opening displacement.

Fracture Surface Anlayses

Fracture surfaces were anlayzed using a JEOL 35SM SEM equipped with a PGT Microanalysis System. Analyses of the fracture surfaces were performed by matching surface fractography, as well as by quantification of the area fraction and particle size distributions of SiC_p present on the fracture surface. Observations were made at magnifications of 1000X-5000X. Particular emphasis was placed on determining the relative amounts of SiC_p fracture vs decohesion, as affected by changes in matrix microstructure, while quantification of the data was accomplished with a Zeiss Videoplan 2 equipped with a statistical package.

RESULTS

Microstructures

Figures 1a and 1b present SEM micrographs of the composites containing the coarse (i.e. Avg. size 13 μm) and fine (i.e. Avg. Size 5μm) SiC. Some fractured particles were observed in the material containing the coarse SiC.

TEM of the as-extruded material has been reported elsewhere (4) while Figs. 2a and 2b illustrate the microstructures of the UA and OA materials, respectively. The SiC_p/matrix interfaces were free of precipitates in the UA material as shown in Fig. 2a. In contrast, the OA material exhibited extensive precipitation in the matrix as well as at the SiC_p/matrix interface. EDAX analyses revealed that the interface precipitates contained both Mg and Zn, representative of the equilibrium precipitate, $MgZn_2$. Neither carbides nor corase intermetallic particles were found at the SiC_p/matrix interface. Precipitate free zones were not observed at the SiC_p/matrix interface in the OA alloys, although they were observed at grain boundaries in the matrix of the OA material.

Mechanical Properties

Table I summarizes the uniaxial tensile properties for the UA and OA materials. It is shown that the yield strengths, ductilities, and work hardening exponents, n, were relatively similar for the conditions tested. Also included in Table I are the results of quantitative fractography, which reveals a large difference in the amount of fractured SiC_p present on the fracture surfaces. The UA material containing 13μm average size SiC (i.e. UA-13μm) exhibited a preference for fracture of the SiC, while the UA-5μm material exhibited a preference for matrix failure and failure near the SiC/matrix interfaces. Both of the OA materials (i.e. Average Size 5μm, 13μm) exhibited a preference for failure near the SiC/matrix interfaces and in the matrix. The fracture surfaces of the material containing 13μm SiC exhibited larger dimples than did the 5μm material. The differences in fracture morphology are illustrated in Figures 3a-d for the four conditions tested. TEM of foils taken below the fracture surfaces of the two materials (i.e. UA vs OA) reflected the differences observed on the fracture surfaces, with a predominance of matrix failure, and near the SiC/matrix interface failure in the OA material (4,10). Fractured SiC_p was observed in the foils taken from the fractured UA material containing 5μm particles.

The surfaces of the smooth bend specimens similarly exhibited a difference in the fracture micromechanisms. The UA-13μm material exhibited both fractured inclusions and fractured SiC adjacent to the fracture surface, as shown in Figure 4a, while the UA-5μm material exhibited less fractured SiC and failure predominantly in the matrix and near the SiC/matrix interface, as shown in Figure 4b. Multiple fracture of elongated SiC was often observed for the UA 13μm material, as illustrated in Figure 4a. OA specimens exhibited preferential failure in the matrix and at the SiC/matrix interfaces as shown in Figure 4c. Quantitative analyses revealed that the large particles in the size distribution fractured preferentially, as did particles whose major axes were parallel to the loading direction. The sequential loading experiments revealed that fracture typically occurred first at large inclusions and elongated SiC particles.

Table II summarizes the notch toughness data obtained on the materials tested. The UA materials consistently provided higher notch toughness values in comparison to the OA material, although little improvement was obtained in the material containing the finer particle (i.e. 5μm) size. The notch toughnesses obtained were similar to results obtained on short rod toughness specimens (4) and to those obtained on precracked specimens(23).

DISCUSSION

Significant differences in fractography were noted between UA and OA tensile specimens possessing equivalent strength. A preference for SiC_p fracture was observed in the UA-13 μm specimens, in contrast to the UA-5μm and the OA specimens which exhibited failure at and near the SiC_p/matrix interfaces. It is interesting to note that despite the large differences in fractographic details, little effect of matrix microstructure was obtained on the macroscopic measures of ductility, such as reduction in area (i.e. RA) or fracture strain. Although a major effect of matrix microstructure on the failure mechanisms have been observed in the present works (4,10) on 7XXX alloys and 6XXX alloys (17) reinforced with SiC_p, other recent work on 2XXX alloys (18) failed to reveal such effects of matrix microstructure on the micromechanisms of failure. One possible source of the differences noted in the present works (4,10,13) and previous work (19) may relate to the details of the processing techniques as well as to the differences in aging treatments and response for the various studies. For example, in materials processed via super-solidus consolidation (17,18) coarse intermetallic particles have often been found to initiate failure in smooth tensile specimens, while material processed using sub-solidus conslidation (4,10,13) often reveals that fracture preferentially initiates in regions of clustered SiC_p both in smooth tensile specimens as well as in notched bend bars. Although clear differences have been noted between the types of fracture initiating features in sub- versus super-solidus processed materials (4,10,13), the details of crack propagation in recent work (18) did not reveal a significant difference in failure mechanism in UA versus OA 2XXX alloys. Recent work on other alloy systems (e.g. 7XXX, 2XXX, 6XXX) have shown a preference for SiC_p fracture in UA materials with failure at or near the interface in OA alloys (4,10,13,17,19). The source of these discrepancies may relate to subtle differences in the aging treatments employed in the various works as well as to the differences in the aging response of the various alloys. Work is continuing to investigate the possible reasons for the observed differences in fracture behavior between various materials.

In the present work on 7XXX alloys, as well as in continuing work on 2XXX alloys (19) consolidated below the solidus temperature, both SEM and TEM observations have revealed a preference for SiC_p fracture in UA-13μm materials, with a preference for interface, or near-interface failure in the OA 7XXX alloys. The smooth bend specimens clearly revealed the sequence of events leading to failure in these composite materials. In the UA-13μm materials, fracture initiation occurred at low strains via fracture of elongated SiC, large SiC, and at inclusions. Multiple fractures in the SiC were observed at higher strains. The UA-5μm materials revealed fracture initiation via inclusion fracture, with additional fracture occurring in the large SiC particles in the matrix and near the SiC/matrix interface. OA specimens similarly revealed inclusion fracture, although subsequent fracture events were always dominated by failure near the SiC/matrix interfaces, and failure in the matrix, regardless of particle size. Previous work (4,10) has indicated that fracture may initiate preferentially in clustered regions.

Although the macroscopic tensile properties did not reveal a substantial effect of matrix microstructure on yield strength, work hardening exponent (i.e. n), or ductility, the notched toughness experiments revealed that OA materials failed at lower nominal bending stresses (and K_Q) than did the UA materials. As in the smooth tensile specimens, quantitative fractography revealed a preference for SiC_p fracture in UA-13μm materials, with interface or near interface failure preferred in OA materials, and in the UA-5μm material. Thus, although a fracture mode transition from particle cracking to matrix and interface failure was observed on going from 13 μm to 5μm particle size UA material, no significant effects of particle size on notch toughness were observed. However, it should be noted that average particle sizes larger than 13μm (in similar matrix material) have been shown to produce significantly lower notch toughnesses and short rod toughnesses (20) for an equivalent level of reinforcement. Work is currently being conducted on fatigue precracked specimens to determine the effect of notch radius on measured toughness values.(23)

The present results indicate that interfacial regions may play a dominant role in the fracture behavior of particulate reinforced MMC's. It is shown that interface and near interface failure in OA alloys accompanies a large drop in notch toughness although the macroscopic tensile properties are relatively unaffected by the aging treatments employed. While few systematic studies have been performed to investigate the details of fracture initiation and propagation in these materials, the present results indicate that both the matrix microstructure (i.e. aging condition, particle size) and the details of the reinforcement/matrix interface may exert major effects on the mechanical behavior of these materials. Furthermore, a particle size effect has been observed (4,10) whereby the large SiC_p fracture during testing in preference to SiC_p at the low end of the size distribution. This has manifest itself in the present work via the transition from SiC fracture in the UA-13μm material to failure predominantly in the matrix and near the interface in the UA-5μm material. Similar observations where the fracture path appears to sample large particles in composite material has been reported by Ritter (21) and by Ritchie (22).

CONCLUSIONS

1. Heat treatments have been designed to systematically vary the matrix microstructure and SiC_p/matrix interface in P/M 7XXX Aluminum alloys reinforced with either 13μm or 5μm SiC particles at the 20 volume percent level. Equivalent smooth tensile properties were obtained for the conditions tested.

2. Despite the relative similarity in macroscopic tensile properties and ductilities between the UA and OA materials, quantitative fractography revealed a preference for SiC_p fracture in UA-13μm materials, with a predominance of interface or near interface failure in the OA materials and the UA-5μm material. Quantitative fractography also revealed that the large SiC in the size distribution are preferentially sampled in the fracture process.

3. Smooth bend specimens sequentially loaded to failure were successfully used to detail the stages of fracture for the conditions tested. Multiple fracture of the individual SiC_p was often observed for the UA-13μm material, with a preference for failure in the matrix and near the interface in the UA-5μm material.

4. Notch toughness values were higher in the UA materials. No significant difference in properties was obtained in the finer particle size material.

ACKNOWLEDGEMENTS

The authors would like to thank W.H. Hunt, Jr. for the supply of material and the provision of a one-year fellowship for one of the authors (CL). Additional support has been supplied by an AIME Research Initiation Grant (JJL) with experimental support from DARPA-ONR-N00014-86-K-0773.

REFERENCES

1. A.K. Dhingra, S.G. Fishman and S.D. Karmarkar, J. Met., 33, 12 (1981).
2. J.E. Shoutens, J. Met., 37, (1985).
3. A.K. Dhingra, J. Met., 38, 17 (1986).
4. J.J. Lewandowski, C. Liu, and W.H. Hunt, Jr., in Powder Metallurgy Composites, M. Kumar, K. Vedula, and A.M. Ritter (eds.), TMS-AIME, Warrendale, PA 1987, pp. 117-137.

5. D.M. Schuster, M. Skibo, and F. Yep, J. Met., 39, 60 (1987).
6. J.F. Knott, Fundamentals of Fracture Mechanics, Butterworths (1973).
7. J.J. Lewandowaski and A.W. Thompson, Met. Trans. A,, 17A, 461 (1986).
8. G.G. Garrett and J.F. Knott, Met. Trans. A, 9A, 1187 (1978).
9. J.J. Lewandowski and J.F. Knott, Proc. ICSMA-7, H. McQueen et al, eds., Pergamon, Oxford, Vol. 2, 1193 (1985).
10. J.J. Lewandowski, C. Liu and W.H. Hunt, Jr., Mat. Sci and Eng., in press (1988).
11. D.L. McDanels, Met. Trans. A, 16A, 1105 (1985).
12. A. Flom, Ph.D. Thesis, University of Maryland, 1988.
13. C.Liu, S. Pape and J.J. Lewandowski, in Interfaces in Polymer, Ceramic, and Metal Matrix Composites, Proc. ICCI-II, (H. Ishida, ed.), Elsevier, 1988, pp. 513-25.
14. J.J. Lewandowski and A.W. Thompson, Met. Trans. A, 17A, 1769 (1986).
15. J.J. Lewandowski and A.W. Thompson, in Proc. ICF6, S.R. Valluri, et al eds., Pergamon, Oxford, Vol. 2, 1515 (1984).
16. J.J. Lewandowski , C.A. Hippsley, and J.F. Knott, Acta Met., 35, 593 (1987).
17. A. Voekel, Unpublished results, Carnegie-Mellon University, (1988).
18. C.P. You, A.W. Thompson, and I.M. Bernstein, Scripta Met., 21, 181 (1987).
19. C. Liu and J.J. Lewandowski, unpublished research, (1988).
20. W.H. Hunt, Jr. Personal Communication (1988).
21. A.M. Ritter, in Powder Metallurgy Composites, M. Kumar, K. Vedula, and A.M. Ritter, eds., TMS-AIME, Warrendale, PA 1987, pp. 59-79.
22. J.K. Shang and R.O. Ritchie, Mat. Sci. Eng., submitted.
23. M. Manoharan and J.J. Lewandowski, unpublished results, (1988).

TABLE I

MECHANICAL PROPERTIES

MAT'L	σ_y (MPa)	UTS (MPa)	n	% RA	% SiC on Surface
UA-13μm	381	502	0.16	7.7	18
OA-13μm	406	460	0.16	7.4	11
UA-5 μm	379	496	N/A	8.3	12
OA-5 μm	418	470	N/A	7.3	10

TABLE II

NOTCH TOUGHNESSES*

MAT'L	K_Q
UA-13μm	24 MPa√m
OA-13μm	16.5 MPa√m
UA-5 μm	24 MPa√m
OA-5 μm	15 MPa√m

* NOTCH ROOT RADIUS 50-60 μm

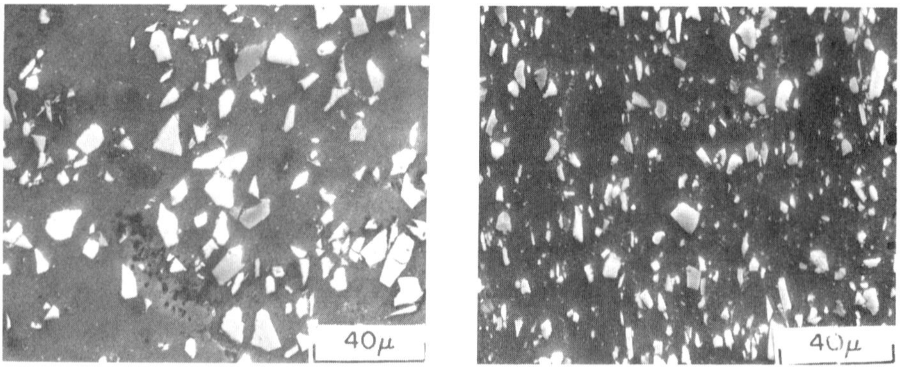

a) b)

Fig. 1 SEM micrographs of the a) 13μm and b) 5μm SiC containing material.

a) b)

Fig. 2 TEM micrographs of the a) UA and b) OA microstructures.

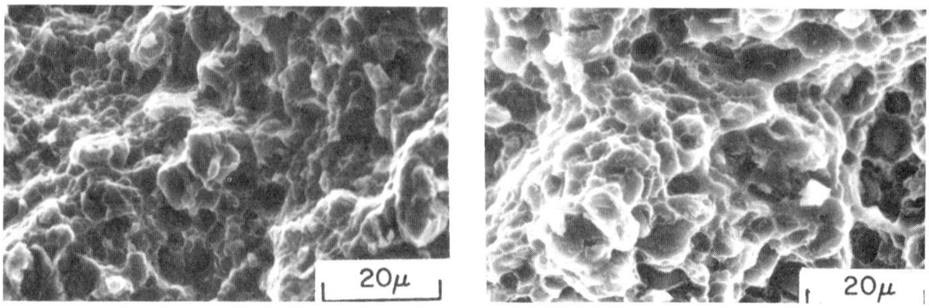

a) b)

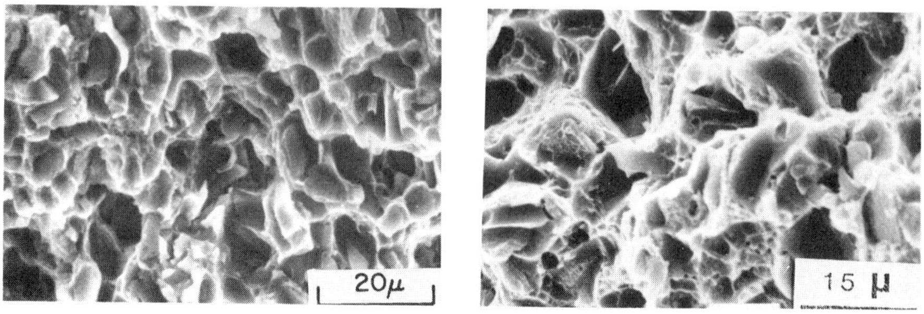

c) d)

Fig. 3 SEM fractographs of UA and OA materials containing either 5μm or 13μm SiC a) UA-5μm, b) OA-5μm, c)UA-13μm, d)OA-13μm.

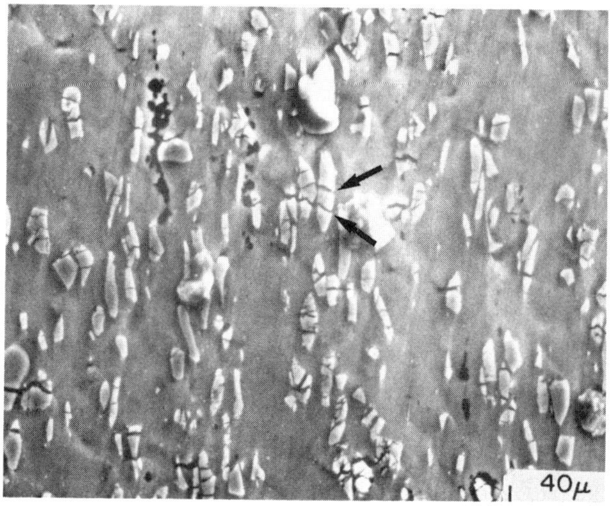

a)

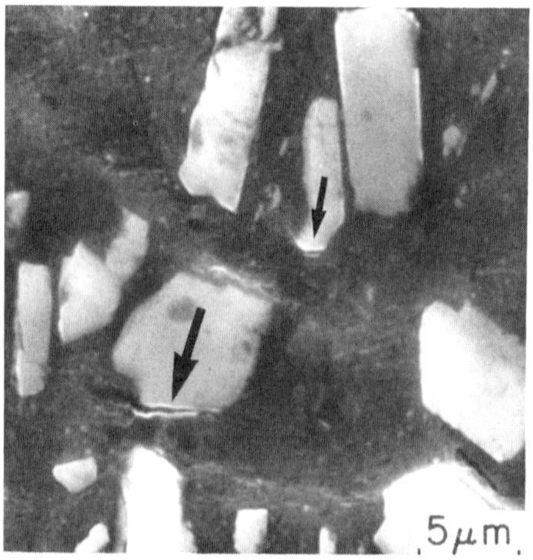

b)

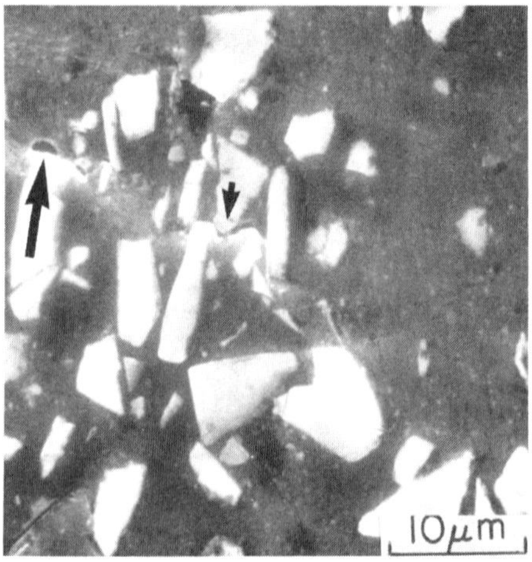

c)

Fig. 4 SEM views of surfaces of strained bend specimens.
a) UA-13µm - Multiple fractures observed
b) UA-5 µm - Matrix and near interface failure observed.
c) OA-13µm - Matrix and near interface failure observed.

HIGH TEMPERATURE DEFORMATION
OF METAL MATRIX COMPOSITES

C. Demetry, J.T. Beals, F.R. Tuler

Worcester Polytechnic Institute,
Worcester, MA

ABSTRACT

Compression tests were performed on as-cast 6061 aluminum with 10 and 20 volume percent silicon carbide particulate over a range of temperatures from 300°C to 550°C and strain rates from 10^{-3} to 1 sec^{-1}. Deformation behavior was characterized using both a thermodynamic modeling technique and constitutive equations relating the temperature and strain rate dependence of the flow stress. Microstructural analysis was conducted to determine the operative metallurgical processes. Flow softening by the mechanism of dynamic recovery is predominant; however, there is also some evidence of dynamic recrystallization.

KEYWORDS

Metal matrix composites; high temperature deformation; dynamic material modeling.

INTRODUCTION

Metal matrix composite materials (MMC) have been receiving increased interest as a result of their outstanding mechanical properties and wear resistance, both of which make them attractive for the aerospace and other high technology applications. MMC's are currently being produced by a modified ingot metallurgy process followed by conventional metalworking and heat treatment. Usually hot working must be carried out within a certain "window" in the strain rate and temperature field in order to produce defect-free parts with controlled dimensions, microstructures, and mechanical properties. In order to optimize the fabrication process, the relationships among flow stress, microstructure, temperature, and strain rate need to be studied. This paper describes the results of such a study investigating the deformation behavior of cast 6061 aluminum containing 10 and 20 volume percent silicon carbide particulate reinforcement.

Constitutive Equations

One method of characterizing deformation behavior is through the use of constitutive equations relating flow stress to temperature, strain, and strain rate. At constant strain and temperature the dependence between flow stress and strain rate can be represented by

$$\sigma = A\,\dot{\epsilon}^m$$

where m is the strain rate sensitivity. The parameter m is a measure of the strain rate dependence of the flow stress and determined experimentally by finding the slope of a log σ versus log $\dot{\epsilon}$ plot.

The temperature dependence of the flow stress at constant strain and strain rate generally can be represented by the constitutive equation

$$\sigma = C\,\exp(Q/RT)$$

where Q = an activation energy for plastic flow, R = universal gas constant, and T = test temperature in K. If this expression is obeyed, a plot of ln σ versus 1/T yields a straight line with a slope of Q/R. The magnitude of the activation energy often can indicate the type of metallurgical process occurring; however, the apparent activation energy may also be a complex result of more than one process.

Dynamic Material Modeling

The dynamic material modeling approach, developed by Prasad, Gegel, Malas and others(1984, 1988), models hot working in terms of the management of irreversible thermodynamic processes. Their assertion is that the intrinsic ability of a material to dissipate power is the most important criterion to be used in optimizing process control. The power dissipation capacity of the workpiece is expressed in terms of the efficiency of dissipation, η, which is related to the portion of available energy used for microstructural evolution (rather than for conversion to heat).

Dynamic material modeling also uses two criteria that determine the stability of material flow. For stable material behavior the rate of energy input cannot exceed the amount which the material can dissipate while avoiding fracture or plastic instability. The mechanical stability criterion requires the system to lower its total energy continuously. The material stability criterion stipulates that the workpiece should store entropy at least as fast as the entropy production rate of heat. The efficiency n and both stability criteria can be readily calculated from the slopes of log flow stress versus log strain rate and flow stress versus inverse temperature curves. Results are typically shown in the form of deformation mechanism maps showing efficiency regimes and stable and unstable regions.

EXPERIMENTAL PROCEDURE

The composite materials were supplied by Dural Aluminum Composites Corporation in the as-cast condition using aluminum alloy 6061 and SiC particles with an average size of 12 microns. Cylindrical specimens, 1 cm in diameter and 1.5 cm long, were machined from the ingots and tested in uniaxial compression on a servohydraulic machine modified to give constant true strain rate conditions. Testing was conducted at seven strain rates between 0.001 sec^{-1} and 1 sec^{-1} and six temperatures between 300 and 550°C. The deformation was continued to a true total strain of 0.45. All specimens were quenched in water in less than five seconds after deformation was completed. Metallography of the deformed specimens was accomplished with diamond grinding wheels and an automatic polishing system using diamond paste and colloidal silica suspension.

RESULTS AND DISCUSSION

Typical stress-strain curves for the 10 volume percent composite at high and low strain rates at 500°C are shown in Fig. 1. In

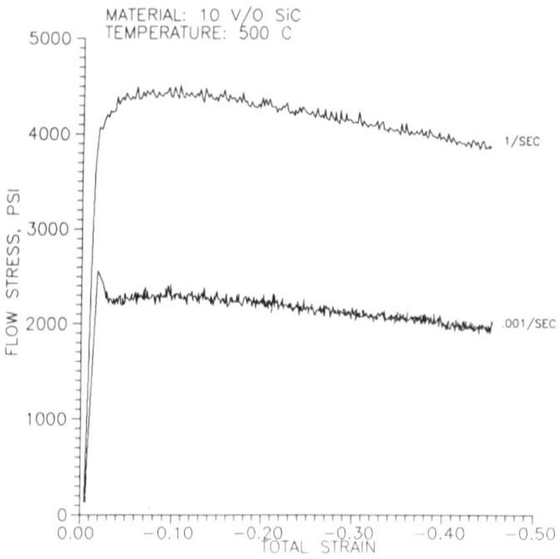

Fig. 1. Typical stress-strain curves at 500°C for 10 vol % SiC/Al.

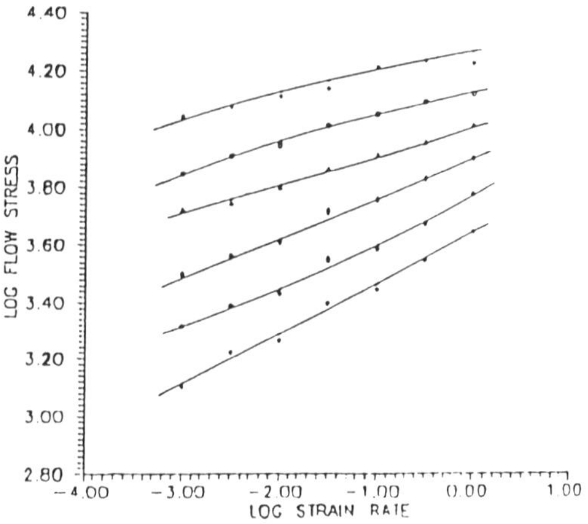

Fig. 2. Effect of strain rate on flow stress for 10 vol % SiC/Al.

all cases flow stress decreases with increasing temperature, and the flow stress is greater for the 20 volume percent composite than for the 10 volume percent material at equivalent temperature and strain rates. Flow softening occurs at all temperatures and strain rates, but to different extents. Tests done at high temperatures and low strain rates showed a single peak in the stress-strain curve which may be indicative of dynamic recrystallization.

Figure 2 shows the increase in flow stress with increasing strain rate. As discussed in a previous section, the slopes of these logarithmic plots give the strain rate sensitivities, m, shown in Table 1. Values of m for the composite materials range from 0.06 at 300°C to 0.17 at 550°C, comparable to those for Al alloy 6061 but much less than those for most metals.

Table 1 Strain Rate Sensitivity

Temperature	10 vol%	20 vol%
300°C	0.039	0.055
350°C	0.062	0.071
400°C	0.113	0.094
450°C	0.137	0.119
500°C	0.178	0.161
550°C	0.178	0.161

The activation energy of the microstructural mechanisms controlling deformation are shown on three-dimensional maps in Fig. 3 as a function of temperature and strain rate. It is clear that more than one deformation mechanism is governing the behavior of the composites over the ranges of temperature and strain rate examined. The magnitudes of these activation energies are less than half those accepted values for self-diffusion and dynamic recrystallization in pure aluminum. In addition, the energies for the 20 volume percent composite are consistently less than those for the 10 volume percent composite. Both of these observations suggest that silicon carbide particles act as nucleation sites for softening mechanisms. It should also be noted that the highest activation energies occur at high temperatures and low strain rates where the single peaks in the stress-strain curves were observed.

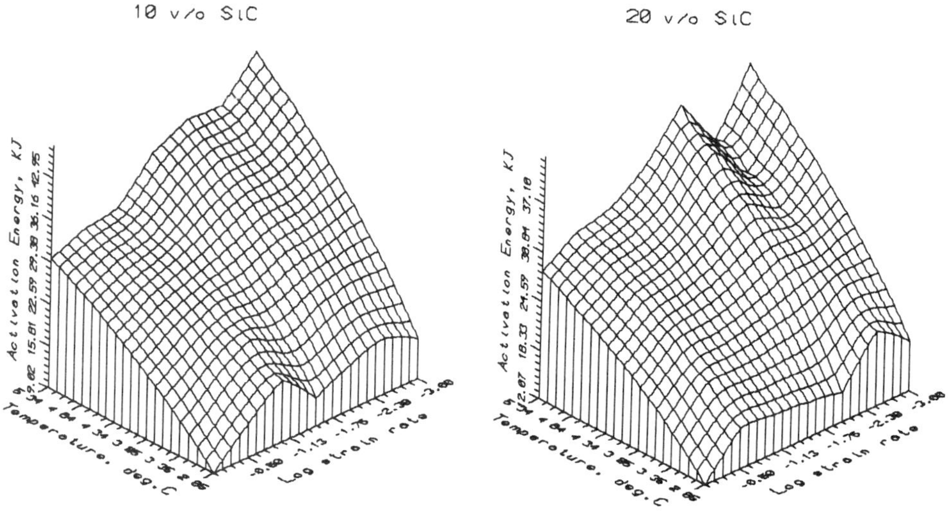

Fig. 3. Activation energy maps for 10 and 20 vol % composite.

The dynamic material modeling technique was relatively easy to implement using our stress-strain data. The countour map in Fig. 4 shows the "efficiencies" over the range of temperatures and strain rates for the 10 volume percent composite. The map confirms that the deformation process is relatively insensitive to strain rate. As one would expect, efficiency increases with temperature up to a maximum of 30%.

All of our data satisfy the criterion for mechanical stability, but some combinations of temperature and strain rate, indicated by the shaded area in Fig. 5, are unstable according to the criterion for material stability. It is not intuitively obvious that these regions should be unstable, and the microstructures in these regions show no obvious defects or evidence of damage mechanisms.

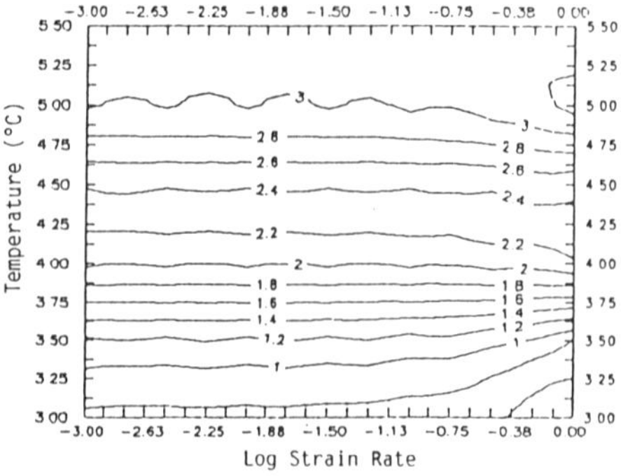

Fig. 4. Efficiency map for 10 vol % composite.

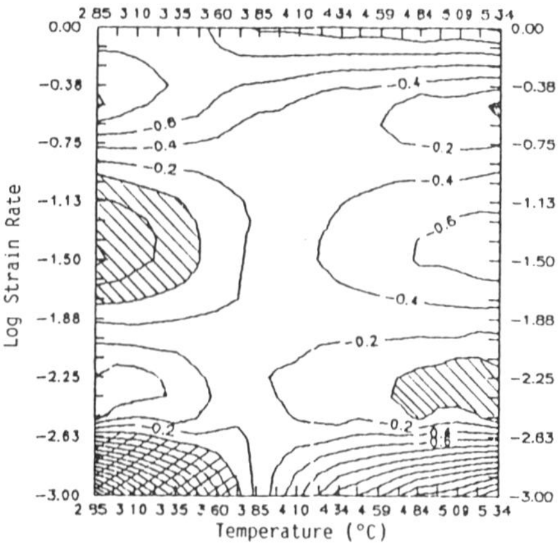

Fig. 5. Stability map for 10 vol % composite. Shaded regions indicate material instability.

Figure 6 shows a micrograph of an unetched, deformed specimen of the 20 volume percent composite. When etched, many of the

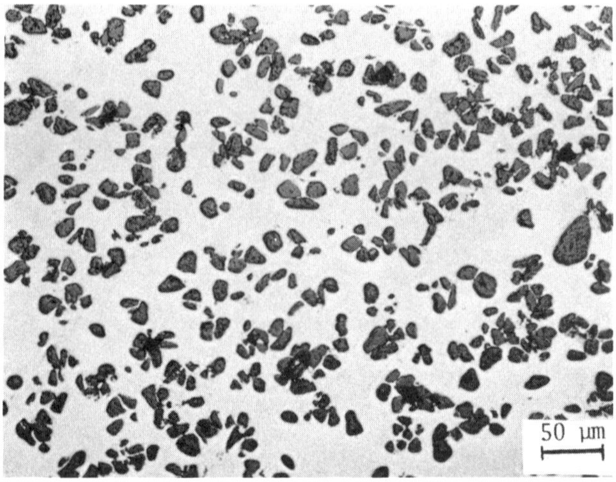

Fig. 6. Unetched microstructure of 20 vol % composite deformed to strain = 0.45 at T = 450°C and strain rate = 0.03 sec^{-1}.

microstructures show the initial cast structure with no well-defined grain structure retained after deformation, as shown in Fig. 7. However, at high temperature and low strain rate, a well-recovered structure is evident, with an occasional region of recrystallization, as shown by the micrograph in Fig. 8. Further testing at higher strain rates was conducted for Alcan International's Kingston Research and Development Centre. This work shows that recrystallization occurs at high temperatures and high strain rates and most often occurs at clusters of silicon carbide particles (Tuler and others).

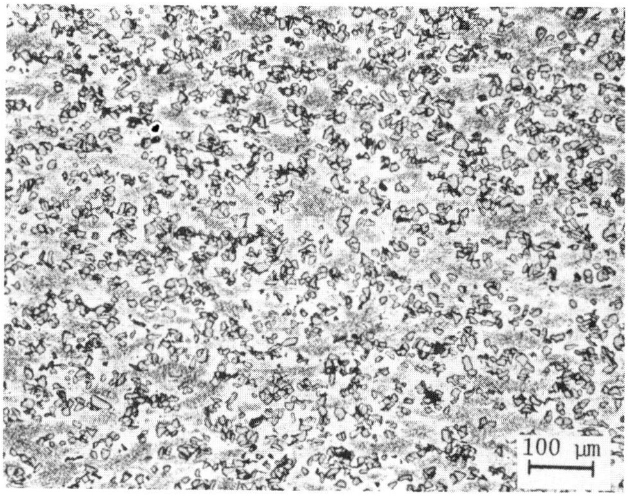

Fig. 7. Quenched microstructure of 20 vol % composite deformed at 300°C and 0.03 sec^{-1}.

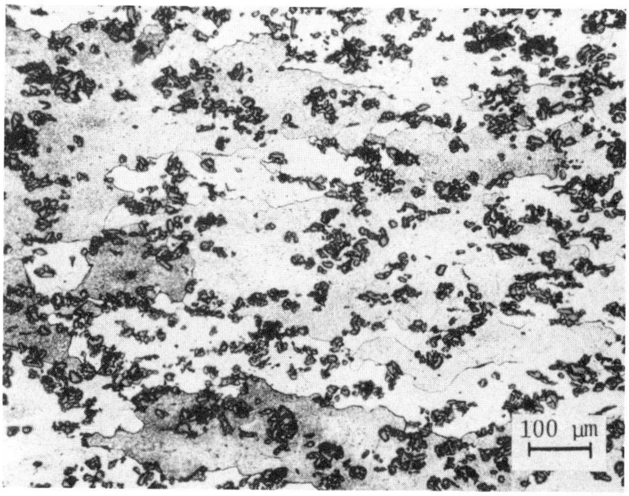

Fig. 8. Quenched microstructure of 10 vol % composite deformed at 550°C and 0.03 sec^{-1}.

CONCLUSIONS

The flow stresses and microstructures of as-cast 10 and 20 volume percent SiC particle-reinforced 6061 aluminum composites have been determined over a broad range of temperatures and strain rates. Several conclusions concerning the deformation of these materials can be drawn from the results:

1) The deformation of the composites shows low strain-rate sensitivity;
2) Stress-strain curves show that flow softening is predominant during deformation, and microstructures confirm that dynamic recovery and, in a few cases, dynamic recrystallization are the operative metallurgical processes for the range of test conditions examined;
3) The variation of activation energy of deformation over the range of test conditions indicates that more than one microstructural mechanism is controlling the deformation in this composite;
4) Trends in activation energies and microstructure analysis suggest that SiC particles act as nucleation sites for recovery and recrystallization.

ACKNOWLEDGEMENTS

The authors are grateful to Alcan International Limited and Dural Aluminum Composites Corporation for permission to publish this work and for their support of composite materials research at Worcester Polytechnic Institute.

REFERENCES

Gegel, H.L., J.C. Malas, S.M. Doraivelu, and V.A. Shende (1988). Modeling Techniques Used in Forging Process Design. Metals Handbook, Vol.14-Forming and Forging. 9th ed.

Prasad, Y., H. Gegel, S. Doraivelu, J. Malas, J. Morgan, K. Lark, and D. Barker (1984). Modeling of Dynamic Material Behavior in Hot Deformation: Forging of Ti-6242. Metallurgical Transactions A, 15A.

Tuler, F.R., J.T. Beals, C. Demetry, D. Zhao, D.J. Lloyd (September 1988). International Symposium on Advances in Cast Reinforced Metal Composites. ASM International Metals Park, Ohio.

The Effects of Pressure on the Fracture of Composite Materials

Frank Zok and J.D. Embury
Dept. of Materials Science & Engineering
McMaster University
Hamilton, Ontario, Canada

ABSTRACT

The use of tensile tests with superimposed hydrostatic pressures enables a variety of combinations of equivalent stress and mean stress to be obtained. In conventional metallic materials the use of tests under superimposed hydrostatic pressure is of value in the study of processes such as ductile fracture which involve dilatent events such as void nucleation and growth. Quantitative models of ductile fracture and fracture mechanism maps have been developed based on the use of tests under hydrostatic pressure (D. Teirlinck, F. Zok, J.D. Embury, M.F. Ashby, Acta Metallurgica 36, 1213-1228 (1988)).

We have recently extended the technique to metal matrix composites, Al alloys containing particulate SiC, Al alloys with SiC whiskers and directionally solidified Al-Ni alloys. It was observed that in all the materials the ductility increased with an increase in superimposed pressure.

In the Al (2014) SiC particulate materials the increase in ductility was up to 20 times. At low pressures fracture was initiated by particle fracture but at high superimposed pressure particle matrix decohesion also occurred. In the materials containing whiskers or aligned rods of Al_3Ni an alternative mode of fracture was observed at higher pressure. Strain localization occurred by a combination of dilatent events such as interface debonding and particle fracture and by strain softening due to fibre rotation. These processes result in shear failure at high pressures.

The use of superimposed pressure tests provides valuable information in the area of plastic response, damage accumulation and control of fracture events in metal matrix composites. The paper presented both experimental and theoretical work on the role of hydrostatic stress. A published account of portions of the work will be available in the Proceedings of the Conference on Interfacial Phenomena in Composites held by Brown University in June 1988, and in the Proceedings of the 9th Riso Conference on Composites in September 1988.

SOLIDIFICATION PROCESSING OF METAL - MATRIX COMPOSITES

P. K. Rohatgi and R. Asthana

Materials Dept., Univ. of Wisconsin - Milwaukee,
Milwaukee, WI 53211, U. S. A.

ABSTRACT

The design and synthesis of fibre reinforced and particle filled metal matrix composites produced by melt metallurgy techniques have been described. The various conventional and modern solidification processing techniques including permanent mold casting, pressure die casting, centrifugal casting and squeeze casting have been discussed. The various engineering properties and applications of these composite materials produced by casting techniques have been described.

KEYWORDS

Metal matrix composites; wettability; casting techniques; microstructures; selective reinforcements; fluidity.

INTRODUCTION

Cast metal matrix composites produced by solidification processing techniques consist of continuous or discontinuous fibres, whiskers and particles distributed in a desired spatial pattern in the matrix alloy which solidifies in the restricted spaces between the reinforcing phase to form the bulk of the matrix. By controlling the relative amounts, shape, size and the distribution of the reinforcing phase, the metal-reinforcement interface, as well as by controlling the solidification conditions cast metal matrix composites can be imparted a tailored set of useful engineering properties which are not realized with conventional monolithic materials. In addition the cast microstructure of the matrix gets refined and modified due to the presence of particles and fibres indicating a possibility of controlling macrosegregation, microsegregation and grain size in the matrix. This represents an opportunity to develop new matrix alloys.

Due to the poor wettability between molten metals and most ceramics which are generally used as reinforcements in composites (Delannay and others, 1987) intimate contact between the two can only be promoted by using techniques to artificially induce wettability or by using external pressure to overcome the interfacial t tensions and viscous drag of fluid flow. The techniques generally used for introducing and uniformly mixing a discontinuous ceramic phase in a metallic melt make use of various kinds of mechanical impellers, injection guns, centrifugal acceleration, electromagnetic stirrers, hydrostatic pressure, and ultrasonic irradiation of the melt (Rohatgi & coworkers, 1986). The basic steps involved in solidification processing of composites are outlined in Fig. 1.

CASTING TECHNIQUES

The choice of the casting technique and the configuration of the mold are of central importance to the quality (defect content, particle distribution) of the composite casting owing to the buoyancy driven particle movement, particle pushing by solidifying interfaces, reduced fluidity and flocculation tendency of particles. Some of the commonly used casting techniques for producing metal matrix composites are described below.

Sand and Die Casting

In slowly cooled sand castings the dispersed particles like graphite, mica and talc which are lighter than most of the commonly used Al alloy melts, tend to segregate near the portion of the casting leaving behind a particle impoverished zone near the bottom of the casting. Heavier particles like zircon, silicon carbide, silica, and glass tend to settle to the bottom part of the casting. The distribution of the particles can be made more homogeneous by using permanent molds with or without water cooling. Figures 2 through 5 show typical microstructures of permanent mold castings of Al alloys containing acceptably uniform distribution of particles of talc, zircon, natural graphite, and rice husk ash respectively.

Pressure Die Casting

Pressure die casting of composites allows larger sized and more intricate component shapes to be rapidly produced at relatively low pressures and equivalent capital expenditure. The pressure die cast particle composites exhibit lower bulk and interfacial porosities, more uniform particle distribution, improved particle-matrix bonding, less particle flocculation and occasional exfoliation/fragmentation of soft particles like graphite, with melt penetrating into fine exfoliation cleavages. Over 60 wt % zircon particles and 7 wt % graphite particles have been successfully dispersed in Al alloys by pressure die casting to produce several engineering components. Figure 6 shows a photomicrograph of a pressure die cast Al-zircon particle composite where uniform distribution of particles was achieved in the matrix.

Centrifugal Casting

Due to centrifugal acceleration on particles in rotating molds, the lighter particles like graphite, mica and porous alumina segregate near the axis of rotation thereby producing high particle concentration surfaces for bearing applications. Upto 8% by weight mica and graphite, and upto 30% by weight zircon particles have been incorporated in selected zones of Al alloys by this technique. Figure 7 shows a photomicrograph of a centrifugal casting of Al-graphite particle composite where the matrix region rich in graphite can be clearly distinguished from matrix regions depleted of graphite. Local modification of eutectic silicon was noted in particle rich regions of centrifugal castings of Al-graphite. Figures 8 and 9 show sections of centrifugal castings of Al-graphite and Al-zircon particle composites respectively.

Compocasting

Particulates and discontinuous fibres of SiC, alumina, TiC, silicon nitride, graphite, mica and glass have been incorporated into vigorously agitated partially solid alloy slurries by compocasting technique. The discontinuous ceramic phase is mechanically entrapped between the proeutectic phase present in the alloy slurry which is held between its liquidus and solidus temperatures. Under mechanical agitation such an alloy slurry exhibits "thixotropy" in that the viscosity decreases with increasing shear rate, and appears to be time dependent and reversible. Increasing the residence time of particles or short fibres in the slurry promotes wetting and bond formation due to chemical reaction at the interface. This semi-fusion process allows near net shape fabrication by extrusion or forging since the deformation resistance is considerably reduced due to semi fused state of the composite slurry.

Squeeze Casting

Squeeze casting or liquid forging of metal matrix composites involves unidirectional pressure infiltration of fibre preforms or powder beds to produce void free, near net shape castings of composites. The processing variables in decreasing order of importance with respect to their effect on structure of the matrix are : Fibre preheat temperature, Interfibre spacing, Infiltration pressure, Infiltration speed, Metal superheat temperature (Masur, 1988). If the metal or fibre temperature is too low, poorly infiltrated or very porous castings are obtained. If the temperatures are too high, excessive fibre-metal reaction causes degradation of casting quality. Finally if the plunger speed is too fast, then the preform is deformed on infiltration. A threshold pressure is required to initiate liquid metal flow through fibrous preform or powder bed. This pressure has its genesis in the thermodynamic surface energy barrier due to non wetting. Since the wetting is a time dependent function the threshold pressure increases with increasing infiltration rate. An additional pressure is required to overcome the frictional forces arising from the viscous drag of fluid flowing through interfibre channels. Al alloy composites containing silicon carbide and alumina powder, saffil fibres and silicon nitride whiskers have been fabricated by this method.

Investment Casting & Vacuum Infiltration

In investment casting, filament winding or prepreg handling procedures developed for fibre reinforced plastics are used to position and orient the proper volume fraction of continuous fibres within the casting. The layers of reinforcing fibres are stacked in proper sequence and orientation and the preform thus produced is infiltrated with liquid metal under pressure. Continuous graphite fibre reinforced Mg alloy composites have been produced by this method (Katzman, 1983; Goddard, 1984). In vacuum infiltration process handleable tapes of fibre yarns are made using a fugitive binder and tapes are suitably positioned in the mold. These fibre tapes are then infiltrated with liquid metal under vacuum while the fugitive binder burns away. Best castings in terms of soundness and mechanical properties are obtained if the mold assembly is preheated and evacuated prior to liquid metal infiltration.

MICROSTRUCTURES

Al alloys have been chiefly used as matrix materials in a wide variety of particle filled (Rohatgi & Coworkers, 1986) and fibre reinforced (Mortensen & Coworkers, 1986) metal matrix composites. Microstructures of these composites show that primary Al in hypoeutectic alloys tends to avoid (or "push") the discontinuous ceramic particles and fibres and tends to nucleate in the interfibre spaces unless special surface modification techniques are used to promote heterogeneous nucleation on the fibre surfaces. However, primary silicon and eutectic silicon in Al-Si alloys tend to concentrate on fibre or particle surfaces as shown in fig. 10. The discontinuous particles also tend to modify or refine the structure, e.g. eutectic silicon in Al-Si alloys gets modified whereas primary silicon is refined when solidification occurs in the presence of high volume fraction of dispersed particles. Figures 11 and 12 show the normal and modified eutectic silicon in Al-Si alloys without and with graphite particles present in the alloy respectively. Figure 13 shows a photomicrograph of an Al-Si-Ni alloy containing graphite particles where the graphite particle is seen pushed in the last freezing eutectic zone. Microstructures in fibre reinforced composites can be modulated in a predetermined manner by controlling the interfibre spacing as well as the cooling rate during processing. If the cooling rate is sufficiently high or fibre volume fraction sufficiently low, the matrix alloy solidifies uninfluenced by the fibres. At sufficiently slow cooling rates, however, matrix can be rendered free of microsegregation as a result of restricted diffusion in the solid state and dendrite coalescence during solidification.

PROPERTIES AND APPLICATIONS

Both particle and fibre reinforced metal matrix composites are potentially important materials for a wide variety of applications including aerospace, automotive industries and consumer goods industries. Fibre reinforced metals like Mg/Graphite can achieve zero thermal expansion coefficient until very high temperatures and are ideally suited for various space structure applications where dimensional integrity over wide temperature ranges is important. Particle filled composites show considerable improvements in tribological (friction, wear, galling) properties making them useful antifriction and antiabrasion materials for applications in bearings, bushings, pistons, cylinder liners, machine shrouds, high temperature impellers etc (Prasad & Rohatgi, 1987). The use of graphitic aluminium alloys in automobile engine parts considerably reduces the wear of cylinder liner as well as improves fuel efficiency and engine horse power at equivalent cost. The room and elevated temperature damping capacity of Al and Cu alloys is enhanced significantly by graphite additions and can exceed that of special vibration insulating metals and alloys. Use has been made of the unique capability of casting techniques to selectively reinforce regions in castings with special ceramic fibres to commercially produce diesel engine parts. Toyota diesel piston is selectively reinforced with ceramic fibre by a special squeeze casting method, and shows enhanced wear resistance and higher use temperature at equivalent cost. Connecting rods have also been produced from fibre FP-Al-SiC composites by casting techniques. The use of cheaper particulates in metal castings have considerably reduced the cost of these materials compared to similar fibre reinforced metals and these are being made available in consumer oriented industries (Schuster & Coworkers, 1987).

CONCLUSIONS

Considerable progress has been made in introducing particles and fibres in metal matrices by solidification processing techniques to produce a wide variety of cast metal matrix composites. The availability of squeeze casting has provided a potentially promising route for manufacturing defect free, near net shape composite castings, thereby making feasible selective reinforcement of castings with ceramic fibres in a predetermined manner. A variety of problems of both basic and engineering origin still need to be solved to improve the quality of the composite castings. Some of these problems are: Improved modeling and tailoring of metal ceramic interfaces, modeling of solidification morphology on reinforcing particles and fibres, interactions between particles and solidifying interfaces, elimination of porosity and freeze choking during pressure infiltration, uniformity of distribution of reinforcing particles within a casting and its reproducibility in batches of castings in large scale manufacture.

REFERENCES

Delannay, F., L. Frozen, and A. Deruyttere (1987). J. Mater. Sci., 22,
Goddard, D. M. (1984). Metal Progress, April, 49.
Katzman, H. A. (1983). Carbon fibre reinforced metal matrix composites, U.S. Patent No. 4, 376, 808.
Mortensen, A., M. N. Gungor, J. A. Cornie, and M. C. Flemings (1986). Journal of Metals, 30.
Masur, L. J. (1988). Infiltration of Fibrous Preforms by a Pure Metal, Ph.D. Thesis, Dept. of Mater. Sci. & Engineering, M. I. T., Mass.
Prasad, S. V. and P. K. Rohatgi (1987) Journal of Metals, 22.
Rohatgi, P. K., R. Asthana, and S. Das (1986). Int. Metals Reviews, 31, 115.
Schuster, D. M., M. Skibo, and F. Yep (1987). Journal of Metals, 60.

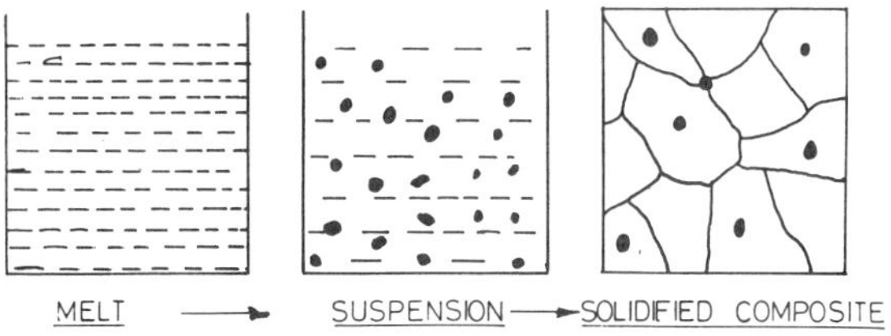

FIG.1. STEPS INVOLVED IN SOLIDIFICATION PROCESSING OF CAST COMPOSITES

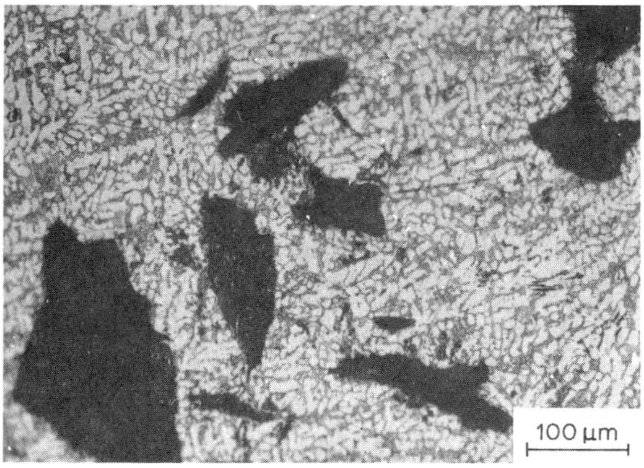

Fig. 2. Gravity die cast Al-Si-Talc particle composite.

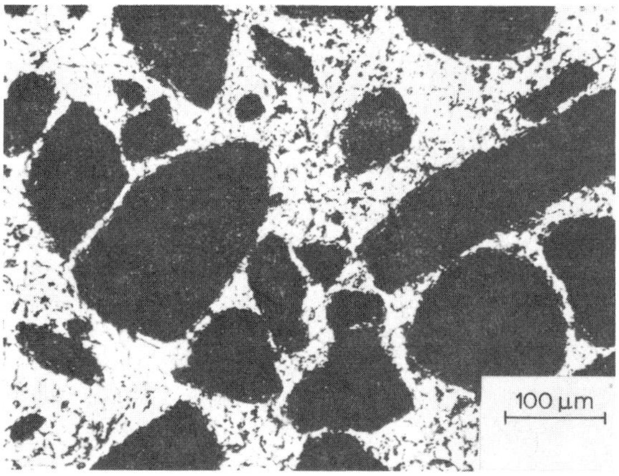

Fig. 3. Gravity die cast Al-Si-Zircon particle composite.

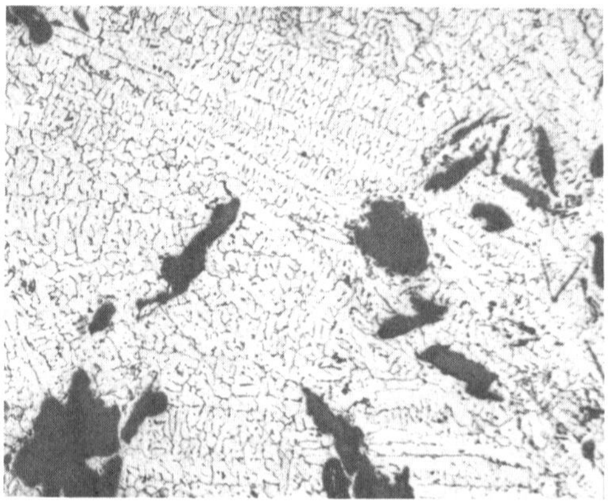

Fig. 4. Gravity die cast Al-Si-Natural Graphite particle composite.

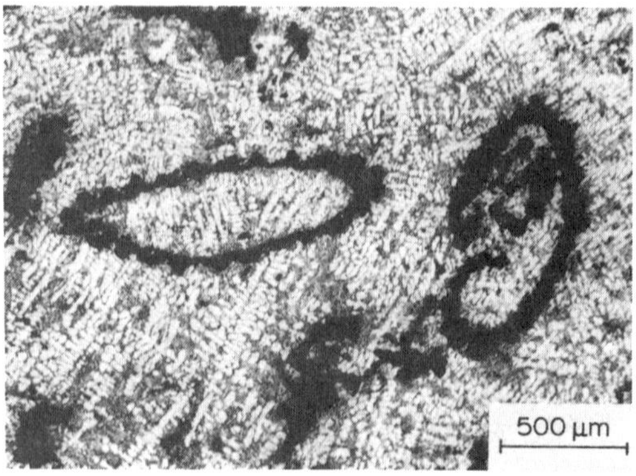

Fig. 5. Gravity die cast Al-Si-Rice husk ash particle composite.

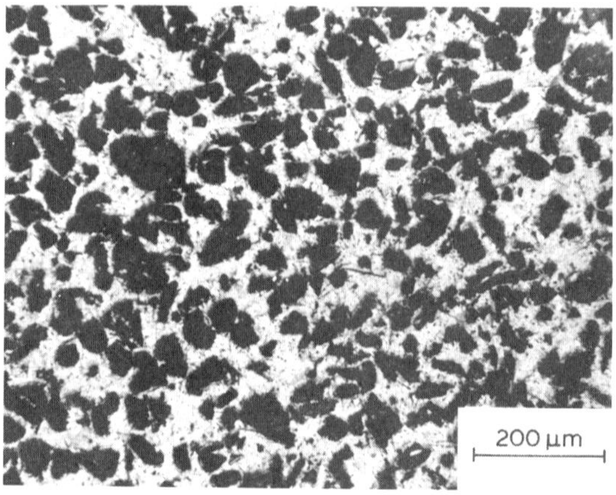

Fig. 6. Pressure die cast Al alloy-60% zircon particle composite.

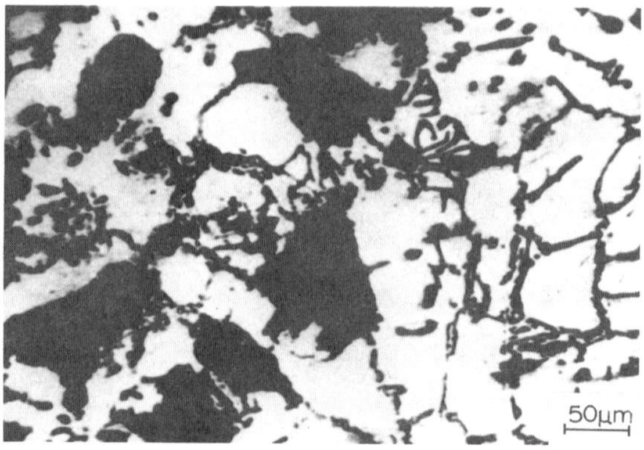

Fig. 7. Modified eutectic Si in centrifugally cast Al-Si-Graphite particle composite.

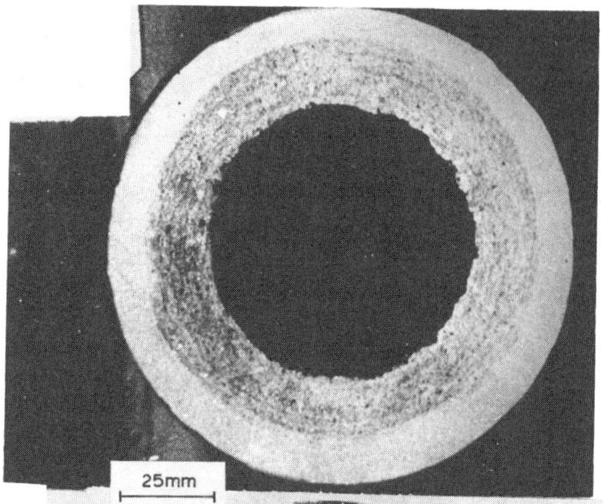

Fig. 8. A section of centrifugally cast Al-Si-Graphite composite bearing.

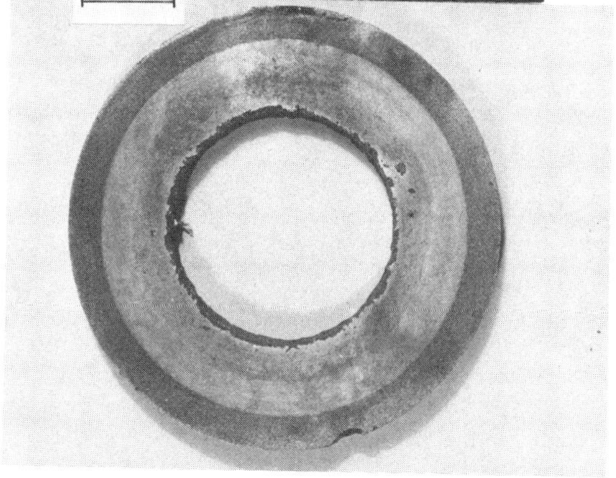

Fig. 9. A section of centrifugal casting of Al-Si-Zircon particle composite.

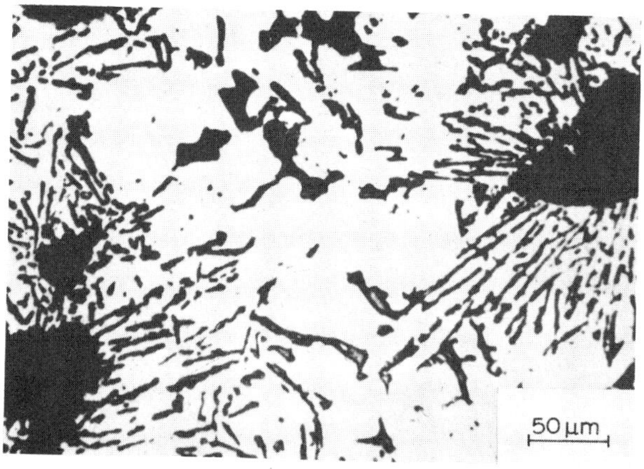

Fig. 10. Preferential nucleation of eutectic Si on graphite surface.

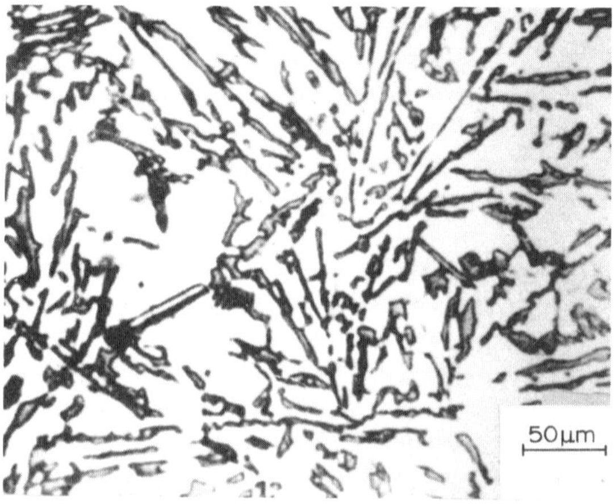

Fig. 11. Eutectic silicon needles in cast Al-11.8 Si alloy.

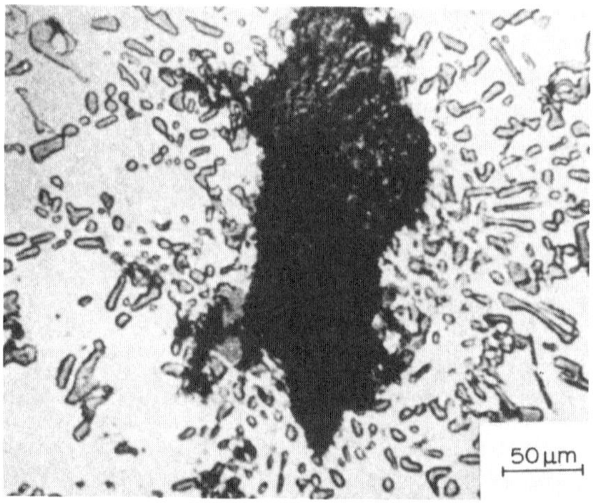

Fig. 12. Modified eutectic silicon in Al-11.8 Si-3% graphite particle composite.

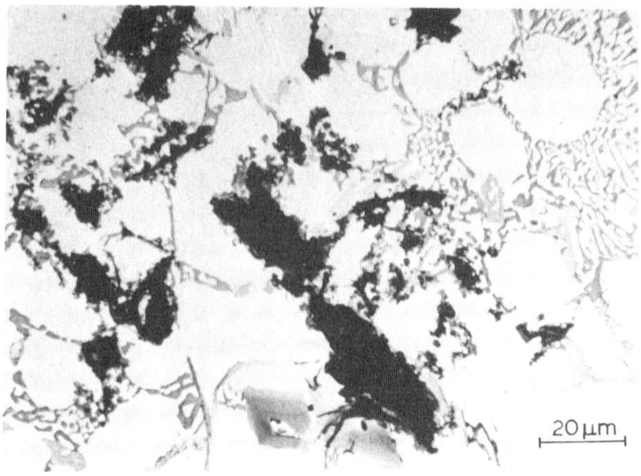

Fig. 13. Photomicrograph showing a graphite particle in the interdendritic region of an Al-Si-Ni alloy.

EXTRUSION OF ALUMINUM METAL MATRIX COMPOSITES

R.W. Hains[+], P.L. Morris[*], P.W. Jeffrey[*]

[+] Alcan Smelters and Chemicals Limited, Kingston
[*] Alcan International Limited, Kingston Laboratories, P.O. Box 8400, Kingston, Ontario K7L 5L9

ABSTRACT

Some properties of particulate metal matrix composites are presented and surface condition and die wear associated with MMC extrusion are discussed.

KEY WORDS

Aluminum; composite; properties; extrusion; surface defects.

INTRODUCTION

Particulate reinforced metal matrix composites (MMC's) are one of the promising advanced engineering materials offering property combinations unattainable in conventional materials. While the degree of improvement in particulate reinforced materials is less than with whiskers, these materials can be produced as pig, or as fabrication ingot for subsequent processing into a wide variety of products by casting, extrusion, rolling or forging, without property loss due to fracture of the reinforcement. Lower manufacturing costs provides a potential for large scale production of a competitively priced product.

Dural Aluminum Composites Corporation, a fully owned subsidiary of Alcan, has developed the necessary technology to produce a range of ceramic particle reinforced MMC's by a molten metal route. Their products have been used to produce cast shapes, forgings and a wide range of extruded solid and hollow shapes.

This paper discusses the effects of ceramic reinforcement of aluminum alloys and their impact on the extrusion process, emphasizing those areas in which future development will produce added gains.

The main thrust of extrusion development has focussed on 6061 and 2014, reinforced either with silicon carbide or alumina at 10 - 20 volume percent. Of principal significance is an increase in elastic modulus but also strength increases are achieved.

Table 1 lists the expected properties of some of these combinations, and Table 2 lists values for 6061 T6 and 2014 T6 extrusions (ref ASTM B221-85). These demonstrate that ceramic reinforcement raise the modulus by up to 40% and minimum tensile property levels of the order of 20%.

TABLE 1 Minimum Expected Tensile Properties for DURAL MMC - Extrusions

	Tensile Strength Ksi Min.	Yield Strength Ksi Min.	Modulus Msi Typ.
6061 - 15 v/o Alumina - T6	49	43	13.3
2014 - 10 v/o Alumina - T6	71	66	13.0
2014 - 15 v/o Alumina - T6	73	67	14.7

TABLE 2 Minimum Guaranteed Properties for 6061 and 2014 Extrusions

	Tensile Strength Ksi Min.	Yield Strength Ksi Min.	Modulus Msi Typ.
6061 - T6	38	35	10.0
2014 - T6*	64	58	10.6

* 0.500 - 0.749" Thick

The ease and rapidity with which these materials enter the market depends on production costs, the strength of market pull, the extent to which they can be produced on existing equipment, and the cost and complexity of process changes necessary to produce acceptable quality.

Dural composites have been manufactured by hot, direct extrusion (lubricated and un-lubricated) and also by hydrostatic extrusion. Hot direct extrusion without lubrication is the simplest, cheapest and most widely used method, therefore this paper will emphasize the effects of this processing method on surface integrity and appearance, extrusion speed and tooling wear.

SURFACE INTEGRITY AND APPEARANCE

Early extrusion trials demonstrated that 6061 and 2014 based MMC's behave very differently from the un-reinforced alloys. Considerable surface tearing occurred at the front end of extrusions and, at low extrusion speeds, persisted throughout the entire extruded length. The intensity of tearing usually reduced with increasing extrusion speed and, on occasions, disappeared completely at intermediate speeds. Further increase in speed resulted in the onset of edge and surface cracking differing from low speed cracking;

appearing similar to a type previously reported with conventional aluminum extrusion. Figure 1 is a low magnification image of such high speed fracture and shows a distinct fracture path. Figure 2 is a high magnification image of the fracture surface, which shows a rounded appearance assumed to be the result of liquation of the interfaces. The defect appears similar to speed cracking, seen in un-reinforced aluminum alloys (see Metal Handbook, 8th edition, volume 10 page 67).

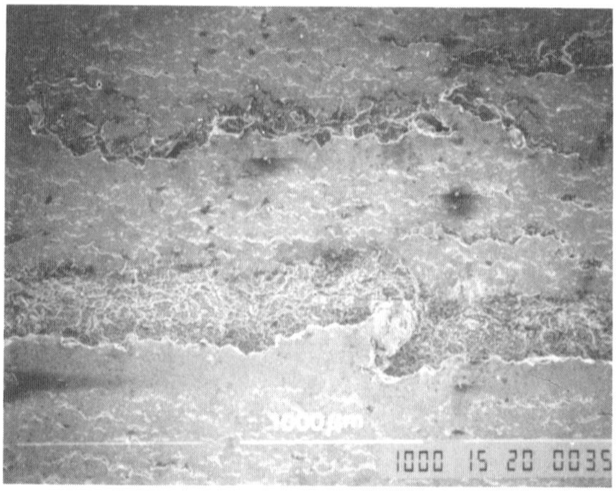

Fig. 1. Surface cracks occuring at high extrusion speeds.

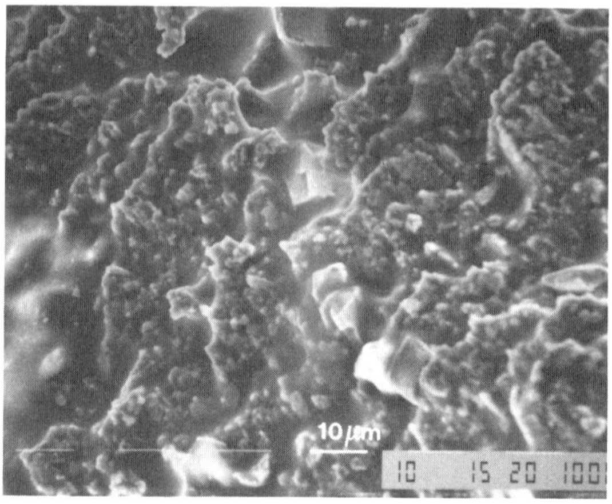

Figure 2. Localized melting.

Reiso (1984) demonstrated that speed - limiting cracking of Al-Mg-Si alloys can occur by two separate mechanisms.

1. Type 1, initiated at the die-extrudate interface, and occurring because the matrix is not strong enough to withstand frictional force at the die.

2. Type 2, initiated at subsurface weaknesses, and assumed to be due to incipient melting.

Reiso (1988) also demonstrated that precipitated magnesium silicide in Al-Mg-Si alloys caused type 2 tearing. Changing the pre-extrusion heating process resulted in a 30-60% increase in extrusion speed. This was attributed to elimination of magnesium silicide precipitates prior to extrusion.

Both types of tearing mechanisms are temperature dependent, and are initiated in a region close to the die-extrudate interface, therefore factors affecting the temperature of extrudate in this region are of interest.

It is known that:

1. In excess of 96% of the mechanical energy of a metal working process is converted to heat.

2. Much of the energy consumed in extrusion is concentrated into shear zones.

   (a) adjacent to the container wall as a result of friction between billet and container; this zone may consume 20-30% of the total extrusion energy.

   (b) in the boundary layer between the dead metal and deformation zones. Akeret (1972) showed in excess of 50% of the energy expended in the deformation zone is concentrated in a layer having a thickness approximately 10% of the billet diameter.

3. Metal from the shear zones (described above) flows into areas at, or close to, the surface of the extrudate.

4. Transport of heat from the shear zone to the surroundings reduces shear zone temperature, and is time dependent. Time for heat transfer from the shear zone decreases with increasing extrusion speed.

MMC's require higher extrusion pressures than un-reinforced material resulting in higher extrudate temperatures at a given extrusion speed. This can result in disintegration of the surface by either type 1 or type 2 cracking; a condition which can be cured by reduction in extrusion speed to permit excess heat to dissipate from the heavily worked shear zone.

Type 2 cracking occurs at a lower temperature than type 1, and therefore becomes the speed controlling factor. This underscores the importance of alloy design and adequacy of pre-extrusion thermal treatments to minimise the quantity of low melting point phases.

High speed cracking, both type 1 and type 2, occur in both monolithic alloys and MMC's. In contrast low speed tearing (Figure 3, LHS) which will be designated Type 3, appears to be a feature found only in MMC's. This sample is from the front end of a .75" x 3.00" rectangular section extruded through an H13 steel die (common to extrusion of un-reinforced aluminum alloys).

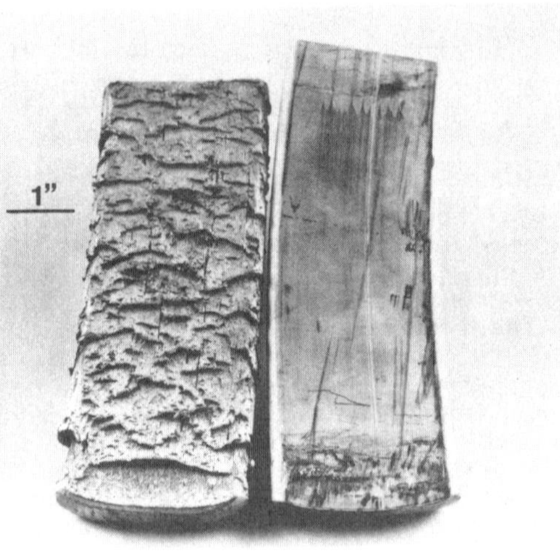

Fig. 3. Extrusion front end.

A probable explanation of the mechanism which results in gross tearing at low speeds and which attenuates, or disappears, as extrusion velocity is increased is as follows.

During the initial build-up of extrusion pressure, the billet is upset to fill the container. Simultaneously the metal spanning the die opening is pushed into the aperture. During this stage the extrudate moves slowly across the bearing surface.

Akeret & Strehmel (1988) suggested that plastic flow of material passing through a converging aperture is mainly due to compression acting normal to the bearing surface (magnitude is of the order of the yield stress of the extrudate), and, in the axial direction, forces are compressive, decreasing to zero (or are even tensile) at the exit side of the die. These conditions produce high frictional forces at the bearing surface which, in earlier work, Akeret (1983) suggested was based on a dynamic equilibrium between adhesion and removal of adhered aluminum particles.

In the initial stages of extrusion the die is colder than the extrudate. Metal adhering to the bearing during this phase is immediately chilled and its flow stress increased. This may create conditions which are energetically favourable for sub-cutaneous fracturing, so that the material in contact with the bearing remains stationary, while the subsurface material continues its forward motion.

Further increase in pressure breaks the bond between the bearing and adhered material; this material then moves across the bearing surface. The entire periphery does not move in unison, with different segments breaking away at different times. New material adheres to the bearing surface and remaining debris. The process is repeated in the classic stick-slip mode until conditions are energetically favourable for continuous movement of material across the bearing surface.

This proposed mechanism is compatible with the findings of Sheppard and Clode (1988) who hypothesised that surface formation during extrusion is a process of generation and modification. The surface is generated at the entry to the die bearing, while the exit side only serves to deteriorate the surface, by imposed ironing and galling of the extruded product.

Additional support to this hypothesis is provided by the following observations.

1. Low magnification SEM images of type 3 fractures indicate a smeared appearance (Fig. 4)

2. The fractures are ductile (Fig. 5), therefore eliminating a mechanism based on localised melting.

3. MMC's have a greater propensity than un-reinforced material to stick to H13 steel. This has been observed frequently during extrusion trials.

4. Front end tearing was eliminated by removing choke at the die entry and reducing bearing length. The sample on the right in Fig. 3 was produced under these conditions.

5. Recurrence of type 3 tearing with the die of reduced bearing length occurred after extrusion of several billets. Deterioration in surface, first occurred for a few inches at the front of the billet but extended to several feet by the end of the extrusion run. Subsequent examination showed that the inlet to the die aperture had eroded during the run, producing a converging bearing. There was also a considerable amount of composite adhering to the exit portion of the bearing.

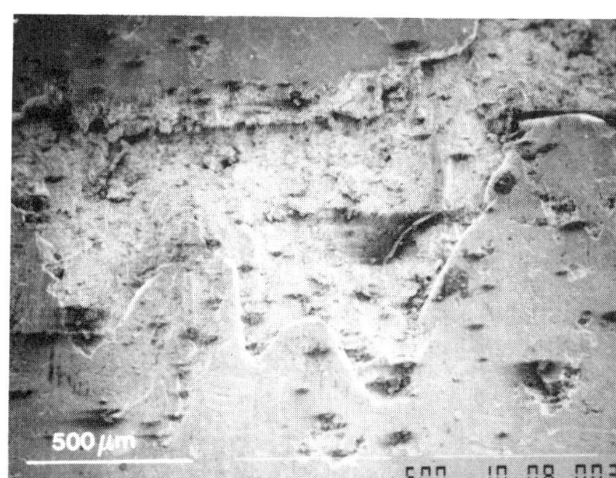

Fig. 4. Type 3 tearing. Surface peeled.

DIE WEAR

Reinforcing agents commonly used to produce MMC's are used, industrially, as abrasives. Knoop hardness values are of the order of 2000 - 2500, which is considerably higher than H13 nitrided steel (900 Knoop) the material most frequently used in dies for aluminum extrusions. It is therefore not surprising that the nitrided layer is completely removed from H13 dies, after the extrusion of less than 1000 ft. of composite. Figure 6 shows such a die. The wear pattern is also very nonuniform and appears to be a function of the extrusion ratio.

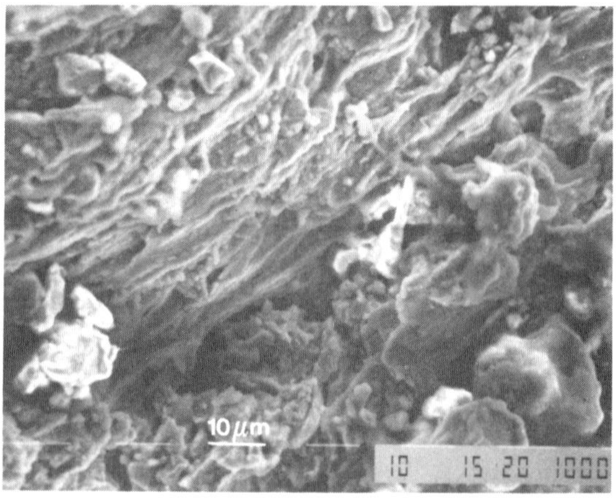

Fig. 5. Type 3 tearing. Ductile fracture.

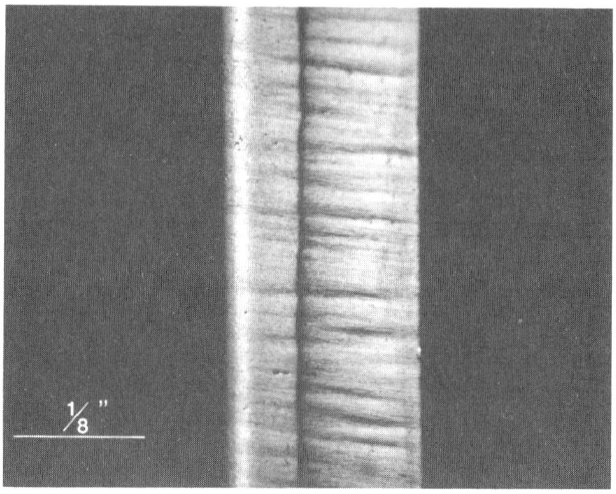

Fig. 6. Die bearing showing severe wear. Nitrided H13 steel. Entry at left.

At low extrusion ratios wear predominantly occurs at the entry of the die, while at higher ratios additional wear occurs part way along the bearing. This appears to follow positions of maximum pressure on the bearing to a first approximation. High surface hardness is an essential characteristic for MMC extrusion dies. Most die materials with hardness of the same order of magnitude as the reinforcing ceramic, are brittle and difficult to machine. This restricts their application to simple shapes; frequently they are available only for extrusion of round shapes.

Die materials for more complex profiles must combine wear resistance with machinability, preferably by wire EDM. This may be achieved by application of wear resistant coatings to softer, machinable dies, or alternatively with ceramics which have been "doped" to improve electrical conductivity to a level at which EDM machining can be employed.

SUMMARY

Particle reinforced MMC's offer the potential for volume production of low priced advanced engineering materials. At this early stage of their evolution, a wide range of extrusions have been produced using conventional aluminum extrusion equipment and tooling. These experiences have shown that future process improvements will result from developments in die materials and design.

Most of the work conducted to date has centered on existing Aluminum alloys with appropriate additions of various ceramic reinforcements. The basis alloys were developed to meet certain market needs, recognising the inherent properties of the un-reinforced material.

Addition of ceramic particles changes the properties of the material and also its behaviour during hot working. It therefore follows that the question of appropriate composition of the basis alloy needs to be re-evaluated, and alloys designed to optimise metal matrix composites.

The structure and nature of precipitates needs to be examined because of their potential for initiation of type 2 cracking, with a consequent reduction in productivity.

The issue of temperature-rise during extrusion also needs careful examination. It is likely that production of composites by indirect extrusion will be beneficial by eliminating container shear, with consequent reduction in pressure demands and associated reduction in temperature rise during extrusion. Liquid nitrogen cooling of the tooling should also prove beneficial by increasing heat transfer from the deformation shear zones.

Die Steels traditionally used for extrusion of monolithic alloys are worn rapidly by ceramic-reinforced composites. Suitable abrasion resistance may be achieved using ceramic or cermet bearing inserts, or by abrasion resistance coatings applied to the bearings of softer more readily machinable materials.

ACKNOWLEDGEMENT

The authors would like to recognise the contribution to this work by Dewi Evans who initiated the studies on extrusion and to Bryn Chamberlain, Nicole Chauvette, Paul Fortin, Christopher Jowett and Pierre Achim for their enthusiastic support and assistance with this work.

REFERENCES

Akeret, R., (1972), Rev. Met., 69, (9), 635-640.
Akeret, R., (1983), Aluminum, 59, (9), E276-280 and 59 (10), E355-360.
Akeret, R., and Strehmel, W., (1988), Proc. 4, Int. Ext. Tech. Sem., 2, 357-367.
Reiso, O., (1984), Proc. 3 Int. Ext. Tech. Sem., 1, 31-40.
Reiso, O., (1988), Proc. 4 Int. Ext. Tech. Sem., 2, 287-295.
Sheppard, T., M. P. Clode (1988), Proc. 4, Int. Ext. Tech. Sem., 2, 329-341.

CONTINUOUS SILICON CARBIDE FIBER REINFORCED METAL MATRIX COMPOSITES

Melvin A. Mittnick
John McElman

Textron Specialty Materials
Lowell, MA 01851

ABSTRACT

Continuous silicon carbide (SiC) fiber reinforced metals (FRM) have been successfully applied on numerous aerospace development programs fulfilling primary design objectives of high specific strength over baseline monolithic materials. This presentation will review the current state-of-the-art in silicon carbide fiber reinforced metals through discussion of their application. Discussion will include a review of mission requirements, program objectives and accomplishments to date employing these FRMs.

KEYWORDS

Continuous silicon carbide (SiC) fiber; reinforced metals (FRM); aerospace; specific strength; application.

INTRODUCTION

The term composite ususally signifies a combination of two or more constituent elements (in this case a high-performance, low-weight filler embedded in a metal) to form a bonded quasi-homogeneous structure that produces synergistic mechanical and physical property advantages over that of the base elements. Theoretically there are three types of composites: (1) a particulate based material formed by the addition of small granular fillers into a binder that generally derives an increase in stiffness but not strength; (2) a whisker/flake filler that realizes a greater proportion of the filler strength due to its higher aspect ratio and hence a greater ability to transfer load; and (3) a continuous fiber system (i.e., Fiber Reinforced Metal, FRM) that , due to fiber continuity, derives the full properties of the high-performance fiber (strength and stiffness).

There are various metal matrix systems commerically available today that conform to the three types of composites discussed above, the mechanical properties of these various systems vary significantly. The major import is the clear distinction between continuous fiber systems and the discontinuous (particulate and whisker/flake) systems. In the latter, only greater stiffness is generally realized, whereas in the former, both the modulus and the strength of the continuous reinforcing fiber are fully translated into the composite. In practice, if all of the filler strength is to be translated into the composites, then (a) the filler must have sufficient aspect ratio to transfer the load from "filler" to "filler" through the matrix (b) the filler particles must be in close proximity to each other to avoid large matrix distortions (high volume loading), and (c) a large number of "filler termination sites" placed in close proximity must be avoided to preclude stress concentrations that overload the filler and matrix. In practice it is very difficult to satisfy all of the requirments, and hence achieving the theoretical potential of the filler is close to impossible.

Although the mechanical properties of the continuous fiber composite system (FRM) are superior to those of the discontinuous systems, the ease of providing material isotrophy is a significant advantage, for there are many stiffness-controlled applications that benefit from discontinuous reinforced material. These applications generally involve complex geometries where it is difficult to position continuous fibers during the fabrication process. However, where FRM systems will usually provide the most efficiency (weight savings).

Silicon Carbide Fiber

Since, the advent of high-strength, high modulus, low-density boron fiber, the role of fibers produced by chemical vapor deposition (CVD) in the field of high-performance composites has been well established. Although best known for its use as a reinforcement is resin-matrix composites[1,2] boron fiber has also received considerable attention in the field of metal matrix composites[3,4,5]. Boron/aluminum was employed for tub-shaped truss members to reinforce the Space Shuttle orbiter structure, and has been investigated as a fan blade material for turbo-fan jet engines. There are drawbacks, however, rapid reaction of boron fiber with molten aluminum[6] and long-term degradation of the mechanical properties of diffusion-bonded/aluminum at temperatures greater than 480°C (900°F) preclude its use both for high-temperature applications and for potentially more economically feasible fabrication methods such as casting or low-pressure, high-temperature pressing. These drawbacks have led to the development of the silicon carbide (SiC) fiber.

Silicon Carbide Fiber Production Process

Continuous SiC filament is produced in a tubular glass reactor by CVD. The process occurs in two steps on a carbon monofilament substrate which is resistively heated. During the first step, pyrolytic graphite (PG) approximately 1 μm thick is deposited to smooth the substrate and enhance electrical conductivity. In the second step, the PG coated substrate is exposed to silane and hydrogen gases. The former decomposes to form beta silicon carbide (β SiC) continuously on the substrate. The mechancial and physical proeprties of the SiC filament are:

Tensile Strength = 3400 MPA (500 ksi)
Tensile Modulus = 400 GPa (60 msi)
Density = 3.045 g/cm^3 (0.11 lb./in^3)
Coefficient of Thermal Expansion
CTE = 1.5 x 10^{-6}/°F) (2.7 x 10^{-6}/°F)
Diameter = 140 mm (0.0056 in.)

Various grades of fiber are produced, all of which are based on the standard SiC deposition process described above where a crystalline structure is grown onto the substrate. The β SiC is present as such across all of the fiber cross-section except for the last few microns at the surface. Here, by altering the gas flow in the bottom of the tubular reactor, the surface composition and structure of the fiber are modified by, first, an addition of amorphous carbon that heals the crystalline surface for improved surface strength, followed by a modification of the silicon-to-carbon ratio to provide bonding with the metal.

Processing Considerations

As in any vapor deposition or vapor transport process, temperature control is of utmost importance in producing CVD SiC fiber. The Textron process calls for a peak deposition temperature of about 1300°C (2370°C).

Temperature significantly above this temperature cause rapid deposition and subsequent grain growth, resulting in a weakening of tensile strength. Temperatures significantly below the optimum cause high internal stresses in the fiber, resulting in a degradation of metal matrix composite properties upon machining transverse to the fiber(7).

Substrate quality is also an important consideration in SiC fiber quality. The carbon monofilament substrate, which is melt-spun coal tar pitch, has a very smooth surface with occassional surface anomalies. If severe enough, the surface anamoly can result in a localized area of irregular deposition of PG and SiC which is a stress-raising region and a strength-limiting flaw in the fiber. The carbon monofilament spinning process is controlled to minimize these local anamalies sufficiently to guarantee routine production of high-strength $>$3450 MPa ($>$ 500 ksi) SiC fiber.

Another strength-limiting flaw which can result from an insufficiently controlled CVD process is the PG flaw(8). This flaw results from irregularities in the PG deposition. Two causes of PG flaws are: (1) disruption of the PG layer due to an anomaly in the carbon substrate surface and (2) mechanical damage to the PG layer prior to the SiC deposition. PG flaws often cause a localized irregularity in the SiC deposition, resulting in a bump on the surface. Poor alignment of the reactor glass can result in mechanical damage to the PG layer by abrasion. A series of PG flaws results in what is called a "string of beads" phenomenon at the surface of the fiber. The mechancial properties of such fiber are severely degraded. These flaws are minimized by careful control of the PG deposition parameters, proper reactor alignment, and the minimization of substrate surface anomalies.

The surface region of Textron's SiC fibers is typically carbon rich. This region is important in protecting the fiber from surface damage and subsequent strength degradation. An improper surface treatment or mishandling of the fiber (e.g., abrasion) can result in strength-limiting flaws at the surface. Surface flaws can be identified by an optical examination of the fiber fracture face. These flaws are minimized by proper process control and handling of the fiber (minimizing surface abrasion).

Typical mechanical properties of the Avco CVD SiC fiber consist of average tensile strength of 3790 to 4140 MPa (550 to 600 ksi) and elastic moduli of (400 to 415 GPa) (58 to 60 msi). A typical tensile strength histogram shows an average tensile strength of 4000 MPa (580 ksi) with a coefficient of variation of 15%.

Fiber Variation

The surface region of the SiC fiber must be tailored to the matrix. SCS-2 has a 1 μm carbon-rich coating that increases in silicon content as the outer surface is approached. This fiber has been used to a large extent to reinforce aluminum. SCS-6 has a thicker (3 μm) carbon-rich coating in which the silicon content exhibits maxima at the outer surface and 1.5 μm from the outer surface. SCS-6 is primarily used to reinforce titanium.

SCS-8 has been developed as an improvement over SCS-2 to give better mechanical properties in aluminum composites transverse to the fiber direction. The SCS-8 fiber consists of 6 μm of very fine-grained SiC, a carbon-rich region of about 0.5 μm, and a less carbon-rich region of 0.5 μm.

Cost Factors

From an economic standpoint, SiC is potentially less costly than boron for three reasons: (1) the carbon substrate used for SiC is lower cost than the tungsten used for the boron; (2) raw materials for SiC (chlorosilanes) are less expensive than boron trichloride, the raw material for boron; and (3) deposition rates for SiC are higher than those for boron, hence more product can be made per unit.

Composite Processing

The ability to readily produce acceptable SiC fiber reinforced metals is attributed directly to the ability of the SiC fiber to (a) readily bond to the respective metals and (b) resist degradation of strength while being subjected to high-temperature processing. In the past, boron and Borsic™ fibers have been evaluated in various aluminum alloys and, unless complex solid-state (low-temperature, high-pressure) diffusion bonding procedures were adopted, severe degradation of fiber strength has been observed. Likewise in titanium, unless fabrication times are severely curtailed, fiber/matrix interactions produce brittle intermetallic compounds that again drastically reduce composite strength.

In contract, the SCS grade of fibers has surfaces that readily bond to the respective metals without the destructive reactions occurring. The results is the ability to consolidate the aluminum composites using less complicated high-temperature casting and low-pressure (hot) molding. Also for titanium composites, the SCS-6 filament has the ability to withstand long exposure at diffusion bonding temperatures without fiber degratation. As a result, complex shapes with selective composite reinforcement can be fabricated by the innovative superplastic forming/diffusion bonding (SPF/DB) and hot isostatic pressing (HIP) processes.

In the following discussion, further details of fabrication techniques will be discussed; however, first the production if intermediary products such as preforms and fabrics used in the component fabrication are described. These are required to simplify the loading of fibers into a mold and to provide correct alignment and spacing of the fibers.

Composite Preforms and Fabrics

"Green tape" is an old system consisting of a single layer of fibers that are collimated/spaced side by side across a layer, held together by a resin binder, and supported by a metal foil. This layer constitutes a prepared (in organic composite terms) that can be suquentially "laid up" into the mold or tool in required orientations to fabricate laminates. The laminate processing cycle is then controllled so as to remove the resin (by vacuum) as volatilization occurs. The method normally used to wind the fibers onto a foil-covered rotating drum, overspraying the fibers with the resin, followed by cutting the layer from the drum to provide a flat sheet of "prepreg".

"Plasma-sprayed aluminum tape" is a more advanced "prepreg" similar to "green tape" but replaces the resin binder with a plasma-sprayed matrix of aluminum. The advantages of this material are (a) the lack of possible contamination for resin residue and (b) faster material processing times because of the hold time required to ensure volatilization and removal of the resin binder is not required. As with the green tape system, the plasma-sprayed preforms are laid sequentially into the mold as required as pressed to the final shape.

"Woven fabric" is perhaps the most interesting of the preforms being produced since it is a universal preform concept that is suitable for a number of fabrication processes. The fabric is a uniweave system in which the relatively large-diameter SiC monofilament are held straight and parallel, collimated at 100 to 140 filaments per inch and held together by a cross-weave of a low-density yarn or metallic ribbon. There are now two types of looms that can be specially modified to produce the uniweave fabric. The first is a single-arm Rapier-type loom capable of producing continuous 60 in. wide fabric with the SiC filament oriented in the "fill" (60 in. width) direction. The other is a shuttle-type loom in which the SiC monofilaments are oriented in the continuous direction with the light-weight yarn a metal ribbon in the "fill" axis. The shuttle loom can weave fabric up to 6 in. wide. Various types of cross-weave materials have been used, such as titanium, aluminum, and ceramic yarns.

Processing Methods

"Investment casting" is a fabrication technique that has been used for many years but is still universally accepted as a very cost effective method for producing complex shapes.

The aerospace business has some time rejected aluminum castings due to the low strengths that are typically achieved; however, with a material that is now fiber dependent and not predominately matrix controlled, significant structural improvements have been derived so as to revive the interest in this low-cost procedure. The investment casting technique, sometimes called the "Lost Wax" process, utilizes a wax replicate of the intended shape to form a porous ceramic shell mold where, upon removal of the wax (by steam heat) from the interior, a cavity for the aluminum is provided. The mold includes a funnel for gravity pouring, with risers and gates to control the flow of the aluminum into the gage section. A seal is positioned around the neck of the funnel, allowing the body of the mold to be suspended into a vacuum chamber. By a combination of gravity and vacuum (imposed through the porous walls of the shell mold), the total cavity is filled with aluminum.

The SiC fibers are installed in mold using the fabric described above by either first placing the fabric into the wax replica or simply splitting open the mold and inserting the fabric into the cavity after the wax has been removed. At present, the latter approach is usually used due to contamination and oxidation of the fibers during wax burnout. At some future date, the necessary techniques for including the fiber in the wax (thereby reducing the processing costs) will probably be developed.

"Hot molding" is a term coined by Textron to describe a low-pressure hot pressing process that is designated to fabricate shaped SiC aluminum parts at significantly lower cost than the typically diffusion bonding, solid-state process. As stated previously, the SCS-2 fibers can withstand molten aluminum for long periods; therefore, the molding temperature can now be raised into the liquid-plus-solid region of the alloy to ensure aluminum flow and consolidation at low pressure, thereby negating the requirement for high-pressure die molding equipment.

The best way of describing the hot molding process is to draw an analogy to the autoclave molding of graphite epoxy where components are molded in an open-faced tool. The mold in this case is a self-heated, slip-cast ceramic tool embodying the profile of the finished part. A plasma sprayed aluminum preform is laid into the mold, heated to a near molten aluminum temperature, and pressure consolidated in a autoclave by a "metallic" vacuum bag. The mold can be profiled as required to produce near net shape parts including tapered thicknesses and section geometry variations.

"Diffusion bonding of SiC/titanium" is accomplished by hot pressing (diffusion bonding) technology, using fiber preforms (fabric) that are stacked together between titanium foils for consolidation. Two methods are being developed by aircraft and engine manufacturers to manufacture complex shapes. One method is based on the HIP technology, and uses a steel pressure membrane to consolidate components directly from the fiber/metal preform layer. The other method requires the use of previously hot pressed SCS/titanium laminates that are then diffusion bonded to a titanium substrate during subsequent super-plastic forming operations.

This is typical of the first fabrication procedure noted above. The fiber preform is placed onto a titanium foil. This is then spirally wrapped, inserted, and diffusion bonded onto the inner surface of a steel tube using a steel pressure membrane. The steel is subsequently thinned down and machined to form the "spline attachment" at each end. Shafts are also being fabricated for other engine fabricators without the steel sheath.

The concept developed for superplastic forming of hollow engine compressor blades. Here the SCS/titanium lamiantes are first diffusion bonded in a press. These are then diffusion bonded to form monolithic titanium sheets, with "stop-off" compounds selectively positioned to preclude bonding in desired areas. Subsequently, the "stack-up" is sealed into a female die. By pressurizing the interior of the "stack-up", the material is "blown" into the female die to form the desired shape, stretching the monolithic titanium to form the internal corrugations.

These processes typically require long times at high temperature. In the past, all of the materials used have developed serious matrix-to-fiber interactions that seriously degrade composite strength, SCS-6, however, due to its unique surface characteristics, delays intermetallic diffusion and retains its strength up to 7 hours in contact with titanium at 925°C (1700°F).

Composite Properties

Since continuous SiC reinforced metals have been in existence for a relatively short period of time, the property data base has been developed sporadically over this period depending on funded applications.

SiC/Aluminum

The most mature of the SiC reinforced aluminum (SiC/Al) consolidation approaches is hot molding, and therefore the greatest mechanical property data base has been developed using this material. The design data base for hot molded SCS-2/6061 aluminum includes static tension and compression properties, in-plane and interlaminar shear strengths, tension-tension fatigue strengths (SN curves), flexure strength, notched tension data, and fracture toughness data. Most of the data have been developed over a temperature range of -55°C to 75°C (-65°F to 165°F) with static tension test results up to 480°C (900°F). As can be seen from these data, the inclusion of a high performance, continuous SiC fiber in 6061 aluminum yields a very high strength 1378 MPa (+200 ksi) high-modulus 207 GPa (30 msi) anisotropic composite material having a density just slightly greater, 2.85 g/cm^3 (0.103 lb/in^3) than baseline aluminum. As in organic matrix composites, cross or angle plying produces a range of properties useful to the designer.

The property data developed to date for investment case SCS/aluminum have been limited to static tension and compression. Fiber volume fractions are lower (40% maximum) than the hot molded laminates (47% typical) due to volumetric constraints in dry loading the shell molds; however good rule-of-mixture (R.O.M.) tensile strengths are excellent compression strengths (twice the tensile strength) are being achieved.

The use of 6061 aluminum as the matrix material and the capability of the SiC fiber to withstand molten aluminum has made conventional fusion melding a viable joining technique. Although welded joints would not have continuous fiber across the joint to maintain the very high strengths of the composite, baseline aluminum weld strengths can be obtained. In addition to fusion welding traditional molten salt bath dip brazing has been demonstrated as an alternative joining method.

An important consideration for emerging materials is corrosion resistance. Testing has been performed on SCS-2/6061 hot molded material at the David W. Taylor Naval Ship R&D Center(9) under marine atmosphere, ocean splash/spray, alternate tidal immersion, and filtered seawater immersion conditions for periods of 60 to 365 days. The SCS/aluminum material performed well in all tests, exhibiting no more than pitting damage comparable to the baseline 6061 aluminum alloy.

SiC/Titanium

SCS-6/Ti 6-4 composites were originally developed at high temperature. There has been a successful program to reinforce the beta titanium alloy 15-3-3-3 with SCS fiber and superior composite properties have been achieved at 1585 to 1930 MPa (230 to 280 ksi) tensile strength(10).

Fabrication of titanium parts has been accomplished by diffusion bonding and HIP. The HIP technique has been particularly successful in the forming of shaped reinforced parts (e.g., tubes) by the use of woven SiC fabric as a preform. The high-strength, high-modulus properties of SCS-6/Ti represents a major improvement over B_4C-B/Ti composites in which the modulus of composite is increased relative to the matrix, but the tensile strength is not as high as would be predicted by the rule of mixture.

SiC/Magnesium and SiC/Copper

SCS-2 has been successfully cast in magnesium(11). Under a recent Naval Surface Weapons Center (NSWC) program(12), development of SiC-reinforced copper has been initiated. At present, about 85% of R.O.M. strengths have been achieved at a volume fraction of 20 to 33%.

Applications

The very high specific mechanical properties of SiC reinforced metal matrix composites have generated significant interest within the aerospace industry, and as a result many research and development programs are now in progress. The principal area of interest is for high-performance structures such as aircraft, missiles, and engines. However, as more and more systems are developing sensitivities to "performance" and "transportation weight", other and less sophisticated applications for these newer materials are being considered. The following paragraphs describe a few of these applications.

SiC/aluminum wind structural elements are currently being developed. Ten foot long "Zee" shaped stiffeners are to be hot molded and then subsequently riveted to wing planks for full-scale static and fatigue testing. Experimental results obtained to date have verified material performance and the design procedures utilized.

SiC/aluminum bridging elements are being developed for the Army to be used for the lower chord and the king post of a 52m assault bridge. Future plans call for development of the top compression tubes of the new Tri-Arch bridge being developed by Fort Belvoir.

Sic/aluminum internally stiffened cylinders are being developed using the previously discussed investment casting process. A wax replica is first fabricated that incorporates the total shape of the shell including internal ring stiffeners and the end fittings. The fabric containing the SiC fibers is then wound onto the inner shell mold, the two halves of the shell are remated and sealed, and infiltration of the aluminum is then accomplished.

SiC/aluminum fins for high-velocity projectiles are in the process of evaluation.

SiC/aluminum missile body casings have been fabricated utilizing a unique variation of filament winding. An aluminum motor case is first produced in the convetional manner; this time, however, with significantly less wall thickness than normally required. The casing is then overwrapped with layers of SiC fibers, where each layer is sprayed with a plasma of aluminum to build up the matrix thickness. No final consolidation of the 90% dense system is required, for the hydrostatic internal pressure on the circular body imposes no (or very minimal) shear stresses on the matrix. It is hoped that further development of this technique will permit full consolidation of the matrix by vacuum bagging the total section and hot isostatic pressing.

SiC-titanium drive shafts are being developed and fabricated by the hot isostatic pressing process described previously. These are generally for the core of an engine, requiring increased specified stiffness to reduce unsupported length between bearings and also to increase critical vibratory speed ranges. SiC-Ti tubes up to 5 ft. in length have been fabricated and have incorporated into their ends a monolithic load transfer section for ease of welding to the splined or flanged connections.

SiC discs for turbine engines are currently under development. Initially discs were made by winding SiC-Ti monolayer over a mandrel followed by hydrostatic consolidation (hot isostatic pressing). The concept now being developed utilizes a "doily" approach where single fibers are hoop wound between titanium metal ribbons to be subsequently pressed together in the axial direction, reducing the breakage of fibers and simplyfing the production of tapered cross-sections.

Selectively reinforced SiC-titanium hollow fan blades are being developed.

SiC/copper materials have been fabricated and tested for high-temperature missile applications. Also, SiC/bronze propellers have been cast for potential Navy applications where more efficient/quiet propellers are required.

FUTURE TRENDS

The SiC fiber is qualified in aluminum, magnesium, and titanium. Copper matrix systems are under development and reasonably good results have been obtained using the higher temperature titanium aluminides as matrix materials. The SCS-6 fiber demonstrates high mechanical properties to above 1400°C (2550°F). It is natural, then, to project systems such as SiC-nickel aluminides/iron aluminide/super-alloys, etc., all of which, on an R.O.M. basis at least, project very useful properties for "engine" and hypersonic vehicle applications. Work required in this area includes diffusion barrier coatings and matrix alloy modifications to facilitate high-temperature fabrication processes. Also required is the detailed investigation of any detrimental thermal/mechanical cycling effects that may occur as a result of the mismatch in thermal expansion coefficients between matrix and fiber.

REFERENCES

1. DeBolt, H., "Boron and Other Reinforcing Agents," in Lubin, G., ed., Handbook of Composites, Van Nostrand Reinhold Co., New York, 1982, Chapter 10.

2. Krukonis, V.J., "Boron Filaments," in Milweski, J.V., and Katz, H.S., ed., Handvook of Fillers and Reinforcements for Plastics, Van Nostrand Reinhold Co., New York, 1977, Chapter 28.

3. McDaniels, D.L., and Ravenhall, R., "Analysis of High-Velocity Ballistic Impact Response of Boron/Aluminum Fan Blades," NASA TM-83498, 1983.

4. Salamme, C.T., and Yokel, S.A., "Design of Impact-Resistant Boron/Aluminum Large Fan Blades," NASA CR-135417, 1978.

5. Brantley, J.W., and Stabrylla, R.G., "Fabrication of J79 Boron/Aluminum Compressor Blades," NASA CR-159566, 1979.

6. Wolff, E., "Boron Filament, Metal Matrix Composite Materials," AF33 (615)3164.

7. Suplinskas, R.J., "High Strength Boron", NAS-3-22187, 1984.

8. Aylor, D.M., "Assessing the Corrosion Resistance of Metal Matrix Composite Materials in Marine Enviornments," DTNSRDC/SMME-83/45, 1983.

9. Kumnick, A.J., Suplinskas, R.J., Grant, W.F., and Corine, J.A., "Filament Modification to Provide Extended High Temperature Consolidation and Fabrication Capability and to Explore Alternative Consolidation Techniques," N00019-82-C-0282, 1983.

10. Corine, J.A., and Murty, Y., "Evaluation of Silicon Carbide/Magnesium Reinforced Castings," DASG46-80-C-0076, 1983.

11. Marzik, J.V., and Kumnick, A.J., "The Development of SCS/Copper Composite Material," N60921-83-C-0183, 1984.

STABILITY OF SiC IN MOLTEN ALUMINIUM

D.J. LLOYD AND E. DEWING

Alcan International Limited
Kingston Research and Development Centre
P.O. Box 8400, Kingston, Ontario, K7L 5L9, Canada

ABSTRACT

In metal matrix composites the stability of the reinforcement in the matrix is critical for optimum properties. This is particularly important in those situations where the reinforcement is exposed to molten metal, where reaction rates to form intermetallic compounds can be very high. In this paper the stability of unprotected SiC particulate in contact with molten aluminium alloys is considered. SiC can react with molten aluminium to form aluminium carbide, which results in a deterioration in the microstructure and properties of the composite. The reactions which occur are considered, together with methods of following the kinetics of the reaction. It is shown that by measuring the liquidus temperature of the composite the kinetics of the reaction can be followed, and this approach gives comparable results to other methods of following the reaction, such as XRD.

KEY WORDS

Composites, aluminium alloys, SiC, particulate, reactivity, kinetics, aluminium carbide, liquidus, XRD.

INTRODUCTION

A very important factor in obtaining the optimum properties in a composite is to control any reaction between the reinforcement and the metal matrix of the composite. This is particularly important in situations, such as molten metal infiltration and composite casting, where the reinforcement is exposed to molten metal. Under these conditions the reaction rate to form intermetallics at the interface between the reinforcement and the matrix can be very high. The degradation in strength that occurs with interaction in Al-C fibre composites has been recently reported by Kohara and Muto (1985). Interaction between the reinforcement and the matrix results in intermetallic compound formation at the interface, and degradation in the reinforcement and interfacial bond strengths. The influence of fibre matrix bond strength on the composite properties has been modelled by Ochiai and Osamura (1985). Therefore, it is important to assess the stability of any reinforcement in the matrix of interest and consider the reactions which can occur.

In this paper untreated SiC particulates in aluminium alloys is considered. Warren and Anderson (1984) pointed out that SiC can react with Al to form aluminium carbide, Al_4C_3,

$$4Al + 3SiC \rightarrow Al_4C_3 + 3Si \quad (1)$$

Aluminium carbide is a brittle compound which has poor corrosion resistance and is expected to degrade the composite. In addition the reaction increases the Si content of the alloy, which modifies the basic metallurgy of the matrix alloy.

Reaction between Al and SiC has been observed in Nicalon SiC fibres by Viala and co-workers (1987), and in SiC by Iseki and co-workers (1984), Cheryshova and Rebrov (1986) and Moshier and co-workers (1987). The extent of the reaction has either been followed by X-ray diffraction methods or by chemical analysis for aluminium carbide after dissolving away the matrix. In the present paper an alternative approach utilising changes in the melting point of the alloy is considered.

EXPERIMENT

Several SiC particulate reinforced aluminium alloy composites have been reheated to above the melting point, held for different times and then resolidified. The extent of the reaction has been examined by three methods:

(1) XRD. The resolidified composites were ground in a pestle and mortar and powder diffraction X-ray patterns taken to measure the intensities of the Si and Al_4C_3 peaks.

(2) Chemical analysis. Using a solution of nitric and phosphoric acids, the matrix is dissolved without attacking the SiC particles and the resulting solution analysed by inductively coupled plasma emission spectroscopy - ICP.

(3) Liquidus measurements. Lloyd and Jin (1988) have suggested that changes in the melting point of the composite can be used to assess the extent of reaction of SiC. From equation (1) it is seen that the reaction results in the formation of Si, which will change the liquidus of the alloy. If the influence of Si on the liquidus of the original matrix alloy is known, the change in the Si content of the matrix resulting from reaction with the reinforcement can be obtained.

RESULTS

Fig. 1 shows the changes in the aluminium carbide and Si XRD peak intensities after holding a 7075 - 15 vol% SiC composite for different times at 675°C. The peak intensities increase very rapidly in the first hour of holding above the liquidus but then the reaction rate drops to a very low level for longer times. This behaviour is also reflected in changes in the melting point of the alloy, as seen from Fig. 2. If, as expected, the changes in the liquidus are due to increases in the Si content of the alloy, the microstructures after holding above the liquidus should show an increase in the Si phases present. To clearly demonstrate this a pure Al-SiC composite was held for 5 minutes at 800°C, solidified and examined micrographically, Fig. 3. The presence of numerous Si plates can be seen in the microstructure, confirming the increase in Si level, together with Al_4C_3 crystals at the particle-matrix interface.

These results show that monitoring the reaction between SiC and the matrix by using XRD and liquidus measurements show the same trends, and since liquidus measurements are relatively easy to do, using standard cooling curve or differential scanning calorimetry methods, it would be useful to confirm that they can give quantitative information. To do this a composite of 6061 - 20 vol% SiC was held for various times at 800°C, solidified, the liquidus measured and the Si content of the matrix analysed chemically. From the liquidus temperature, and knowing the influence of Si on the liquidus from the Al-Mg-Si phase diagram, the Si content of the composite can be calculated (Lloyd and Jin, 1988). For a 20 vol% SiC composite the total Si available from breakdown of the reinforcement is 16.1%. Fig. 4 shows that the liquidus measurements tend to fall slightly below the chemical analysis results but the agreement is quite good. This data also confirms that the reaction rate drops to a very low level after the first hour, and in this 6061 matrix it saturates for the sensitivity available with these analytical methods.

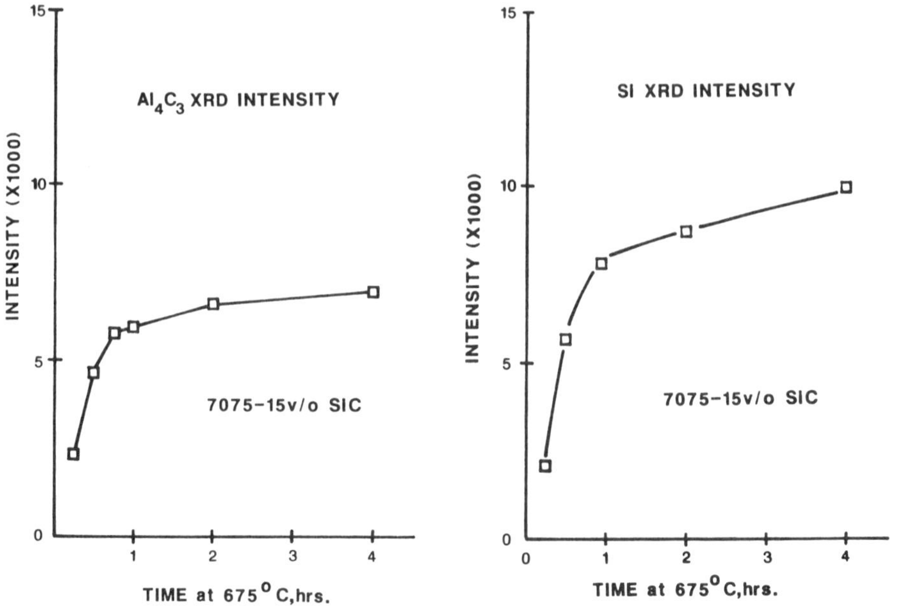

(a) for Al_4C_3 (b) for Si

Fig. 1. Changes in the XRD peak intensities with time in 7075 - 15 v/o SiC.

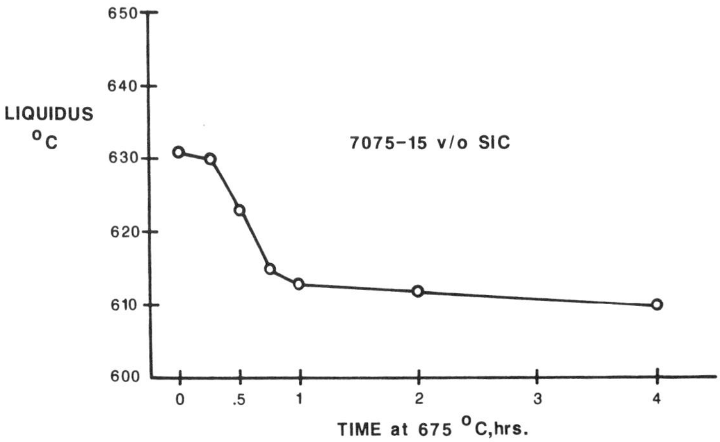

Fig. 2. Changes in the melting point with time at 675°C for 7075 - 15 v/o SiC.

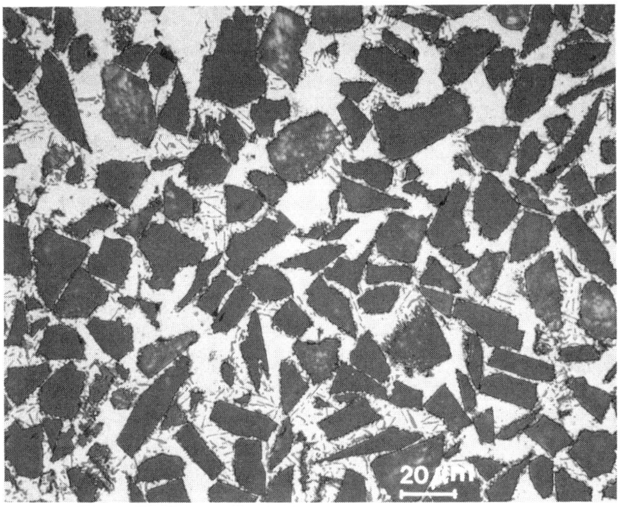

Fig. 3. Microstructure of Al - SiC composite after holding for 5 minutes at 800°C.

DISCUSSION

These experiments show that reaction between SiC and molten aluminium alloys can be assessed by monitoring the changes in the melting point, since the Si content of the alloys increase with the extent of the reaction. The experiments also show that the reaction rate is very high initially, but then decreases to a very low level after about one hour. It is of interest to see if the very low reaction rate at long times is due to the Si level reaching such a high level that aluminium carbide formation according to equation (1) becomes thermodynamically unfavourable.

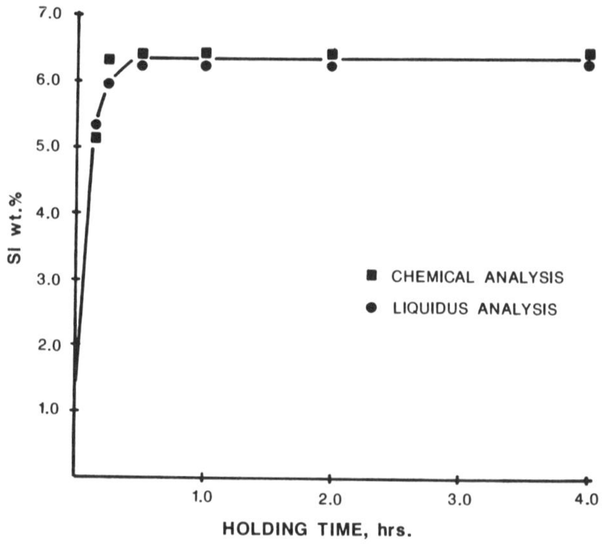

Fig. 4. Increase in the Si content of the matrix for 6061 - 20 v/o SiC composite held at 800°C.

The excess molar free energy for the aluminium-silicon system is given by Murray and McAlister (1984) as

$$\Delta G^{xs} = x(1 - x)\ [A + B(1 - 2x) + C(1 - 6x + 6x^2)]\ \text{J/mol} \qquad (2)$$

where x = atom fraction of Si $\quad A = -10{,}695.4 - 1.823T$

$B = -4{,}274.5 + 3.044T \quad C = 670.4 - 0.460T$

The activity coefficient is given by

$$RT\ln\gamma = \Delta G^{xs}/\partial x \qquad (3)$$

and the partial derivative is

$$\partial\Delta G^{xs}/\partial x = \Delta G^{xs} + (1 - x)(d\Delta G^{xs}/dx) \qquad (4)$$

From equations (2), (3) and (4) the concentration of Si in liquid Al in equilibrium with aluminium carbide and silicon carbide can be calculated at various temperatures. Fig. 5 shows a plot of the results from such a calculation, the equilibrium Si level increasing from about 8.4 wt% at 607°C to 12.8 wt% at 827°C. So if the Si level reaches these concentrations at the different temperatures, the reaction will become thermodynamically unfavourable, and will stop. Fig. 5 also shows the results obtained from holding experiments at different temperatures. For these experiments a holding time of 2 hours was used, which from the previous plots is a sufficiently long time for the reaction to essentially saturate. Clearly, the reaction is not proceeding to a Si level which attains thermodynamic equilibrium. However, it should be noted that this conclusion assumes that the Si is homogeneously distributed through the matrix and ignores the possibility of high local concentrations in the neighbourhood of the SiC particles. Since the matrix is liquid, diffusivities should be sufficiently high to rapidly remove any concentration gradients, but this cannot be independently examined. On subsequent solidification, Si particles are associated with the SiC particulate, but this is inevitable in these alloys due to the nature of the solidification process. The reaction may also be inhibited by the crystals of aluminium carbide formed at the interface by the reaction of the SiC, and large aluminium carbide crystals are present in reacted material (Lloyd and co-workers, 1988). Finally, it should be noted that the kinetics of aluminium carbide formation will depend on a variety of factors in addition to the temperature. As seen from the thermodynamic calculations, the Si level in the matrix will be very important, and at a sufficiently high concentration of Si the reaction will not occur. In addition, the reaction occurs at the surface of the SiC reinforcing particles, so the surface area ie. SiC particle size and volume fraction will have a major effect, the finer the particle size the greater the extent of reaction.

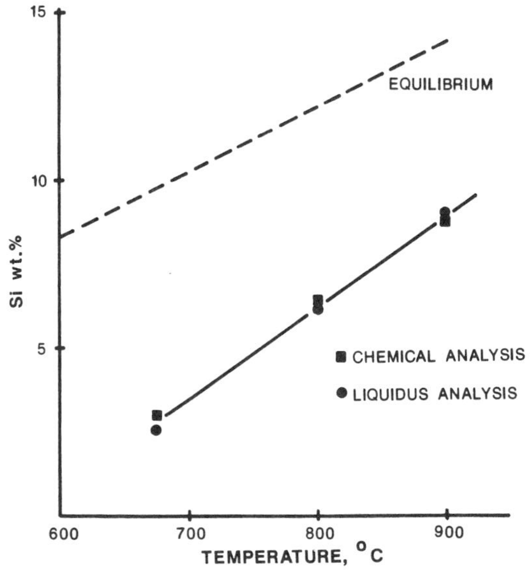

Fig. 5. Plot showing the equilibrium Si content necessary to prevent Al_4C_3 formation in an Al-SiC composite, and the Si content obtained after holding a 20 vol.% composite for 2 hours at different temperature.

SUMMARY

The reactivity of SiC particles in an aluminium alloy matrix has been examined. It has been shown that monitoring the melting point of the composite is a straightforward method of following the extent of the SiC reaction, and gives results in good agreement with other methods, such as chemical analysis of the matrix. At temperatures above the liquidus of the matrix the SiC particles in the composite react to form aluminium carbide and Si, the reaction rate being very high initially, but decreasing to a very low rate after about one hour. This decrease to a very low rate does not appear to be due to the Si concentration of the matrix achieving the equilibrium level necessary to inhibit the reaction. Some other shielding phenomenon, such as the formation of aluminium carbide crystals at the interface, maybe limiting the reaction.

ACKNOWLEDGEMENTS

The authors are grateful to Alcan International for permission to publish this work.

REFERENCES

Chemyshova, T. A., and A. V. Rebrov (1980). *J. of Less-Common Met., 117,* 203-207.

Iseki, T., T. Kameda and T. Maruyama (1984). *J. of Mater. Sci., 19,* 1692-1698.

Kohara, S., and N. Muto (1985). ICCM-V, *The Metallurgical Society, Inc.,* 631-638.

Lloyd, D. J., and I. Jin (1988). *Met. Trans.,* in press.

Lloyd, D. J., H. Lagace, A. McLeod and P. L. Morris (1988). *Brown Symposium on Metal Matrix Composites,* in press.

Moshier, W. C., J. S. Ahearn and D. C. Cooke (1987). *J. of Mater. Sci., 22,* 115-122.

Ochiai, S., and K. Osamura (1985). *Z. Metallkde, 76,* 485-491.

Viala, J. C., F. Bosselet, P. Fortier and J. Bouix (1987). *ICCM-V and ECCM-2, Elsevier-Applied Science,* 2.146-2.155.

Warren, R., and C. H. Anderson (1984). *Composites, 15,* 101-111.

THE INTERACTION BETWEEN VARIOUS Al ALLOYS AND SiC

K. Kannikeswaran and R. Y. Lin
Department of Materials Science and Engineering
University of Cincinnati
Cincinnati, OH 45221-0012

ABSTRACT

The reaction between SiC and Al alloys has been investigated. Composite specimens prepared under vacuum were examined using a Scanning Electron Microscope (SEM) and Energy Dispersive X-ray Analyzer (EDAX) on both the cross section and the SiC surface. Definite reactions were observed to have occurred at the interface between SiC and Al. EDAX scans across the interface showed that silicon penetrates into the Al alloy matrix. An examination of the surfaces of SiC in contact with Al has revealed a considerable modification in the surface morphology of SiC. Alloy systems investigated include 6061, 2024 and 7075 Al alloys and a series of Al-Cu and Al-Si-Cu genetic alloys. The segregation of alloying elements at the interface has been explained using a ternary Al phase diagram. The fact that there is a significant amount of Si observed at the interface between SiC and Al in the vicinity of the SiC surface supports a physico-chemical model for the interfacial reaction.

KEYWORDS

Composite Materials; Interfacial Reactions; Al/SiC Composite; Physical Chemistry; Interface Analysis; Interfacial Reaction Modeling

INTRODUCTION

Compatibility between the matrix and the reinforcement is one of the most important criteria in the development of composite materials. Adequate bonding between the matrix and the fibers is essential to enable maximum loading of fibers. In choosing a reinforcement for a matrix, any potential reaction between the matrix and the reinforcement must be carefully evaluated. Thus, a study of the compatibility between the matrix and the reinforcement, as well as the interface between the two, is essential in developing advanced engineered materials.

During the fabrication of aluminum/silicon carbide composite materials, the reaction at the aluminum and silicon carbide interface depends on such fabrication parameters as temperature, atmosphere, and the chemical composition of both the aluminum matrix and the SiC reinforcement. Iseki and others (1984) examined the interfaces between Al and various types of SiC and found that the reaction between aluminum and silicon carbide with free silicon is not as extensive as the reaction between aluminum and silicon carbide without free silicon. Distinct reaction zone layers were observed through a transmission electron microscope. Using an electron diffraction technique, SiC, Al_4C_3, and Al with Si layers were identified at the interface between aluminum and SiC without free Si. Meanwhile, no Al_4C_3 product was detected when free silicon was present in SiC.

Restall and co-workers (1970) prepared SiC fiber/Al composites by vacuum infiltration of molten aluminum through a silicon carbide fiber bundle. At the matrix-fiber interface, they found a less than 3 ιm-thick reaction zone, even after holding the fibers for 15 minutes in molten aluminum at 680^oC. The microprobe mapping of a cross section around the fibers revealed that, in the reaction zone, both Si and Al coexisted. A specific reaction product was not proposed.

Arsenault and others (1984) examined the fracture surface of Al/SiC whisker composites produced by ARCO SILAG. The samples were fractured and examined in a scanning Auger microprobe. The fresh fracture surface showed 100% Al, even in the area of the whiskers, indicating that the fracture was through the Al layers. After sputtering the sample with an ion jet, Si and C signals were obtained, but there was no indication of SiO_2, Al_2O_3 or Al_4C_3.

The objectives of the composite program in this laboratory are to investigate the interfacial reactions in the composite systems and to develop characterizing techniques in the composite intrfaces. In a previous paper, Kannikeswaran and Lin (1987) examined the characteristics of the interfaces between aluminum and silicon carbide, with an emphasis on the effect of impurity elements in a commercial Al 1100 alloy. It appears that the trace amount of silicon in the 1100 alloy slows down the interfacial reaction. Thermodynamic considerations suggest that although Al_4C_3 may not always be one of the reaction products, SiC surfaces will always be modified in contact with aluminum. This work extends the study of the interfacial reaction between Al and SiC into much more complicated aluminum alloys, such as 2024, 6061 and 7075, as well as genetic Al-Si-Cu alloys.

EXPERIMENTAL TECHNIQUES

Silicon carbide filaments, approximately 1mm x 1mm x 15mm in size, were cut from 12.5 mm diameter by 12.5 mm height silicon carbide buttons using a diamond wafering blade. These silicon carbide buttons, supplied by SOHIO, were sintered in nitrogen and contained less than 0.5 percent boron as the sintering aid material and less than 5.0 percent porosity. Before casting, the filaments were held in position in a graphite mount and then packed around with chips of aluminum alloys in an alumina tube. The entire set up was enclosed in a quartz reaction chamber, which could be evacuated using a mechanical (vacuum) pump connected to it through a series of copper tubes. The chamber was maintained at 0.1 pa throughout the experiment.

The heating was provided with a Marshall furnace using an Eurotherm Model 984 temperature controller. The furnace was preheated to the desirable temperature at a location below the reaction chamber. After evacuating the system overnight, the heated furnace was raised up to a preset position to quickly heat up the specimen. Since the reaction chamber and sample mass are both small, the specimen normally reaches the experimental temperature in about five minutes. The reaction was stopped by quenching the reaction chamber at the end of a preset reaction time by quickly lowering down the furnace. The composite specimen was then removed and prepared for examination with a scanning electron microscope (SEM, Cambridge Stereoscan 600) and Energy Dispersive X-ray Analyzer (EDAX) on the cross sections and SiC side of the interface.

RESULTS

There are three groups of Al alloys used in this study as matrix materials: (A) genetic Al-Cu alloys with SiC, (B) genetic Al-Cu-Si alloys with SiC, and (C) commercial alloys. The composition of these alloys are shown in Table 1.

Table 1. Compositions of Aluminum Alloy Matrix

Alloy	Cu	Mg	Mn	Si	Cr	Zn	Al
2024	4.5	1.5	0.6	---	---	---	bal.
6061	---	1.0	---	0.6	---	0.2	"
7075	1.6	2.5	---	---	0.23	5.6	"
Al-1Cu	1.0	---	---	---	---	---	"
Al-2Cu	2.0	---	---	---	---	---	"
Al-4Cu	4.0	---	---	---	---	---	"
Al-5Cu	5.0	---	---	---	---	---	"
Al-1Si-1Cu	1.0	---	---	1.0	---	---	"
Al-1Si-3Cu	3.0	---	---	1.0	---	---	"
Al-1Si-4Cu	4.0	---	---	1.0	---	---	"

All samples were processed at 700°C for one hour in vacuum.

A. Al-Cu Alloys/SiC Composites. Fig. 1 shows the cross section SEM micrographs of the composites with the Al-Cu alloys as the matrix material. The interface between SiC and pure Al without Cu is characterized by good bonding and the absence of intermetallic compound at the interface. The effect of processing temperature and time on the extent of interfacial reaction is outside the scope of this study and will be reported elsewhere.

With Cu additions to aluminum as the matrix material, the interface morphology is gradually changed by the presence of a third phase at the interface region when the copper content exceeds a threshold value. With the matrix containing just 1% copper and no other alloying elements or impurities, the interface is similar to that between pure Al without Cu and silicon carbide. There is no distinct intermetallic phase observed. As the copper content rises to 2%, small particles of intermetallics appear at the interface. It is also found that the ones near the silicon carbide surface contain silicon along with aluminum and copper. When the copper content increases to 4%, a greater extent of copper segregation at the interface is observed. With 5% copper in the matrix, the segregation of copper is in the form of layers or large chunks at the interface.

B. Al-Cu-Si Alloys/SiC Composites. With 1 wt.% Si addition to the matrix alloy, the third phase observed at the interface differs significantly from those without Si. As shown in Fig. 2, even with only 1 wt.% Cu, there exist distinct thin layers of intermetallic compounds at the interface in place of the particle cluster type morphology. These intermetallic compound layers consist of copper, aluminum and silicon. The thickness of these layers increases with the copper content in the matrix alloys.

C. Commercial Alloys. Three types of commercial aluminum alloys were used in this study: 2024, 6061 and 7075. No intermetallic compound was observed at the interface between SiC and 6061 Al alloy (Fig. 3). Since there is no copper present in the 6061 alloy, this observation is consistent with those for Groups (A) and (B) above. At the interface between SiC and 2024 Al alloy (Fig. 4), however, a third phase layer containing copper along the interface was observed. As shown in Table 1, copper is one of the primary alloying elements in this alloy. Although the copper content in this alloy is only about 4.5%, the amount of copper observed at the interface is significant.

For the 7075 Al-SiC system (Fig. 5), a layer of the third phase was again found along the metal ceramic interface; but the thickness of the layer is smaller than that for the 2024 aluminum alloy. This observation is in line with those of Fukunaga and Goda (1985) who found that the thickness of the copper-rich layer at the interface increased with increasing copper content in the matrix alloy. The morphology of SiC surfaces in contact with molten Al alloys is shown in Fig. 6. All SiC filaments appear to have been attacked by molten aluminum alloys.

DISCUSSION

The observed interfacial reaction in the Al alloy/SiC composites may be summarized as follows. At the beginning of the casting process, SiC dissolves in molten aluminum alloys. The extent of the dissolution depends on the processing temperature, the silicon content in the melt, the diffusion rate of silicon in the melt, and other thermodynamic properties of the melt containing Al, Si, C and any other elements. Impurities present in SiC, e. g. free silicon, will also affect the interfacial reaction (Iseki, 1984). As the solidification process commences near the end of the composite preparation, heterogeneous nucleations at the SiC surface and homogeneous nucleations in the melt away from the SiC surface may occur, depending on the concentration of alloying elements in the melt. The silicon concentration distribution in the melt plays a significant role during this solidification stage. Finally, in the very last stage of casting, due to the thermal expansion coefficient difference between the solidified matrix and SiC, microgaps may be created at the interface causing the residual ternary eutectic melt to be sucked into microgaps and to quickly solidify as a thin layer of ternary eutectic compound. As shown in the ternary liquidus diagram of the Al-Si-Cu system (Fig. 7) from Chang and co-workers (1979), for an aluminum-rich melt containing greater than 20 wt.% Si and greater than 2 wt.% Cu, pure silicon will first be precipitated. Towards the end of the casting process, the final melt composition will be at the ternary eutectic point with roughly 26 wt.% Cu and 5 wt.% Si. The amount of such ternary eutectics is proportional to the copper content in the original alloy as may be estimated by the lever rule.

CONCLUSION

Using a special casting design, the interfacial reaction between aluminum alloys and silicon carbide at certain temperatures and time has been studied. The results from a series of alloys containing varying amounts of Cu and Si reveal that SiC dissolves at the beginning of the reaction and third phase precipitates form at later stages of the reaction as the concentration of silicon in the melt, through the diffusion process, increases to a certain level. The phase diagram of multi-component systems may be applied to explain the formation of third phases during the casting of the composites.

ACKNOWLEDGEMENT

The authors would like to express their gratitute to the Ohio Research Challenge Program for the support of this work. Thanks are also due to Mr. Ernest Clark for helping portions of the SEM and EDAX analysis.

REFERENCES

Arsenault, R. J., and C. P. Pande (1984). Scripta Metallurgica, 13, p. 1131.

Chang, Y. A., J. P. Neumann, A. Mikula, and D. Goldberg (1979). Phase Diagrams and Thermodynamic Properties of Ternary Copper-Metal Systems, Incra Monograph Series VI. International Copper Research Asso. Inc., pp. 225-232.

Fukunaga, H., and K. Goda (1985). Bulletin of JSME, Vol. 28(235), pp. 1-6.

Iseki, T., T. Kameda, and T. Maruyama (1984). Journal of Materials Science, 19, p. 1692.

Kannikeswaran, K., and R. Y. Lin (1987). Journal of Metals, 39(9), pp. 17-19.

Restall, J. E., A. Burnwood-Smith, and K. F. A. Walles (1970). Metals and Materials, 4, p. 467.

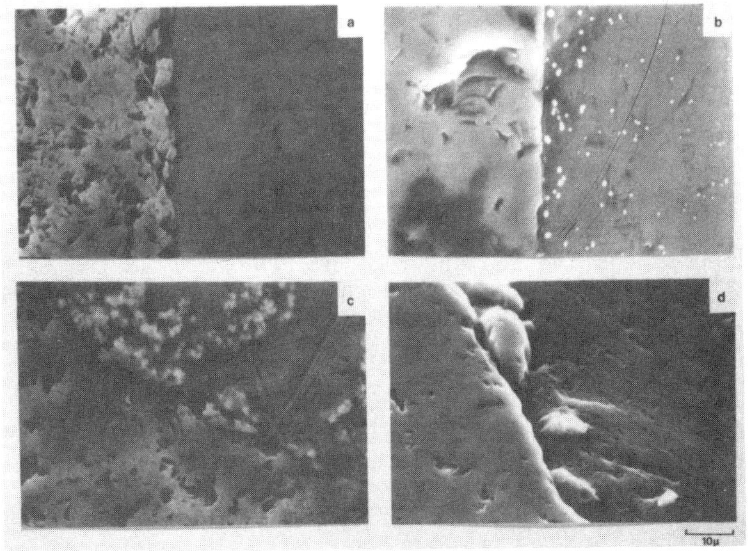

Fig. 1. SEM photos for the cross sections of interfaces between silicon carbide and Al-Cu alloys a) 1% Cu, b) 2% Cu, c) 4% Cu and d) 5% Cu.

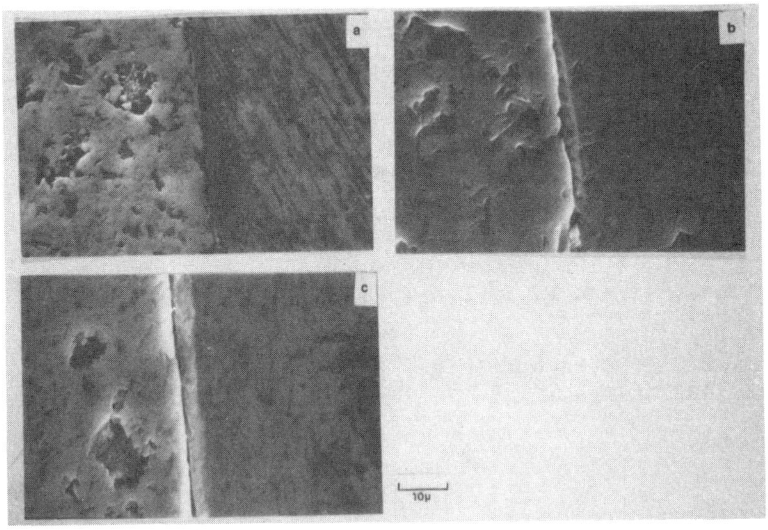

Fig. 2. SEM photos for the cross sections of interfaces between silicon carbide and Al-Cu-Si alloys: a) 1% Cu, b) 3% Cu and c) 4% Cu, all with 1% Si.

Fig. 3. The cross section of the SiC/6061 Al interface,
a) low magnification SEM image,
b) high magnification SEM image,
c) Si x-ray map of the region in (b), and
d) Al x-ray map of the region in (b).

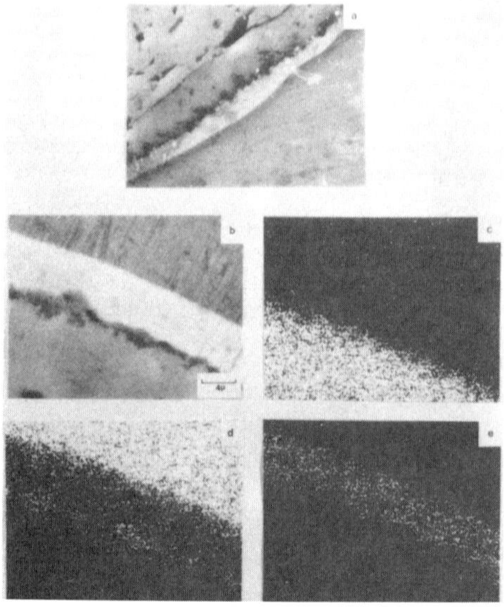

Fig. 4. The cross section of the SiC/2024 Al interface,
a) low magnification SEM image,
b) high magnification SEM image,
c) Si x-ray map of the region in (b),
d) Al x-ray map of the region in (b), and
e) Cu x-ray map of the region in (b).

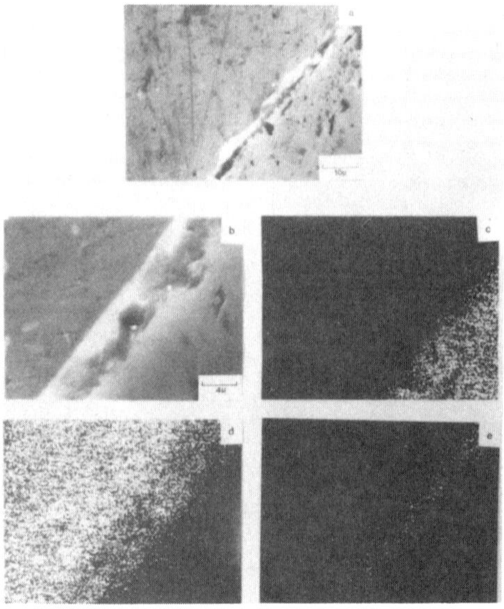

Fig. 5. The cross section of the SiC/7075 Al interface,
a) low magnification SEM image,
b) high magnification SEM image,
c) Si x-ray map of the region in (b),
d) Al x-ray map of the region in (b), and
e) Cu x-ray map of the region in (b).

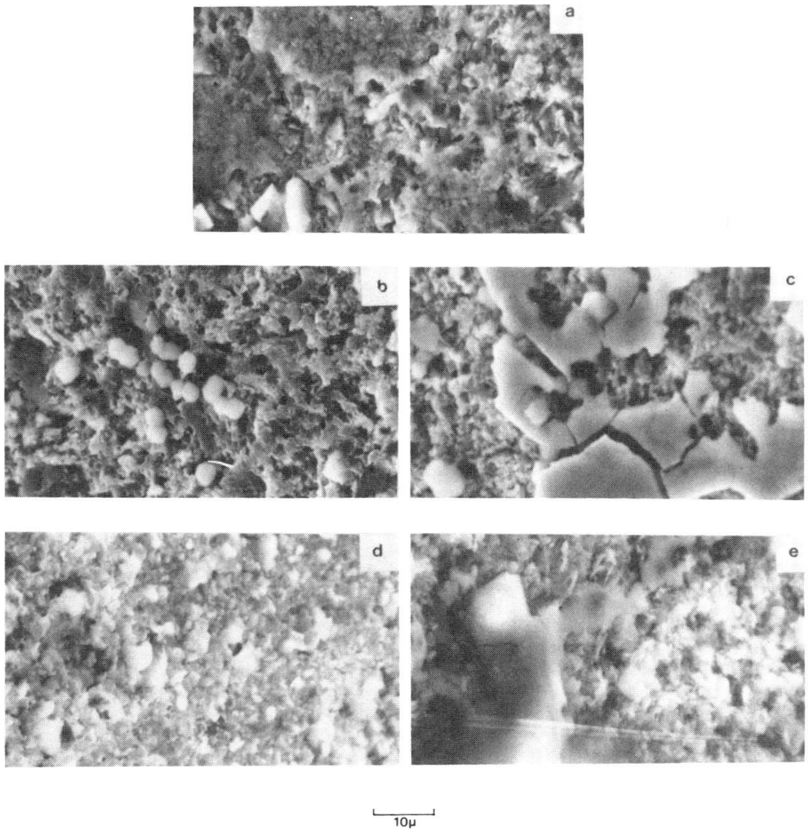

Fig. 6 SiC surfaces. (a) Reacted with 6061 Al; (b), (c) Reacted with 2024 Al; (d), (e) Reacted with 7075 Al.

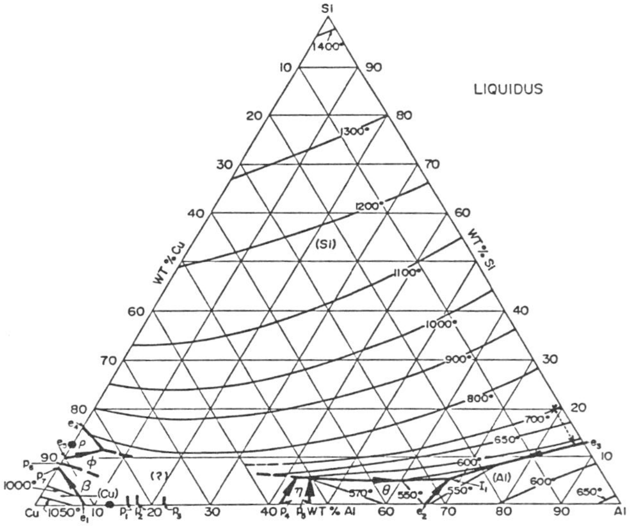

Fig. 7. Ternary liquidus diagram of the Al-Cu-Si system and the anticipated solidification path

CHARACTERIZATION OF INTERFACES IN METAL-MATRIX COMPOSITES

L.M. Dignard-Bailey*, T.F. Malis*, J.D. Boyd* and J.D. Embury**

*Metals Technology Laboratories, CANMET
**Dept. of Materials Science and Engineering, McMaster University

ABSTRACT

The interface microstructure and composition in a SiC particulate reinforced Al 2014 composite have been characterized by transmission electron microscopy (TEM), X-ray microanalysis (EDS), electron energy loss spectroscopy (EELS), Auger electron spectroscopy (AES), and secondary ion mass spectroscopy (SIMS). The investigation concentrated on determining the alloy partitioning, the precipitation of distinct phases and the presence of impurity elements at or near the interface, that results from the processing and subsequent heat treating of the metal matrix composite.

KEYWORDS

Interface; composite; transmission electron microscopy (TEM); X-ray microanalysis (EDS), electron energy loss spectroscopy (EELS), Auger electron spectroscopy (AES), and secondary ion mass spectroscopy (SIMS).

INTRODUCTION

The interface between the ceramic reinforcing phase and the metal (alloy) matrix strongly influences the processing and the mechanical properties of metal-matrix composites (MMCs). The intrinsic surface characteristics of the ceramic and metal control the ability to fabricate a satisfactory MMC by powder metallurgy techniques or solidification processing. Optimum interface strength in the fabricated MMC must be sought to obtain the best combination of yield strength and fracture toughness. Hence, it is important to know the character of the interface between the reinforcing phase and the matrix at all stages of MMC processing, and to develop techniques for controlling these interfaces for optimum MMC properties.

To study the mechanical properties of MMC interfaces, several approaches have been taken, including tensile deformation, fatigue and impact tests, and measurement of the Bauschinger effect (Warner and co-workers, 1988; Fu, Schmerling and Marcus, 1986; Wu, 1988; Vasudevan and co-workers, 1988). There have also been a number of recent studies of interface structure and composition (Nutt and Carpenter, 1985; Nutt, 1986; Warner and co-workers, 1988; Arsenault and Pande, 1984; Fu, Schmerling and Marcus, 1986). In these studies, combinations of transmission electron microscopy (TEM), X-ray microanalysis (EDS), and Auger electron spectroscopy (AES) were used to ascertain the defect structure of the interface, identify interfacial precipitates and/or films, and study chemistry variations across the interface. However, all microanalytical techniques suffer from various limitations, and it is often the case that such data can only be rigorously confirmed by use of several techniques. Accordingly, this paper reports on a comprehensive investigation of the structure and composition of a SiC (particulate)-Al 2014 alloy MMC, employing TEM, EDS, AES, electron energy loss spectroscopy (EELS) and secondary ion mass spectroscopy (SIMS).

EXPERIMENTAL

The material was a commercial MMC consisting of 20 vol percent SiC particulate in Al 2014 alloy (nominal composition: 4.4 wt % Cu-0.8 Si-0.5 Mg-0.8 Mn-93.5 Al), prepared by casting and extruding. It was determined by quantitative image analysis that the average SiC particle length was 13 μm, and the average aspect ratio was 1.23 in the longitudinal plane and 1.06 in the transverse plane (w.r.t. extrusion direction). Samples of the as-extruded material were heat treated as shown in Table 1.

TABLE 1. Details of Heat Treatments

Sample	Treatment
A	as-extruded
B	treated 2 h at 500°C, water quenched
C	solution treated 2 h at 500°C, water quenched, aged 18 h at 160°C
D	solution treated 2 h at 500°C, furnace cooled

The general microstructure was determined by optical and SEM metallography on specimens prepared by diamond polishing and etching in Keller's etchant. Electron microprobe analyses were done with a Cameca camebax-Micro operated at 20 KV. AES was done with a PHI-600 instrument equipped with a fracture stage for in-situ fracture and analysis. SIMS was done with a Cameca IMS-4f instrument, primarily in the microscope mode to record secondary ion images with a lateral resolution of 1 μm. The O_2^+ (12.5 KV) ion source was used to image positive secondary ion images, and the CS^+ (10KV) source to image negative secondary ion images. Thin foils for TEM were prepared by standard ion milling techniques which included prior specimen dimpling to minimize milling time. Foils were examined in a Philips-EM400T TEM-STEM equipped with an EDAX-9100/60 microanalysis system and a GATAN-607 serial EELS spectrometer. Care was taken to measure foil thickness so that EEL spectra would be suitable for quantification and updated quantification factors (cross-sections) were used (Malis and Titchmarsh, 1985).

RESULTS

Figure 1 shows the general microstructure for the different heat-treated conditions. The SEM micrographs were taken with a mix of backscattered electron signal (highlighting the intermetallic phases) and secondary electron signal (high spatial resolution). Alignment of the SiC particles in the extrusion direction is evident, for example in Fig. 1c. Electron microprobe analysis confirmed the presence of three main intermetallic phases: $CuAl_2$, $(Mn,Fe)_3SiAl_{12}$ and $Cu_2Mg_8Si_6Al_5$. The solution heat-treatment effectively dissolves the $CuAl_2$ and $Cu_2Mg_8Si_6Al_5$ phases (Fig. 1b,c), but does not dissolve the $(Mn,Fe)_3SiAl_{12}$ phase. Large particles of the (Mn,Fe) rich phase appeared to have been broken up and aligned in the extrusion direction (e.g., Fig. 1c). Some large, needle-shaped $CuAl_2$ formed in the furnace cooled condition, and precipitate-free zones around the SiC particles were observed (Fig. 1d).

AES analyses of the as-extruded composite were performed on "in-situ" fractured surfaces and on plane polished specimens. The fracture surfaces exhibited transgranular fracture of the SiC particles and ductile fracture in the Al matrix (Fig. 2). AES analyses of the fractured SiC particles detected only Si and C, and confirmed that fracture had occurred through the particles rather than along the interface. In order to evaluate if diffusion of Al into SiC had occurred, AES line profile analyses of the Al-SiC interface were obtained. The line profiles were difficult to measure due to the roughness of the fracture surface, nevertheless a few measurements were obtained on relatively flat interfaces. An Auger line profile of the Si and Al signals for such an interface is shown in Fig. 3. The slope of the concentration profiles were very similar and, within the resolution of the AES profile (0.2 μm) did not indicate diffusion of Al into SiC.

TEM-EDS was used as well to determine if Al gradients existed in the SiC particles. Fig. 4 illustrates the two types of Al-SiC interfaces observed. The majority were relatively smooth (Fig. 4a) but a significant portion showed varying degrees of marked facetting or similar serrations (Fig. 4b). In almost all cases, EDS point analysis traces produced Al profiles similar to those observed by Arsenault and Pande (1984). However, the profile shape varied greatly and, in a few cases, the Al levels were quite low. When EELS was used to check this phenomenon, no Al was detected in SiC up to within 20-30 nm of most interfaces. That is, the concentration of Al was no higher than the minimum detectable limit of about 5 at. % for the Al-K edge in the EEL spectrum. The few exceptions were at serrated interfaces, such as Fig. 4b, where Al contents of up to 25 at. % in the SiC were measured by EELS close to the Al-SiC interface. No corresponding amount of Si was found in the Al matrix across the interface (although a few % Si was detected by EDS uniformly throughout the specimen, a common artefact in ion-milling). The presence of dissolved Al in the SiC varied considerably along the serrated interfaces. EEL thickness limitations precluded extensive mapping of this Al distribution, but there were many segments along such interfaces where the clarity of the interface between the two phases was not sharp at any foil orientation, i.e., suggestive of some form of interaction.

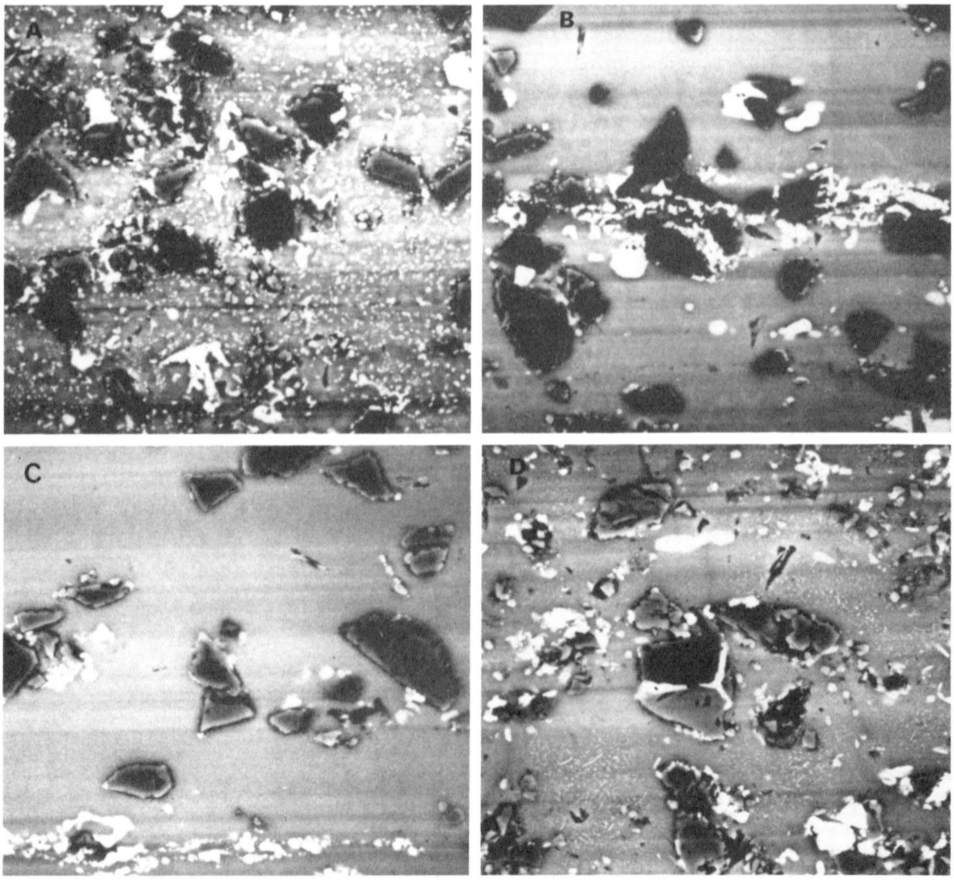

Fig. 1. SEM micrographs taken with 50% BSE-50% SE detection.
(a) sample A, (b) sample B (c) sample C, (d) sample D.

The majority of SiC particles were single crystals with a relatively low defect density (e.g., isolated stacking faults (Fig. 4a). Some heavily-faulted particles were observed, however, which were quite distinct from the occasional mixed-polytype single crystals. Such particles were often divided into subgrains, as shown in Fig. 5. The fault density was quite high, and the faults appeared to emenate from the Al-SiC interface (Fig. 6a). Occasionally, intense perfect dislocation sources on the interface were observed to be operating into the SiC (Fig. 6b), with no corresponding dislocation generation observed into the Al.

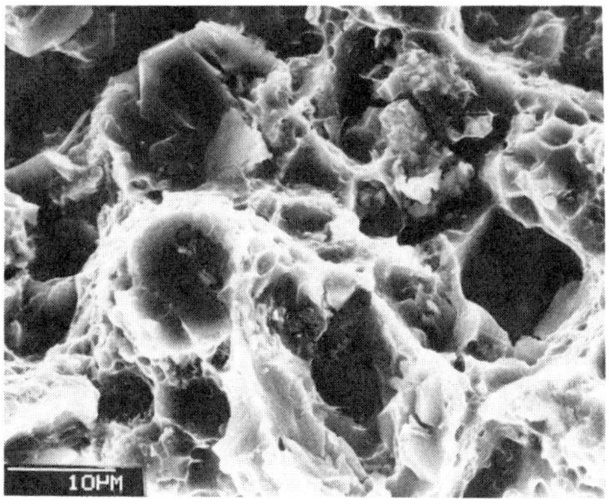

Fig. 2. Fracture surface at sample A fractured in-situ for AES analysis.

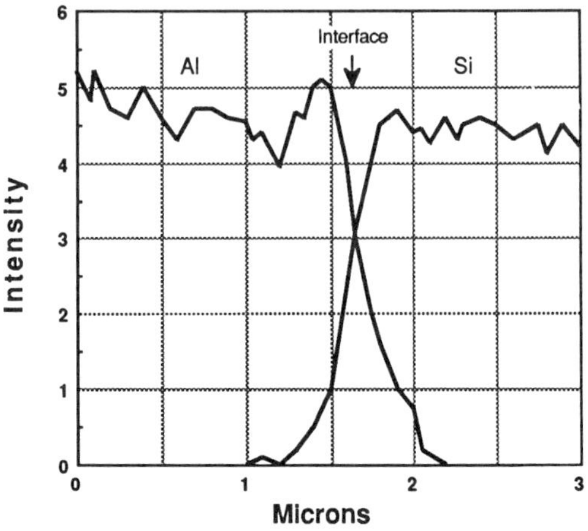

Fig. 3. AES line profile for Al and Si taken across on Al-SiC interface.

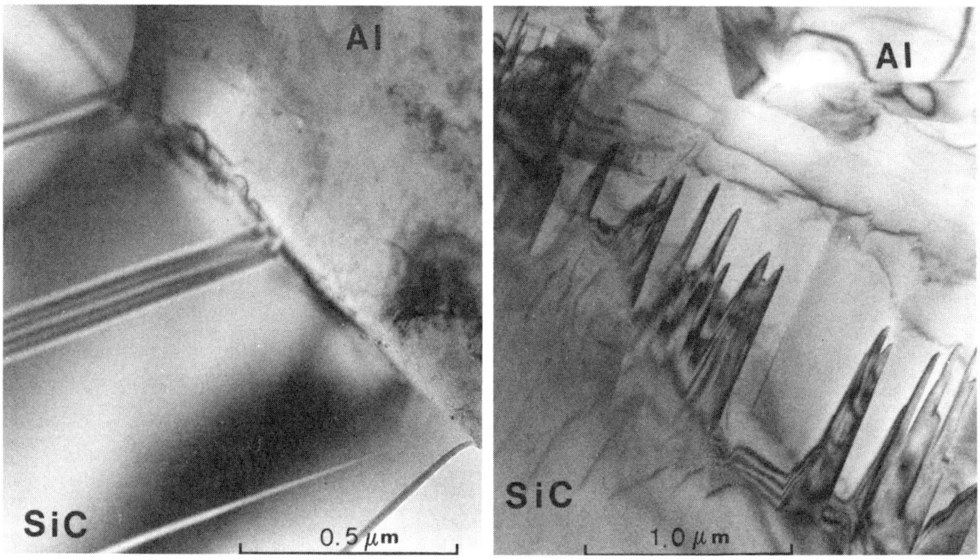

Fig. 4. Morphologies of Al-SiC interfaces. (a) flat, (b) serrated.

An examination of the Al-SiC interfaces in both as-extruded and furnace-cooled material showed some boundary segments which contained MgO precipitation. This was confirmed by EELS in thin regions, followed by EDS "fingerprinting". An example of fine (10-20 nm) MgO precipitates is shown in Fig. 7, but some segments (in both heat treatments) contained MgO precipitates up to ten times larger. Traces of Cu and Cl were detected in some of these precipitates, with the former being visible as co-precipitates in furnace-cooled material. MgO precipitation was observed on segments of intermetallic-SiC interface as well. There were no signs of thin intergranular glassy films, either by Fresnel fringe detection or by use of inelastic dark-field imaging.

The major feature of furnace-cooled material was zones of precipitate depletion around SiC particles and larger intermetallics, particularly of the lath-shaped $CuAl_2$ precipitates (Fig. 8). This was accompanied by extensive re-precipitation of both large and small $CuAl_2$ particles on the Al-SiC interface, as shown in Fig. 8. For serrated interfaces, the $CuAl_2$ often appeared to "fill in" the serrations, as seen in Fig. 9, which also illustrates the degree of complexity involved in interfacial co-precipitation. Both Fig. 9a and 9b show lath-like Al_4C_3 particles (identified by EELS), which were observed on a portion of SiC particles in both extruded and heat-treated material. In Fig. 9a this has been complicated by the $CuAl_2$, and in Fig. 9b, by a large $Cu_2Mg_8Si_6Al_5$ particle as well. (Note that the latter two particles were identified by EDS fingerprinting only). Many of the interface particles contained significant amounts of Cl. In contrast, an extensive examination of all matrix precipitates and intermetallics showed no such Cl enhancement.

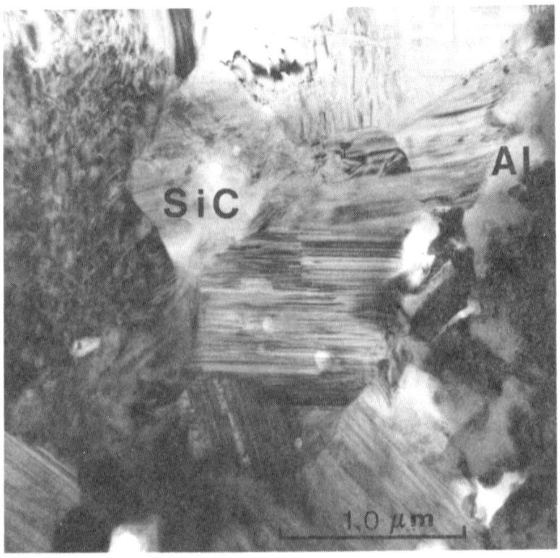

Fig. 5. Heavily faulted polycrystalline particle.

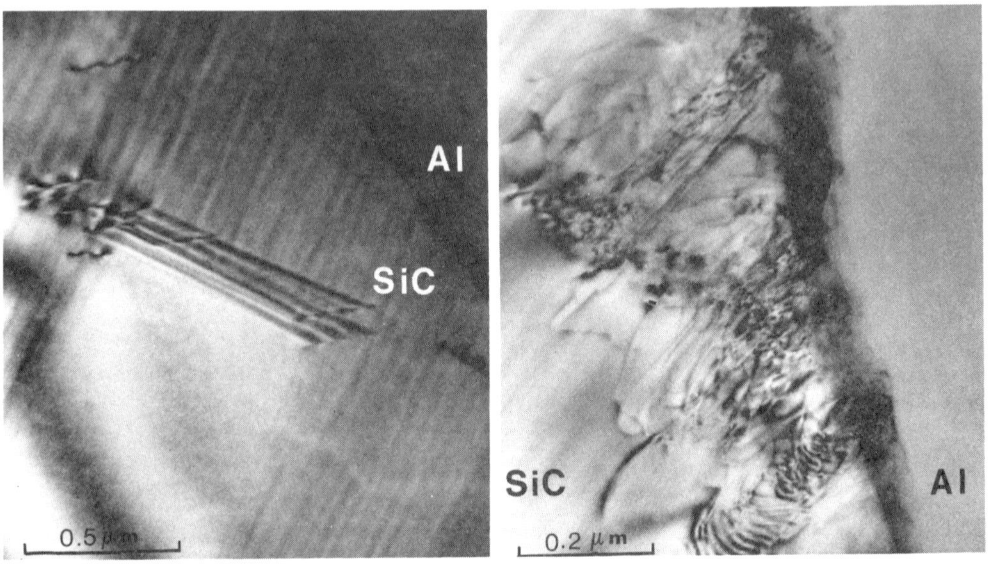

Fig. 6. Generation of stacking faults (a) and dislocations (b) in a SiC particle of sample D, which was deformed during specimen preparation.

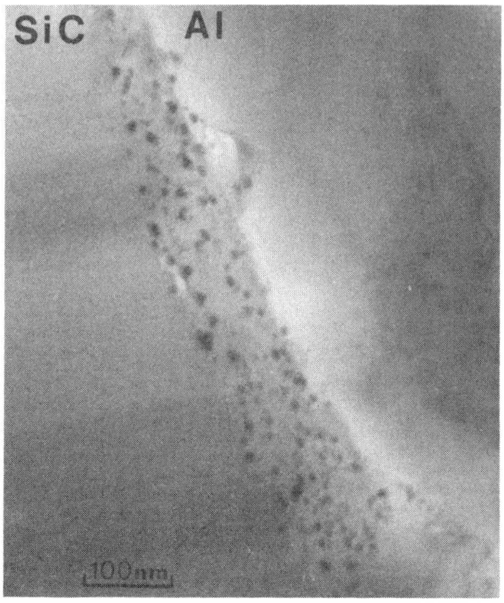

Fig. 7. MgO precipitates on Al-SiC interface.

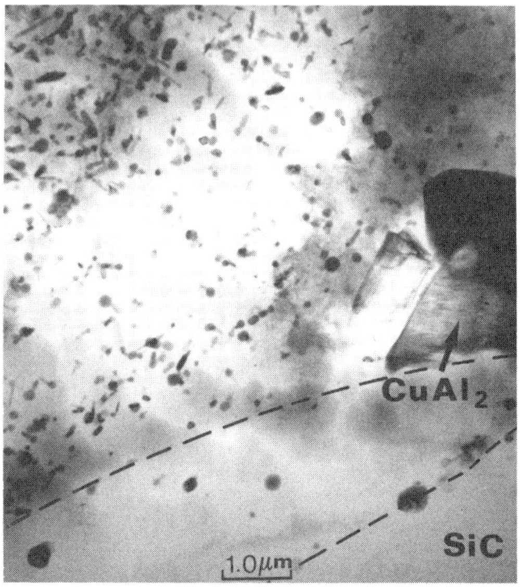

Fig. 8. $CuAl_2$ precipitate-free zone in matrix adjacent to SiC particle in sample D. Dashed lines indicate approximate position of the highly-inclined Al-SiC interface, on which $CuAl_2$ precipitates are observed.

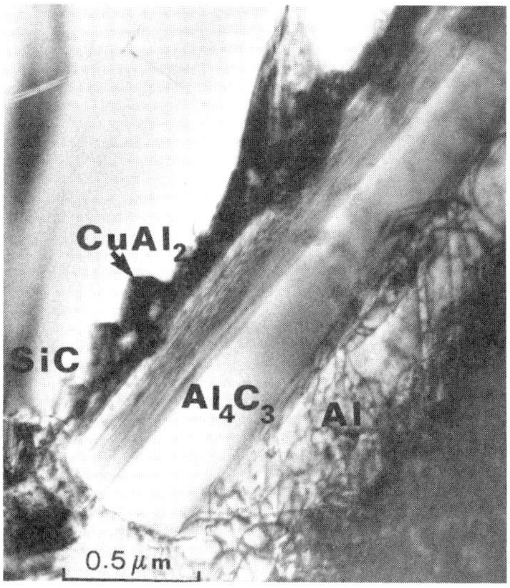

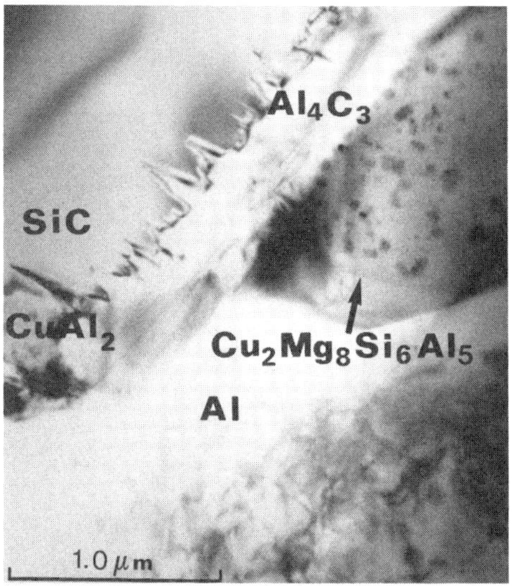

Fig. 9. Precipitation of $CuAl_2$, Al_4C_3, and $Cu_2Mg_8Si_6Al_5$ along a serrated Al-SiC interface in sample D.

SIMS analyses were performed in order to evaluate the distribution of impurity elements present in concentrations too low to be easily detected by the other techniques. Direct ion images indicated the presence of Na, Cl and O at the interfaces. As shown in Fig. 10, the Na^+ image indicated that Na is present in precipitates, and at the Al-SiC interfaces. The Cl^- and O^- images also indicate that these elements are present at the interfaces (Fig. 11). SIMS results do not necessarily imply that a uniform film of Na, Cl and 0 is present at the interfaces, since the lateral resolution of the SIMS is 1μm. It is unlikely that these images are due to surface contamination effects, since careful sample polishing and cleaning, followed by a pre-sputtering period was done. The images as seen in Figs. 10 and 11 remained relatively unchanged with depth profiling down to several μm. Subsequent depth profiling of unreinforced Al 2014 alloys (as extruded and heat treated) showed Cl and Na to be present in similar amounts (similar secondary ion yields), and SIMS imaging showed that Na and Cl were present as precipitates in the matrix.

DISCUSSION

The morphology of the Al-SiC interface is predominantly planar. The diffraction contrast at the interfaces in TEM is similar to that of a high angle grain boundary (Fig. 4a). The serrated interface morphology (Fig. 4b) is less common. It could result from a dissolution/growth reaction between SiC and liquid Al, or it could be present in the SiC starting material.

The present investigation concentrated on determining the alloy partitioning and the precipitation of distinct phases at or near the interface, during processing of the MMC.

All three of the intermetallic phases commonly occurring in Al 2014 [Cu Al_2, $(Mn,Fe)_3SiAl_{12}$ and $Cu_2Mg_8Si_6Al_5$] form at the interface of SiC particles, but not exclusively there. The $(Mn, Fe)_3SiAl_{12}$ forms during solidification or during cooling of the as-cast structure, whereas, the $CuAl_2$ and $Cu_2Mg_8Si_6Al_5$ form during solid-state processing. Specifically, the latter two phases dissolve at the solution treatment temperature (Fig. 1b,c) and precipitate during extrusion or slow cooling from the solution treatment temperature (Fig. 1a,d). The pattern of precipitation in the vicinity of a particle interface is intragranular precipitation, a precipitate-free zone adjacent to the interface, and coarse particles on the interface. This pattern is the same as that generally observed in the vicinity of grain-boundaries in aluminum alloys. From this analogy with grain-boundary precipitation, it is concluded that the solute elements (Cu, Mg, Si) diffuse to the SiC interface during high temperature deformation (extrusion) and slow cooling from the solution treatment temperature, and sustain the growth of large particles at the interface. For these treatments the precipitate-free zone in the matrix adjacent to SiC particles is caused by solute depletion.

MgO precipitates were detected at relatively few interfaces. By comparison, MMCs prepared by compaction of SiC and Al powder have a much larger density of MgO precipitates at SiC interfaces (Nutt and Carpenter, 1985). In the case of powder-processed MMCs, it was proposed that Al_2O_3 on the surface of the metal powder transformed to MgO during liquid-phase sintering.

Significant numbers of Al_4C_3 particles were also detected. This phase forms by the reaction between SiC and molten Al,

$$4Al + 3\,SiC \rightleftharpoons Al_4C_3 + 3\,Si$$

The amount of Al_4C_3 is minimized by increasing the concentration of Si in the melt (reaction is completely suppressed for >8% Si) and reducing the length of time SiC is in contact with molten Al (Lloyd, 1988).

There was no evidence of any amorphous phase at Al-SiC interfaces, as reported by other workers for SiC-2014 MMCs (Warner and co-workers, 1988).

The apparent discrepancy between the EDS and EELS determinations of the Al concentration profile in the vicinity of a SiC interface is attributed to X-ray fluorescence of the Al matrix by Si-K X-rays from SiC. This produces an incorrect detection of Al in the SiC particle near the interface. This phenomenon has been reported recently for intergranular phases in Si_3N_4 (Chadwick and Malis, 1988), and it can be readily verified by several tests. The conclusion, then, is that for the planar interfaces the Si and Al concentration profiles are quite sharp at the interface, and there is no evidence of interdiffusion into either phase. Note however, that the presence of up to 25% Al in the SiC near the boundary of serrated interface could not be caused by fluorescence. This suggests Al diffusion from the molten metal or from a solid state reaction into pre-existing serrations on SiC particles. Further study is needed to confirm this phenomenon.

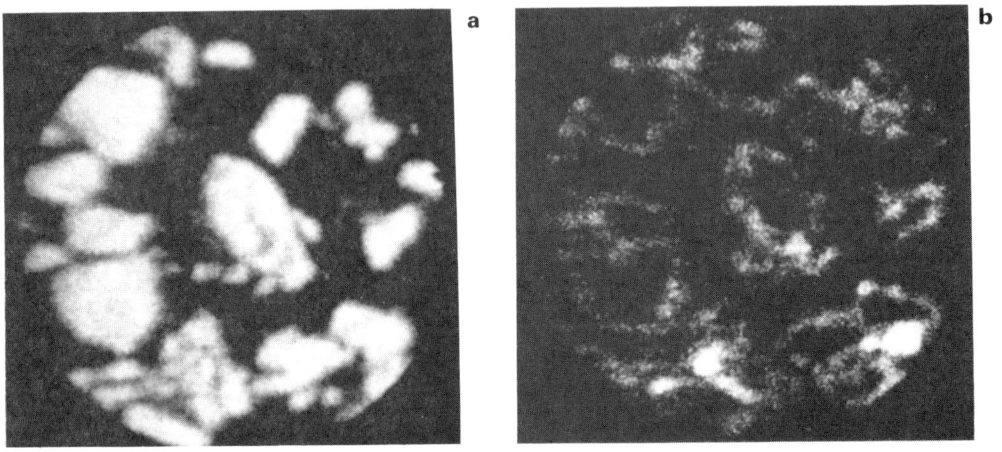

Fig. 10. SIMS direct ion images for (a) $^{28}Si^{+}$, and (b) $^{23}Na^{+}$.

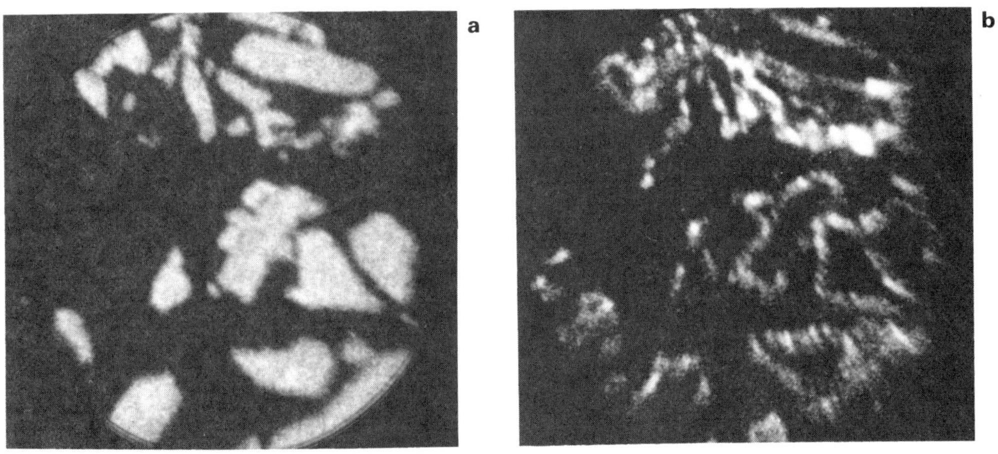

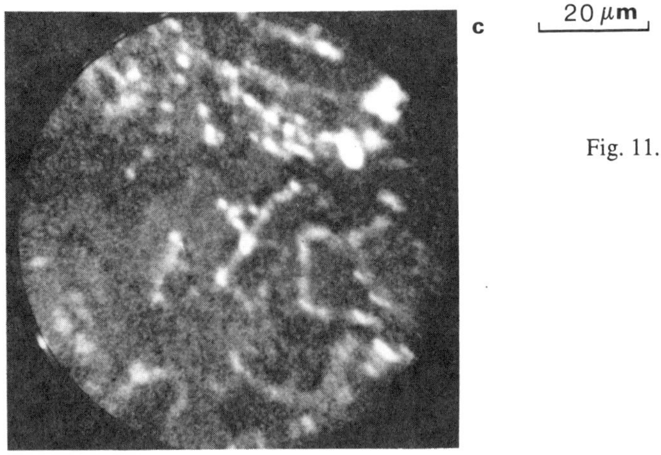

Fig. 11. SIMS direct ion images for
(a) $^{28}Si^{-}$
(b) $^{35}Cl^{-}$
(c) $^{16}O^{-}$.

SIMS analysis detected significant Cl and lesser amounts of Na present at the interface, probably in the form of small precipitates although this is difficult to confirm due to the relatively poor lateral resolution (1 μm) of the SIMS images. TEM-EDS analyses also detected Cl in association with interface precipitates. The TEM foils were prepared by ion thinning, therefore it is unlikely that the presence of Cl is an artifact. TEM-EDS analyses did not detect Na, however Na is difficult to detect by conventional EDS. Previous SIMS analysis of oxidized Al (99.9%) by Degreve and Lang (1985) found that Na was present both as precipitates and at grain boundaries. This suggests that the presence of SiC may act as preferential sites for segregation of impurities, such as Cl and Na. Cl and Na, if present in low concentration, are beneficial in increasing the wettability of SiC and/or decreasing the surface energy of molten aluminum (Goddard, 1987; Wu, 1988). However, without further knowledge of the material processing (pretreatment of SiC and purity of the Al alloy), it is not possible to make any conclusion on the origin of Cl and Na, and their effect on the bonding characteristics of SiC and Al.

CONCLUSION

In summary, the interface between SiC particles and the matrix in the MMC studied exhibits characteristics of a high-angle grain boundary. Particles of MgO and Al_4C_3 can form at the interface during manufacture of the MMC by solidification processing. In addition $(Mn,Fe)_3SiAl_{12}$ particles form at the interface as well as throughout the matrix. During heat treatment, soluble alloy elements (Cu, Mg, Si) diffuse to the interface, form $CuAl_2$ and $Cu_2Mg_8Si_6Al_5$ precipitates there, and give rise to precipitate-free zones in the adjacent matrix. The evidence for interdiffusion between SiC particles and the matrix appears limited to serrated interfaces. There is evidence of segregation of the matrix Cl and Na impurity elements to the Al-SiC interface.

In spite of these various interface precipitation and segregation effects, all indications are that the interface is strongly bonded. During in-situ fracture tests in the present studies,and tensile tests carried out by Vasudevan and co-workers (1988), the interface remains intact during the whole process of matrix yielding, transgranular fracture of SiC and ductile fracture of the matrix. However, the combination of coarse particles at the interface, precipitate-free zone and segregation of Cl and Na suggests that the SiC interfaces could be preferential paths for corrosion and/or stress corrosion cracking.

REFERENCES

Arsenault, R.J., and C.S. Pande (1984), Scripta Metall., 18, 1131-1134.
Chadwick, M., and T. Malis (1988), Submitted to Ultramicroscopy.
Degreve, F., and J.M. Lang (1985), SIMS Characterization of Thin Thermal Oxide Layers on Polycrystalline Aluminum. Mat. Res. Soc. Symp. Proc.,48, 241-262.
Fu, L.-J., M. Schmerling, and H.L. Marcus (1986). Interface Studies of Aluminum Metal Matrix Composites. Composite Materials: Fatigue and Fracture, ASTM STP 907, H.T. Hann, Ed., 51-72.
Goddard, D.M. (1978), J. Mat. Sc., 13, 1841-1848.
Lloyd, D.J. (1988), to be published in: Composite Science and Technology.
Malis, T. and J.M. Titchmarsh (1985), Inst. of Phys. Conf. Series 78, London, 181-182.
Nutt, S.R. (1986), Interfaces and Failure Mechanisms in Al-SiC Composites. Interfaces in Metal-Matrix Composites, Dhingra and Fishman, Ed., 157-167.
Nutt, S.R. and R.W. Carpenter (1985), Mat. Sc. Eng., 75, 169-177.
Vasudevan, A.K., O. Richmond, F. Zok, and J.D. Embury (1988), to be published in: Interfacial Phenomena in Composites: Processing, Characterization and Mechanical Properties.
Warner , T.J., P.J. Withers, J. White, R.M. Jordan, and W.M. Stobbs (1988), Microstructure and Properties of Spray Deposited Al Alloy/SiC Composites. Interfaces in Polymer, Ceramic, and Metal Matrix Composites, H. Ishida, Ed., 537-551.
Wu, R. (1988), Interfacial Behaviors and its Influence on the Mechanical Properties of Carbon Reinforced Aluminum Composites. Ibid, 43-57.

WETTABILITY OF SILICON CARBIDES BY LIQUID METALS

Kiyoshi Nogi* and Kazumi Ogino*

*Department of Materials Science and Processing
Faculty of Engineering, Osaka University, Suita, Osaka, Japan

ABSTRACT

The wettability of silicon carbides (a hot-pressed SiC, a reaction-bonded SiC and a single crystal of α-SiC) by liquid metals was investigated using the sessile drop method.
Measurements were carried out in Ar atmosphere and in vacuo.
Generally, non-wetting contact angles were observed for Pb, Sn, Ag, Ge, and Cu, whereas Si, Fe, Ni, and Co were found to wet.
Wetting behaviours of Cu and Ge depended on the surface properties of SiC. In particular, the liquid Cu showed the hexagonal spreading, which was regulated by the surface property of α-SiC, on the single crystal of α-SiC.

KEYWORDS

Wettability, Contact angle, Silicon carbide, Liquid metal, Joining, Composites

INTRODUCTION

The wettability of ceramics by liquid metals gives information important for the production of metal-matrix composites (MMC) and the joining of metal and ceramics . However, even with the same metal-ceramics system, different results have often been reported, because the wettability is easily affected by the presence of small amounts of impurities.
In the present work, the wettability of the reaction-bonded SiC, the hot-pressed SiC, the carbon treated reaction-bonded SiC whose surface had been carburated, and the single crystal of α-SiC by various liquid metals was investigated.
The wettability was evaluated by the contact angle and the work of adhesion between liquid metals and SiC using the sessile drop method.

EXPERIMENTS

Of the metals used, Pb, Ag, Ge, and Si were of over 99.99% purity, Sn was of over 99.9% purity, and Cu, Fe, Co, and Ni were melted in a reduced pressure or H_2 atmosphere.
Oxygen contents of the metals before and after experiments were 0-3

ppm except for the Fe which was 18 ppm and 4 ppm before and after the experiment, respectively.
For the silicon carbides, the reaction-bonded SiC containing 10% of metallic Si was purchased from Nippon Kagaku Togyo Co. Ltd., the hot pressed SiC was prepared prior to the experiments, and the single crystal of α-SiC was obtained from Pacific Rundom Co. Ltd..
For the single crystal of α-SiC, the experiments were carried out on the surface of a (0001) plane.
Some of the experiments were carried out using the carbon-treated reaction-bonded SiC, whose surface had been carburated before the experiments.
The two kinds of the furnaces used in the experiments are shown in Fig.1. Both of the furnaces consisted of a graphite for heating element and an observation window enabling the shape of the liquid metal to be photographed.

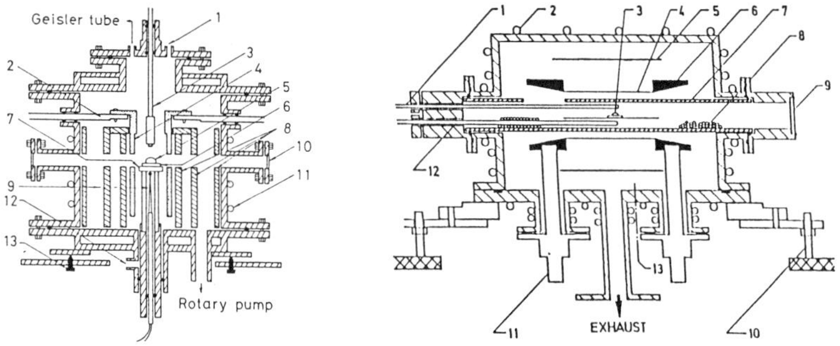

Fig. 1 Apparatuses for measurement
Fig. 1-a: 1) gas inlet, 2) water cooled electrode, 3) metal dropping assembly, 4) graphite heating element, 5) metal drop, 6) SiC plate 7) graphite plate, 8) alumina refractory, 9) thermocouple, 10) observating window, 11) water cooled pipe, 12) gas outlet, 13) levelling screw
Fig. 1-b:1) metal dropping assembly, 2) water cooled pipe, 3) liquid metal drop on SiC, 4) carbon net, 5) Mo radiation shield, 6) graphite ring, 7) alumina pipe, 8) Ti shots, 9)observating window 10) levelling screw, 11) water cooled Cu electrode, 12) thermocouple 13) vacuum gauge

For the experiments in Ar atmosphere, the furnace (Fig. 1-a) was mainly used and the furnace (Fig. 1-b) was used for the experiments under reduced pressure.
Experimental method and procedures have been explained in detail in a previous report (Ogino, 1983).
When the contact angle of the liquid metal was obtuse on SiC, it's value was determined using the tables of Bashforth and Adams (1883). When the contact angle was acute, the value was directly obtained from the shape of liquid metal drop (Hasouna, 1988)

RESULTS

Time dependence of contact angle of liquid metal on SiC depended on the metal and SiC examined. It can be classfied into several types.

1. no or slight time dependence of contact angle of liquid metal was observed on all of SiC and the contact angle was larger than

90° for the liquid Pb, Ag, and Sn/reaction-bonded and hot-pressed SiC systems.

2. no or slight time dependence of contact angle of liquid metal was observed on all of SiC and the contact angle was smaller than 90° for the liquid Ni, Co, and Fe/reaction-bonded and hot-pressed SiC systems.
3. Time dependence depended on the SiC used, and the value of the contact angle reached zero at a certain temperature for the liquid Ge/ reaction-bonded SiC systems.
4. Considerable time dependence was observed, and the value of the contact angle still remained larger than 90° for the liquid Cu/ hot-pressed SiC system.
5. Perfect wetting (contact angle = 0°)was observed at the initial stage for the liquid Si/SiC system.
6. Characteristic wetting was observed for the liquid Cu/ α-SiC system.

Typical examples of each case are shown in Figs. 2-6.

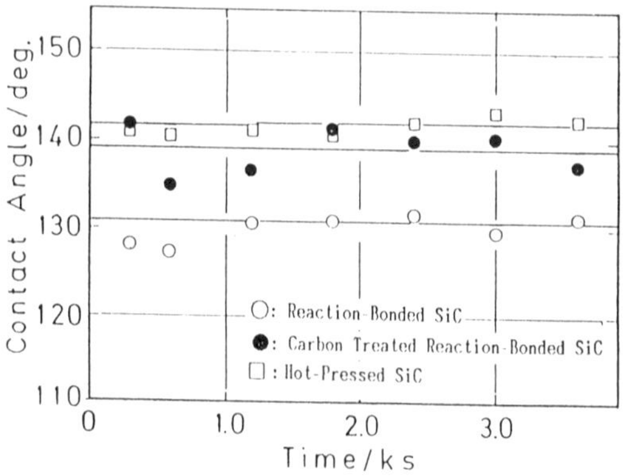

Fig. 2 Time dependence of contact angle of liquid pure Ag on SiC at 1283K

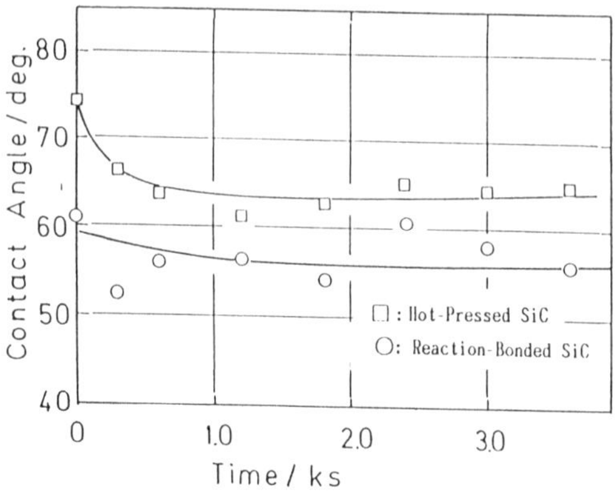

Fig. 3 Time dependence of contact angle of liquid pure Co on SiC at 1803K

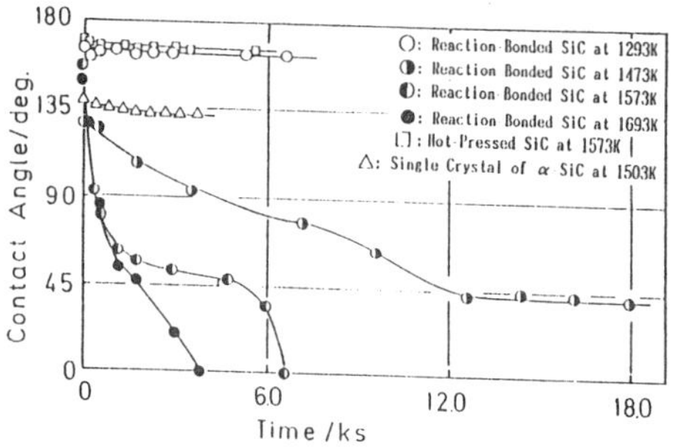

Fig. 4 Time dependence of contact angle of liquid pure Ge on SiC

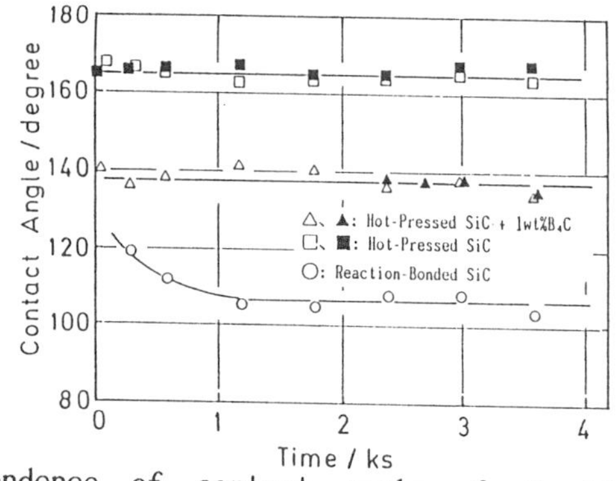

Fig. 5 Time dependence of contact angle of liquid pure Cu on SiC at 1408K

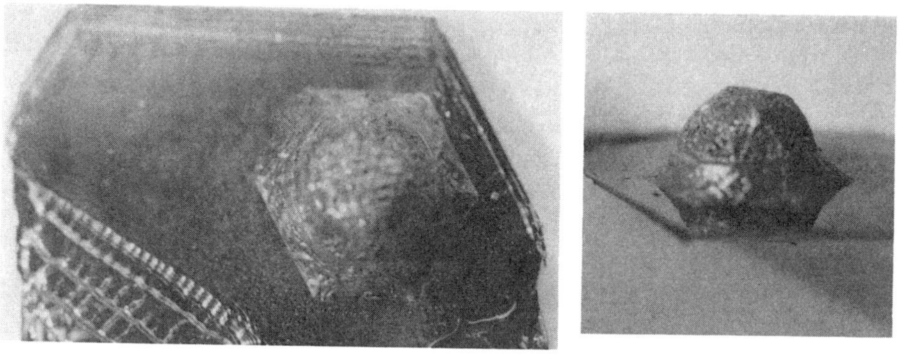

Fig. 6 Solidified shape of Cu drop on a single crystal of α-SiC

For the liquid Si/SiC system, the contact angle was nearly 0° at the dropping moment.
The characteristic wetting was observed for the liquid Cu/α-SiC system under the limited experimental conditions.
Work of adhesion between a liquid and a solid can be obtained using eq.(1).

$$W_{ad} = \gamma_l(1 + \cos\theta) \qquad (1)$$

where, W_{ad} is the work of adhesion between liquid metal and SiC (erg/cm^2), γ_l the surface free energy of liquid metal (erg/cm^2), and θ the contact angle of liquid metal on SiC.
The surface free energy of liquid metals could not be determined in the present work because of the small size of the liquid metal drop. Accordingly, reported values of the surface free energies of the liquid metals (Nizhenko, 1981) were used for the calculation of W_{ad}.
As no values are available for the surface free energy of liquid Ag-Si alloy, the surface free energy of liquid pure Ag (Nogi, 1988) was used for the calculation.
The contact angles and the work of adhesion between liquid metals and SiC are summarized in Table 1.

Table 1 Contact Angle, θ, and Work of Adhesion between Liquid Metals and Various Silicon Carbides and Si Content in Liquid Metals

Metal	SiC	Temp.(K)	θ(°)	W_{ad}(erg/cm^2)	Si(at%)
Sn	H.P.+B_4C	1373	150	64	0
	R.B.	1373	131	163	0.1
	S.C.	1297	126	198	0.3
	S.C.	1473	123	212	0.8
Pb	H.P.	603	167	9	0
	R.B.	603	161	25	0
Ag	H.P.	1283	142	191	0
	R.B.	1283	131	310	4.8
	C.R.B.	1283	138	232	0
Ge	R.B.	1293	164	31	0.9
	R.B.	1473	42	1168	34.5
	R.B.	1573	0	1152	
	R.B.	1693	0	1198	47.6
	H.P.	1573	165	26	0
	S.C.	1503	132	258	0.5
Cu	H.P.	1408	167	36	0
	H.P.+B_4C	1408	134	344	1.1
	R.B.	1408	106	898	4.3
	S.C.		*n.d.		
Fe	H.P.	1873	24	2334	46.0
	H.P.+B_4C	1873	28	2333	43.0
	R.B.	1873	29	2278	48.2
Ni	H.P.+B_4C	1773	72	1936	33.6
	R.B.	1773	68	2033	33.3
Co	H.P.	1803	65	1696	42.1
	R.B.	1803	56	1858	42.1

H.P.; hot-pressed SiC, R.B.; reaction-bonded SiC, S.C.; single crystal, C.R.B.; carburated reaction-bonded SiC,
*n.d.: not determined

DISCUSSION

The values of the contact angle and their time dependence relate to the solubility of Si and C into the liquid metals.
It has been reported that liquid Pb, Ag and Sn, which show no time dependence and large values of contact angle, can dissolve only small amounts of Si and C at the experimental temperature, and in addition these metals and Si do not form silicides (Johnson, 1951; Ruff, 1919; Moissan, 1904; Tamaru, 1909). While liquid Ni, Co, and Fe, which show no time dependence and small values of contact angle, can dissolve large amounts of Si and C compared with liquid Pb, Ag and Sn at the experimental temperature (Boecker, 1912; Epstein, 1936; Ruff, 1914; Lewkonja, 1908; Guertler, 1917; Ruttewit, 1940), liquid Cu and Ge, whose contact angle decrease with time lapse, can dissolve considerable amounts of Si and do not dissolve C at the experimental temperature (Hansen, 1936; Stohr, 1939; Bever, 1946). As is clear from Figs. 2-5 and Table 1, the larger the solubility, the smaller the contact angle, which usually corresponds to a large value of work of adhesion. It can be concluded that a dissolution reaction causes a decrease in the contact angle and an increase in the work of adhesion.
The values of the contact angle of liquid metals on SiC previously reported are summarized in Table 2.

Table 2 Contact Angle of Liquid Metals on SiC

Metal	Temp.(K)	θ(°)	Reference
Sn	1373 1323	135 135	Naidich (1972) Allen (1959)
Ag	1373	128	Allen (1963)
Au	1423	138	
Ge	1323	113	
Cu	1373 1373	60 140	Gnesin (1969)
Si	1703 1723 1723 1723	42 35 33 45	Naidich (1972) Allen (1963) Bevolo (1980) Yupko (1973)

It can be seen from Tables 1 and 2 that the values obtained in the present work agree to a great extent with those in the previous works.
As is clear from Fig.6, a hexagonal spreading was observed in the system of liquid Cu and α-SiC. The hexagonal spreading was always observed in the limited experimental conditions.
It is interesting to find that each rim of the hexagon runs parallel to the edge of α-SiC. This suggests that the hexagonal spreading reflects on the surface property of α-SiC. The wettability of α-SiC by liquid Cu has already been investigated by Naidich (1972) and Gnesin (1969) but the hexagonal spreading has not been reported.
The results obtained are summarized in Table 3.

Table 3 Wetting Behaviour in Liquid Cu/ α-SiC System

Temp.(K)	Atm.(mmHg)	Hex. sp.	Temp.(K)	Atm.(mmHg)	Hex. sp.
1371	10^{-5}	◎	1440	10^{-5}	○
1393	10^{-5}	◎	1473	10^{-5}	×
1393	10^{-4}	◎	1393	10^{-1}	○
1393	10^{-3}	○	1393	760, Ar	×

Hex. sp.; hexagonal spreading
◎ ; clearly observed, ○ ; observed, × ; not observed

It can be seen that the hexagonal spreading was observed at a lower temperature and a lower pressure.
SEM observations were carried out for the surface of α-SiC after removing the Cu using HNO_3 solution. Results are shown in Fig.7

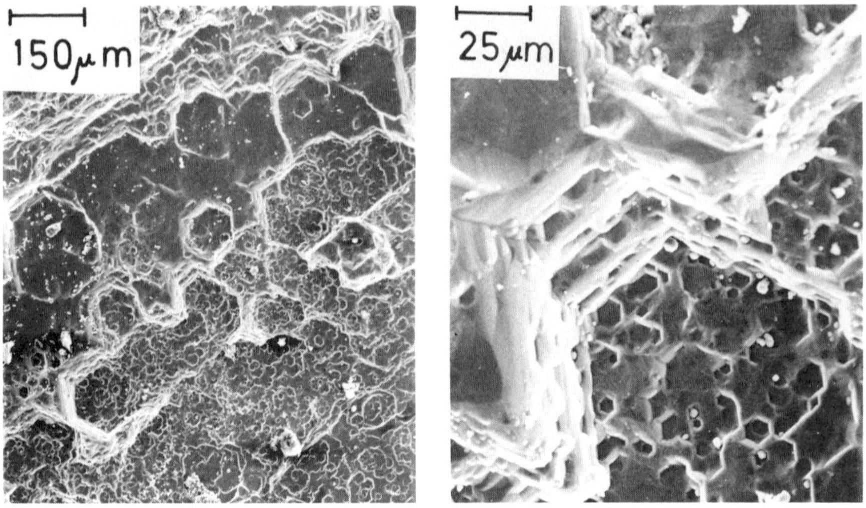

Fig. 7 SEM observation at the surface of α-SiC after removal of Cu

It can be seen from Fig. 7 that small hexagons exist on the whole surface of α-SiC where the liquid Cu existed. When the hexagonal spreading was not observed, the small hexagon could not be found. Hexagonal etch pits resulting from the liquid Cu, whose rims were confirmed to be parallel to edges of α-SiC, may be the reason for the hexagonal spreading.
Detail of the hexagonal spreading has already been described in the authors' previous paper(Nogi, 1988).

CONCLUSION

The wettability of various silicon carbides by liquid pure metals has been investigated using the sessile drop method.

1. Contact angle was larger than 90° in the liquid Cu, Pb, Ag, and Sn/reaction bonded and hot pressed SiC systems.
2. Contact angle was always smaller than 90° in the liquid Ni, Co, and Fe/reaction bonded and hot pressed SiC systems.
3. The contact angle reached zero at higher temperatures in the liquid Ge/SiC systems.
4. Liquid Si covered the whole surface of SiC.
5. Carburation of the surface of reaction-bonded SiC reduced the interaction between liquid metal and the reaction-bonded SiC.
6. Characteristic wetting was observed in the liquid Cu/α-SiC system.
7. The degree of work of adhesion between liquid metals and SiC depends on the solubilities of Si and C in the liquid metals.

ACKNOWLEDGEMENT

The authors are indebted to the Ministry of Education, Science and Culture of Japan and also to the Light Metal Educational Foundation, Inc. for their financial support in the form of a Grant-in-Aid for Scientific Research.

REFERENCES

Ogino, K., and K. Nogi (1983). Trans. ISIJ, 23, 234-239
Bashforth, F., and J. C. Adams (1883). An Attempt to Test the Theory of Capillary Action, Cambridge Univ. Press
Hasouna, A. T., K. Nogi and K. Ogino (1988). Trans. JIM. (in press)
Nizhenko, V. I., and L. I.Floka (1981). Poverkhnostnoe Natiazhenie Zhidkikh Metallov i Splavov, Moskva, Metallurgiia
Nogi, K., K. Oishi, and K. Ogino (1988). Japan J. Inst. Met. 52, 72
Johnson, W. R., and M. Hansen (1951). AF Technical Report 6383,June
Ruff, O. and B. Bergdahl (1919). Z. Anorg. Chem., 106, 91
Moissan, H. and F. (1904). Siemens. Compt. rend., 138, 657-661
Tamaru, S. (1909). Z. Anorg. Chem, 61, 40-42
Boecker, G. (1912). Metallurgie, 9, 296-303
Epstein,S. (1936). The Alloys of Iron and Carbon, vol.1, McGraw-Hill, New York
Ruff, O.(1914). Z. Anorg. Chem., 88, 386-396
Lewkonja, K. (1908). Z. Anorg. Chem., 59, 327-338
Guertler, W. (1917). Metallographie, 1, 658-673
Ruttewit, K. (1940). Z. Metallkunde, 32, 60
Hansen, M. (1936). Der Aufbau der Zweistofflegierungen, Springer-Verlag OHG, Berlin, 621-630
Stohr, H., and W. Klemm (1939). Z. Anorg. Chem., 241, 305-323
Bever, M. B., and C. F. Floe (1946). Trans. AIME, 166, 128-141
Naidich, Ju. V. (1972). Kontaktnie javlenia v metallicheskih rasplavah, Naukova Dumka, Kiev
Gnesin, G. G., and Ju. V. Naidich. (1969). Porosh. Met., 2
Allen, B. C. (1963). Trans. Met. Soc. AIME, 227, 1175
Allen, B. C., and W. D. Kingery (1959). Trans. Met. Soc. AIME, 215, 30
Bevolo, A. J., and co-workers (1980). J. Appl. Phys., 51, 5390
Yupko, V. R., and G. G. Gnesin (1973). Porosh. Met., 97
Nogi, K., and co-workers (1988). Japan J. Inst. Met. 52, 663

LIQUID PHASE SINTERING OF ALUMINA-(Ni+TiC) CERMETS*

Hamid Mostaghaci
National Research Council of Canada
Atlantic Research Laboratory
Halifax, Nova Scotia
Canada, B3H 3Z1

ABSTRACT

Apart from the metal bonded carbides which have been the most successful type of cermet produced to date, other systems, particularly oxide-metal combinations, have been relatively unsuccessful. While carbides are quite compatible with their metallic binders in every aspect, the poor wetting of oxide ceramics by metals, and their lack of solubility in metals have been found to be the major barriers for the fabrication of oxide-based cermets. The objective of this work was to investigate the possibilities of fabrication of alumina-based cermets, using new alloy compositions and different alumina powders. Through theoretical and experimental investigations, a number of Ni-alloys have been found to be suitable for this objective, among which (Ni+TiC) alloys were chosen and mixed with different alumina powders to fabricate a series of alumina-based cermets. The effect of processing parameters on the liquid phase sintering of these systems was studied thoroughly. The lack of homogeneity in the powder mixtures was found to be a major problem in developing structurally uniform cermets through liquid phase sintering.

KEYWORDS

Alumina, Nickel, Titanium carbide, Cermet, Liquid phase sintering.

INTRODUCTION

In the past two decades, considerable progress in materials development has been achieved by applying the methods of liquid phase sintering well known from heavy alloy and hard metal production. The development of cemented carbides (WC-Co and TiC-Ni) (1-4), and the hot pressing of silicon nitride in the presence of a liquid phase (5-8) are the most prominent examples.

Liquid phase sintering is a process by which, in the presence of liquid phase, rearrangement and mass transport mechanisms are enhanced, leading to the accelerated reduction of the total interface free energy of a particulate system. This behaviour is manifested in densification and changes in grain size and

*NRCC No. 29363

shape. However, it seems clear that certain minimum requirements must be met during sintering, regarding wetting of the solid phase by liquid and solubility of the solid in the liquid phase, in order to successfully sinter using a liquid phase.

In the absense of solubility of the solid, which is one of the requirements for an ideal liquid phase sintering when a system is sintered at a liquid phase temperture, the densification process in the system is restricted to viscous flow and particle rearrangement. Since Kingery's theoretical analysis (9), it has been emphasized by various authors that in the absense of such mutual solubility the densification by the rearrangement process alone cannot yield a fully dense product (10-13). The reasons given for this lack of densification are the formation of stable particle bridges (10-12), and a rigid solid skeleton during heat-up and sintering. Shrinkage-porosity resulting from differential shrinkage of the solid and liquid during cooling from the sintering temperature is also considered as a cause for imcomplete densification.

Kingery (9) predicted that complete densification can be attained by rearrangement alone when the volume of the liquid exceeds 35 vol%. For this to occur he prescribed partial solubility of the solid in liquid and complete wetting of the solid by the liquid as essential requirements. This view has been confirmed experimentally (14) in Fe-Cu alloy which is a partially soluable system. However, since this work the importance of solubility has been over-emphasized to achieve complete densification at the 35 vol% limit of liquid. In contrast, it has been shown that the same extent of densification could occur even in the absence of mutual solubility and at a wetting angle greater than zero, provided the system is sufficiently homogeneous (15). This view has been illustrated experimentally in a W-Cu composite which has been obtained from co-reduction of their oxides. It can then be concluded that a zero wetting angle is important so far as the rate of densification is concerned. However, a nearly complete densification can be achieved in co-reduced W-Cu composite at higher wetting angles. Activation energy of densification by rearrangement for this system has been claimed to be less than that of non-homogeneous systems in which the mixtures of W and Cu were prepared by ball milling (15). In this connection, Huppmann and Bauer (16) have also shown that a close relation exists between the microscopically charaterized degree of mixing, and densification, during liquid phase sintering (Figure 1).

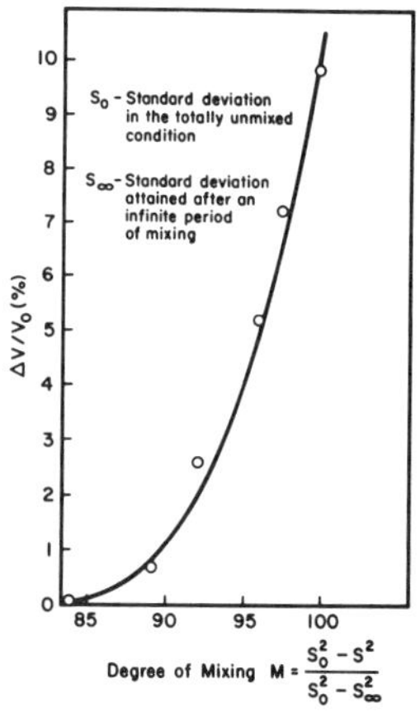

Fig. 1. Volume shrinkage as a function of degree of mixing (16)

In the present work the liquid phase sintering of Al_2O_3-(Ni+TiC) has been investigated. This system was found to have good wetting characteristics but with almost no mutual solubility. Therefore a (W-Cu) model system can be employed for the interpretation of its sintering behaviour. Among different processing parameters studied the lack of homogeneity in powder mixtures was found to be a great barrier in achieving full densification during liquid phase sintering. This will be clearly demonstrated in our experimental results.

EXPERIMENTAL PROCEDURE

For compatibility studies the samples were made in the form of alumina-metal couples. Initially, metallic powders were compacted into small discs (1 cm diameter, 0.3 cm thickness) at a pressure of 200 MPa. These metallic discs then were embedded in ceramic powders which were contained inside polymer bags, and the whole assembly pressed isostatically at a pressure of 125 MPa. Subsequently, the couples were sintered under various processing conditions.

For the fabrication of cermets two different alumina powders were used. The first powder (PUCK, A6)* was a very fine powder but highly agglomerated, while the second was a non-agglomerated powder (A-15SG)+ with a narrow particle size distribution (2.0-2.5μm). Nickel and titanium carbide powders with an average size of 1μm were employed for mixing with these alumina powders. The mixtures of powders in appropriate proportions were milled in a stainless steel mill containing tungsten carbide balls and isopropanol for 24 hours, in order to homogenize adequately. The compacts from these mixtures were sintered in argon at temperatures 1500-1550°C. Characterization of sintered samples were carried out using Scanning Electron Microscopy (SEM) and Electron Probe Miroanalysis (EPMA).

RESULTS AND DISCUSSION

1. Compatibility of Alumina with (Ni+TiC) Alloys

The interaction between alumina and (Ni+TiC) alloys were discussed in an earlier study (17). The wetting of alumina by these alloys was investigated through standard "sessile drop" experiments, and the contact angles were measured for (97% Ni+3%TiC) and (90% Ni+10% TiC) alloy compositions. The results indicated that at 1500°C, in vacuum, these alloys wet alumina at 66° and 52° for alloys containing 3% TiC and 10% TiC respectively. Also, a number of couples made from alumina with the above alloys were fired at 1500°C in flowing H_2. The SEM micrographs of polished sections of typical samples are shown in Figures 2 and 3. The interesting feature of these samples is the formation of a coherent and continuous bond between metal and ceramic. Microprobe analysis of the interface showed the clear presence of Ti.

---

* Produits Chimiques Ugine Kuhlmann, France.
+ Alcoa, CH-1000 Lausanne, Switzerland.

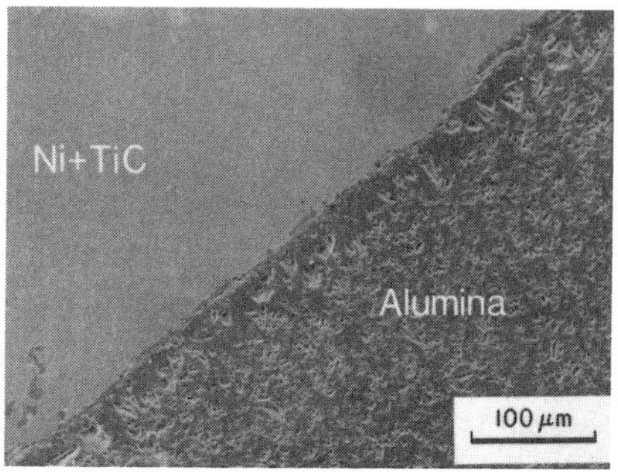

Fig. 2. Scanning electron micrograph of the polished section of an Al_2O_3-(97% Ni+3% TiC) couple, sintered at 1500°C for 1 hour in H_2

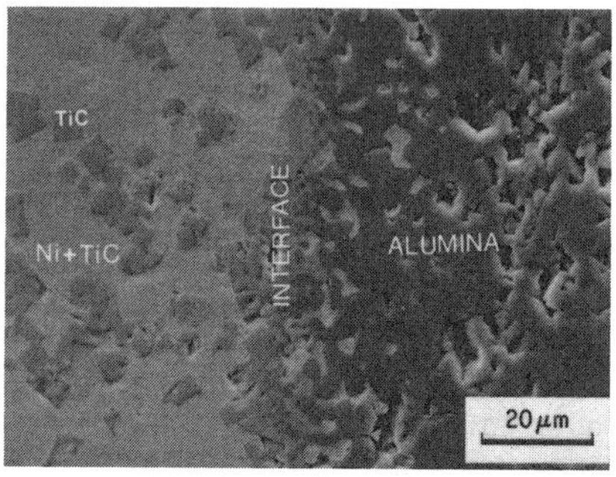

Fig. 3. Scanning electron micrograph of the polished section of an Al_2O_3-(Ni+10% TiC) couple, sintered at 1500°C for 1 hour in H_2

2. Effect of heating rate on the liquid phase sintering of Al_2O_3-(Ni+TiC) alloys

It has been agreed generally that there are at least three stages which can be distinguished during liquid phase sintering (18-19), provided that appropriate wetting has been established to permit sintering. The extent of any one stage will depend on the details of the systems under consideration and particularly whether one is dealing with a completely dispersed or a coalesced liquid-solid system. During the first stage, rearrangement, the liquid wets the solid particles and flows between them to dissolve particle-particle bridges which have formed during heat-up. A more efficient packing is brought about by the capillary forces acting on the particles. The properties which control the extent of sintering during this phase include the magnitude of the interfacial energies, the solubility of the solid in liquid, the amount of liquid present and the particle size and shape. However, in system such W-Cu or Al_2O_3-(Ni+TiC) with no mutual solubility, the liquid is not able to dissolve any of the particle-particle bridges which have formed during heating up. Therefore, a primary requirement for consolidating these systems in the presence of varying amount of liquid phase is the fact that, at the temperature at which liquid forms, the ceramic has not formed a solid skeleton. In that case liquid will be able to flow between the particles causing the ceramic particles to slide freely over reach other, and rearrangement takes place. Having this in mind, an attempt was made to reduce the solid state sintering of compacts as far as possible. This was accomplished by introducing the samples to the sintering temperature for the shortest possible time (∿2 min). The idea was to retard initial stage sintering of alumina until the metal melts completely and flows between the alumina particles. The effect of heating rate on the shrinkage of compacts made from 75% Al_2O_3 and 25% (Ni+TiC) at 1550°C in an argon atmosphere is shown in Figure 4. This Figure shows that no major change has occurred in the shrinkage

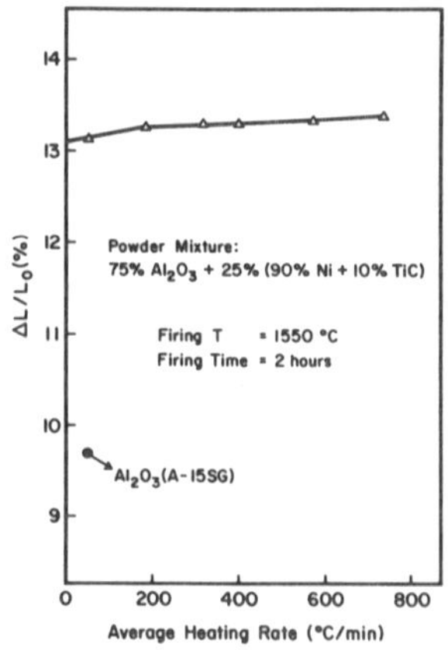

Fig. 4. The effect of heating rate on the shrinkage of Al_2O_3-(Ni+TiC) cermets

of cermets sintered over a span of heating rates from 50°C/min to 750°C/min, although the shrinkage values are considerably higher than that for pure Al_2O_3 sintered under the similar conditions. An almost constant shrinkage over a span of heating rates suggests that even by applying high heating rates (750°C/min), solid state sintering of Al_2O_3 has not been prevented and the material has densified to a considerable extent even after a short stay at temperature before the metal melted. Nevertheless, microstructure development in these samples is indicative of some improvement in liquid phase sintering. In Figure 5, the second phase can be seen at the ceramic grain boundaries to be partially wetting the grains. Scanning electron micrographs from the polished section of a sample

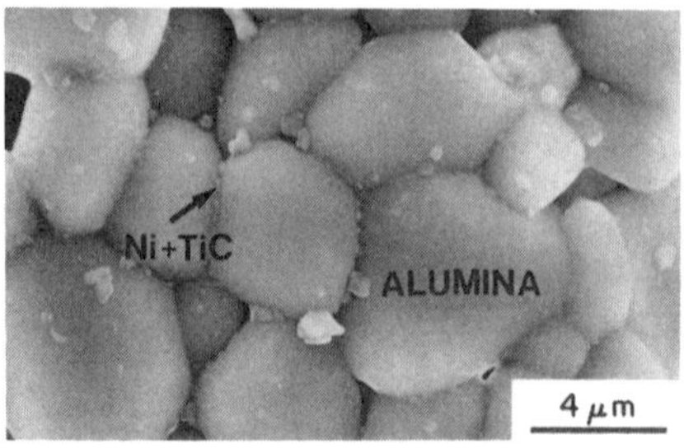

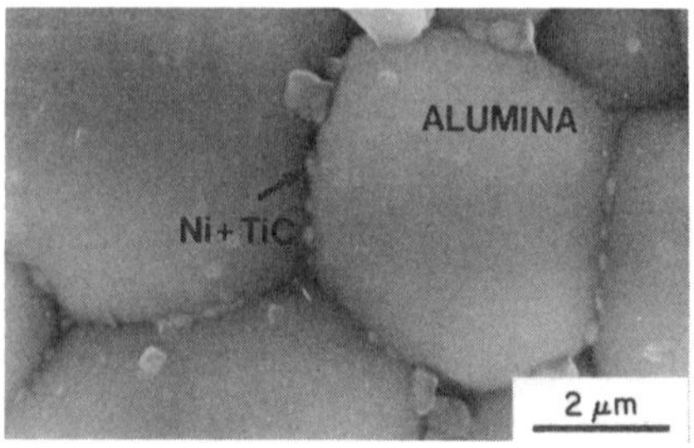

(A) (B)

Fig. 5. SEM micrographs from the fracture surface of an Al_2O_3-(Ni+TiC) cermet, sintered at 1500°C for 2 hours in argon.

of 75% Al_2O_3-25% (Ni+TiC) composition, fired at a rate of 300°C/min, also is indicative of liquid phase sintering in some regions while complete homogeneity has not yet been achieved (Figure 6). Using the EPMA technique, the distribution of different phases at the fracture surface was analysed, and the results are shown in Figure 7. From the spectra shown in this Figure, traces of Ti and Ni can be seen to be accompanied with Al at the grain boundaries. This suggests that Al_2O_3 has dissolved in the (Ni+TiC) melt to a minor extent.

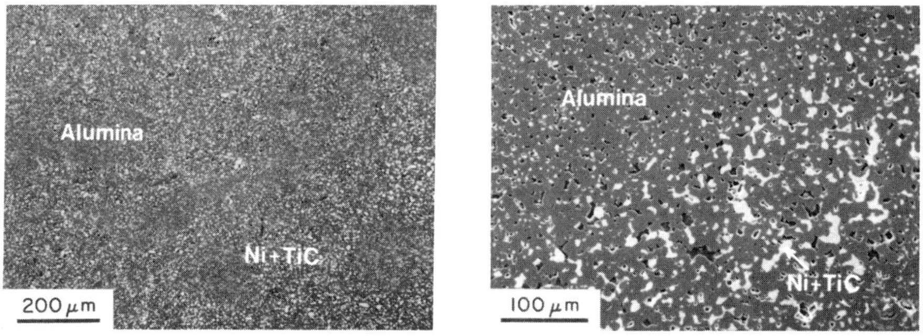

(A) (B)

Fig. 6. SEM micrographs from the polished section of an Al_2O_3-(Ni+TiC) cermet sintered at 1600°C for 1 hour in argon, showing the inhomogeneous distribution of metallic phase within the ceramic matrix.

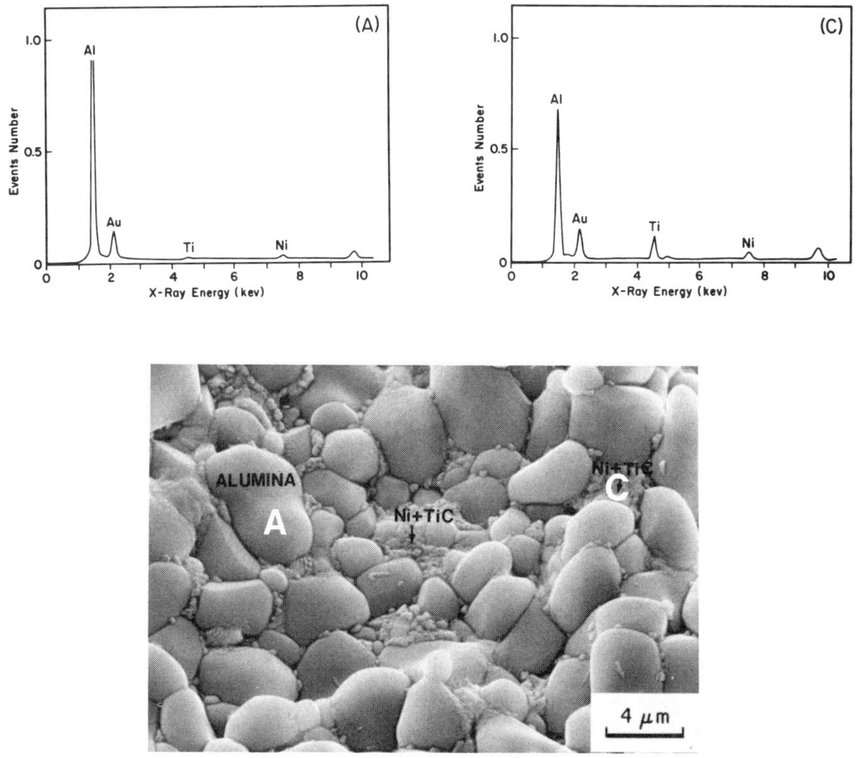

Fig. 7. SEM micrograph from the fracture surface of an Al_2O_3-(Ni+TiC) cermet sintered at 1550°C for 1 hour in argon. Microanalysis of different regions shows different distribution for Al, Ni and Ti.

3. Effect of vol % of Liquid phase during Sintering

The relation between shrinkage and volume percent of liquid in a system which is densified by liquid phase sintering has been investigated by many researchers (20-22). Heady and Cahn (20) formulated the interparticle force as a function of contact angle, volume of liquid, size of initial spherical particles and surface tension. They concluded that the maximum in the F vs V relation occurs at the point V=0, where the interparticle force due to the liquid discontinuity goes from $2\pi R\gamma_{LV}\cos\theta$ to zero (R, the radius of initial particles, θ contact angle). According to Huppmann and Riegger (21) increased wetting angle and increased amount of melt both cause the force F to decrease and this decrease causes the pore growth mechanism to occur. On the other hand, thin liquid film has been shown to be closely related to the solid state (22).

In the present work, two alumina powders with two different characteristics were mixed with various vol% of an alloy having (90% Ni+10% TiC) composition. The results of shrinkage of compacts as a function of the amount of liquid in these samples are shown in Figures 8 and 9. In each case, one sample of pure alumina was sintered at the same conditions at which the other compacts had been fired. This served as a basis of comparison with the samples containing the liquid phase. Alumina powder used for fabrication of the samples in Figure 8 was highly agglomerted powder (PUCK, A6) which at the sintering temperature, contributed considerably to the solid state sintering of compacts. It is evident from Figure 8, that even in the samples containing nearly 50 vol% of liquid phase, shrinkage remained at a level of about 11%. The other alumina powder (Alcoa, A-15SG), a fine, non-agglomerated powder which was used for fabrication of the samples in Figure 9, showed better liquid phase sintering, and the level of shrinkage in the samples from this powder was higher. However, the initial solid state sintering in both powders was rather considerable, although the presence of liquid phase contributed to the attainment of higher shrinkage levels in the compacts. From the trend of shrinkage in the compacts, which remained almost unchanged in relation to the variation of volume percent of liquid phase (Figures 8 and 9), it is suggested that the liquid phase sintering in these samples has taken place by rearrangement only.

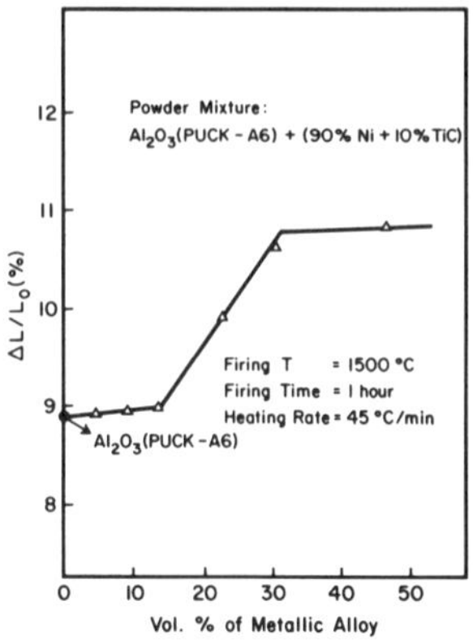

Fig. 8. The effect of vol % of the liquid phase on the shrinkage of Al_2O_3 (PUCK - A6) - (Ni+TiC) cermets

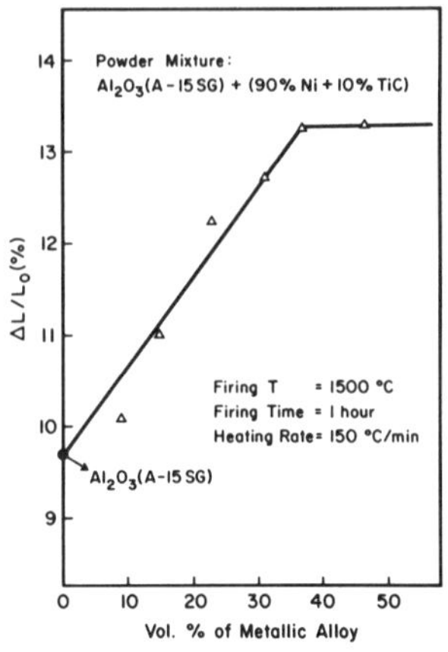

Fig. 9. The effect of vol % of the liquid phase on the shrinkage of Al_2O_3 (Alcoa, A-15SG) - (Ni+TiC) cermets

CONCLUSIONS

The important conclusions from this work are as follows:

(i) Alloys between nickel and titanium carbide wet alumina readily, but the compacts made from alumina and (Ni+TiC) powder mixtures, sintered at the similar conditions, did not densify entirely, through liquid phase sintering.

(ii) Increasing the metallic phase content in Al_2O_3- metal systems up to 40 vol%, enhanced the shrinkage of compacts. However, due to the characteristics of the systems studied in this work (almost mutual insolubility), it is suggested that this enhancement is attained only by the rearrangement mechanism.

(iii) The lack of homogeneity in Al_2O_3-(Ni+TiC) powder mixtures is a major contributing factor to the solid state sintering of alumina, preventing complete densification of the cermets through liquid phase sintering.

ACKNOWLEDGEMENTS

The author greatly appreciates Professor A. Mocellin of the Ceramic Laboratory of EPF-Lausanne, Switzerland, for many helpful discussions and Stellram Société Anonym for the financial support of this work.

REFERENCES

(1) D. J. Miller and J. A. Pask (1983). Liquid Phase Sintering of TiC-Ni Composites. J. Am. Ceram. Soc., 66, 841-846.

(2) D. Moskowitz and M. Humenik, Jr. (1965). Cemented Titanium carbide Cutting Tools. Mod. Dev. Powd. Met., 3, 83-94.

(3) H. Pastor (1984). Fabrication and Usage Properties of Tungsten carbide-Cobalt Cermets. Perspectives in the Cutting Field. Mat. Tech., pp. 433-450.

(4) H. E. Exner, E. SantaMarta, and G. Petzow (1971). Grain Growth in Liquid Phase Sintering of Carbides. Mod. Dev. Powd. Met., 4, 315-325.

(5) S. Wild, et al. (1972). The Role of Magnesia in Hot-pressed Silicon nitride. In P. Popper (Ed.), Special Ceramics 5, BCRA, pp. 377-384.

(6) R. J. Weston and T. G. Carruthers (1973). Kinetics of Hot-pressing α-Silicon nitride Powder with Additives. Proc. Brit. Ceram. Soc. No 22, 197-206.

(7) D. R. Clarke (1980). Densification of Silicon nitride Alloys Using a Eutectic Liquid: An Experimental Test. Met. Sci. Res., 13, 303-310.

(8) R. J. Brook, et al. (1977). Mass Transport in the Hot-pressing of Silicon nitride. In F. L. Riley (Ed.), Nitrogen Ceramics, Noordhoff-Leyden, pp. 383-390.

(9) W. D. Kingery (1959). Densification during Sintering in the Presence of a Liquid Phase. I-Theory. J. Appl. Phys., 30, 301-306.

(10) T. J. Whalen and M. Humenik Jr. (1967). Sintering in the Presence of a Liquid Phase. In G. C. Kuczynski, N. A. Horton, and C.F. Gibbon (Eds.), Sintering and Related Phenomena, Gordon & Breach Science Publisher, N.Y., pp. 715-746.

(11) V. H. Eremenko, Yu. V. Naidich, and I. A. Lavrinenko (1970). Liquid Phase Sintering. Consultant Bureau, N.Y., London, pp. 1-15.

(12) W. J. Huppmann and G. Petzow (1977). Particle Rearrangement during Liquid Phase Sintering of Several Carbide-Metal Combinations. Mod. Dev. Powd. Met., 9, 77-89.

(13) R. J. Nelson and D. R. Milner (1971). Liquid Flow Densification in the Tungsten carbide-Copper System. Powder Met., 14, 39-63.

(14) W. D. Kingery and M. D. Narasimhan (1959). Densification during Sintering in the Presence of a Liquid Phase. II-Experimental. J. Appl. Phys., 30, 307-310.

(15) K. V. Sebastian and G. S. Tendolkar (1979). Experimental Observations on the Densification by Rearrangement in Liquid Phase Sintering. Powder Met. Int., 11, 62-64.

(16) W. J. Huppmann and W. Bauer (1975). Characterization of the Degree of Mixing in Liquid Phase Sintering Experiments. Powder Met., 18, 249-258.

(17) H. Mostaghaci (1988). Compatibility of Alumina with Nickel-Titanium carbide Alloys. J. Can. Ceram. Soc., in press.

(18) W. J. Huppmann (1975). Sintering in the Presence of Liquid Phase. Mat. Sci. Res., 10, 359-378.

(19) R. B. Heady and J. W. Cahn (1970). An Analysis of the Capillary Forces in Liquid Phase Sintering of Spherical Particles. Met. Trans., 1,185-189.

(20) W. J. Huppmann and H. Riegger (1975). Modelling of Rearrangement Processes in Liquid Phase Sintering. Acta Met., 23, 965-971.

(21) Z. Panek (1984). The Optimization of Liquid Amount during Sintering. Science of Sintering, 16, 13-20.

POROSITY DEPENDENCE OF STRENGTH IN CERAMICS

V.D. Krstic

Department of Metallurgical Engineering, Queen's University
Kingston, Ontario, K7L 3N6

ABSTRACT

A quantitative analysis of the strength response of a porous brittle solid is presented and tested against experimental data on borosilicate glass. The analysis is developed from the relationship between the elastic modulus, pore size and spherical pore volume fraction. It is shown that the full characterization of a porous solid in terms of strength requires the knowledge of pore size, pore volume fraction and the annular (radial) flow size to pore size ratio.

KEYWORDS

Stress intensity factor; annular flaw size; strength; pore size; pore volume fraction; crack extension.

INTRODUCTION

Most ceramic materials of technolgical importance contain pores. The presence of pores generates the stress concentration which ultimately leads to fracture initiation. A number of experiments have proven that the strength of glass (Hasselman and Fulrath, 1967, Bertolotti and Fulrath, 1967) and polycrystalline ceramics (Coble and Kingery, 1956, Spriggs and Vasilos, 1963) is strongly dependent on pore size and pore volume fraction. Without taking into account the cavity size, (Bowie, 1956) has shown that the stress intensity factor of cracks emanating from the cavity surface is magnified by the stress concentration due to the presence of a cavity. However, it has been shown (Evans and Tappin, 1972) that in addition to annular flaw size the size of the pore also plays a critical role in controlling the strength level of porous ceramics.

Most recently a complience method (Krstic and Erickson, 1987) was used to relate the elastic modulus to the pore size and pore volume fraction. The analysis was based on the crack (pore) opening displacement in which the pore size is taken to be an integral part of the total crack length consisting of cavity size and an annular flaw size.

In the present manuscript, the energy balance criterion is used to model the porosity dependence of strength in brittle ceramics.

STRESS FIELD INTERACTION EFFECT

Before discussing the effect of pore volume fraction on the strength of brittle solids, it is considered appropriate to briefly discuss the role of the stress field interaction between neighboring pores. In this context, consider a system containing uniformly distributed spherical pores of diameter D, and of interpore separation, d (Fig. 1). If the pores are arranged in a

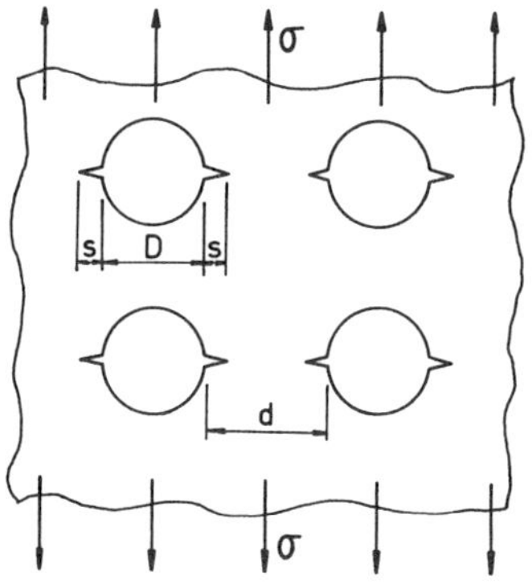

Fig. 1. Array of spherical pores with annular flaws subjected to external stress.

single coplanar row with radial cracks on both sides of each pore, the change in strength with pore separation is given by the expression (Paris and Sih, 1965) (for s<D).

$$\sigma_p = \sigma_f \left[\frac{\pi}{(2\frac{d}{D} + 1)}\right]^{1/2} \left[\tan \frac{\pi}{2(\frac{d}{D} + 1)}\right]^{-1/2} \quad \ldots\ldots (1)$$

where σ_f is the strength of an infinite solid containing a single isolated pore. For a simple cubic system, the relationship between the pore volume fraction, V, the pore separation, d, and the pore size, D, is given by:

$$\frac{d}{D} = \left(\frac{\pi}{6V}\right)^{1/2} - 1 \quad \ldots\ldots (2)$$

Substitution of Eq. (2) in Eq. (1), gives

$$\sigma_p = \sigma_f \left[\frac{3V\pi^2}{4}\right]^{1/6} \left[\tan \left(\frac{3V\pi^2}{4}\right)^{1/3}\right]^{-1/2} \quad \ldots\ldots (3)$$

The predicted change of strength with pore volume fraction from Eq. (3) is illustrated in Fig. 2 along with typical experimental data for borosilicate glass. Evidently, the highest reduction in strength due to the stress field interaction effects of neighbouring pores occurs at high volume fractions when the pores are close to each other. Whereas experiments show just the opposite: the most severe reduction in strength occurs at low pore volume fractions where the stress field interaction effects are minimal. This behaviour clearly demonstrates that the stress field interaction of adjacent pores plays no important role, at least for low pore volume fractions.

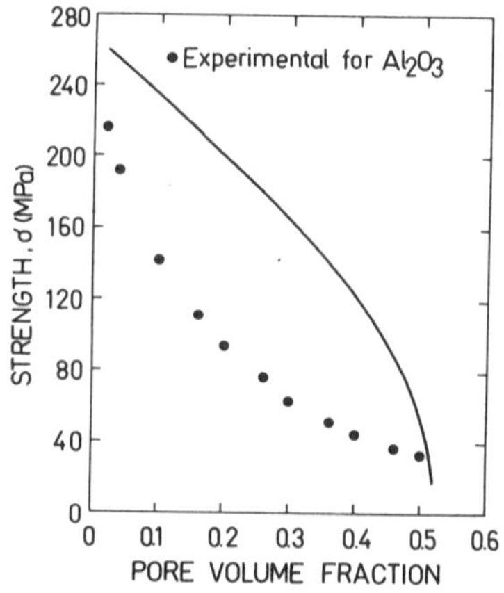

Fig. 2. Comparison between experimentally determined and predicted (based on simple cubic arrangement of pores) variation of strength (from Eq. (3)) in a polycrystalline alumina (from Coble and Kingery, 1956).

STRENGTH-PORE VOLUME FRACTION RELATION

It is well known that when an external stress is applied on a brittle solid, its elastic energy decreases due to the presence of cracks or crack like defects. For a solid containing a single isolated pore with an annular crack emanating from its surface, the reduction in elastic energy stored is (Snedon, 1946).

$$U = -\frac{8(1-\nu^2)\sigma^2}{3E}(R + s)^3 \qquad ....(4)$$

where ν is the Poisson's ratio, σ is the applied stress, R is the pore radius, s is the annular flaw size, and E is the effective elastic modulus of a solid containing pore.

It has been shown that the presence of cracks in a stressed body lowers its Young's modulus, and ultimately leads to stress relaxation (Berry, 1960, Hasselman, 1969). Without taking into account the effect of stress concentration due to the presence of pores the relationship between the Young's modulus and the number of sphercial pores per unit volume (N) is (Krstic, 1987).

$$E_o = E_o\left[1 + \frac{16(1-\nu^2)\ N\ (R+s)^3}{3}\right]^{-1} \quad ....(5)$$

where E_o is the Young's modulus of a pore-free solid, and C is the crack length, C = R+s.

In the following analysis it will be assumed that the interaction between the annular crack tip stress field and the stress field due to the presence of pores is minimal. Based on this, the total energy per unit volume can be expressed as the sum of the elastic energy stored and the surface energy of the crack:

$$V = -\frac{8(1-\nu^2)\sigma^2(R+s)^3}{3E_o}\left[1 + \frac{16N(1-\nu^2)R^3(1+s/R)^3}{3}\right] + 2\left[(R+s)^2 - R^2\right]\pi\gamma \quad ..(6)$$

where γ is the fracture surface energy of a crack, and E_o is the elastic modulus of a pore-free solid. The condition for crack extension is obtained by using the well-known thermodynamic criterion introduced by Griffith, i.e.

$$dV/ds = 0 \quad ....(7)$$

Differention of Eq. (6) with respect to s, followed by rearrangement results in the expression relating the strength and the number of cracks (pore) per unit volume:

$$\sigma_f = \left[\frac{\pi\gamma E_o}{D(1-\nu^2)(1+s/R)}\right]^{1/2}\left[1+32(1-\nu^2)NR^3(1+s/R)^3\right]^{-1/2} \quad ....(8)$$

where D is the pore diameter.

Equation (8) may be expressed in terms of pore volume fraction using the expression which relates the number of pores per unit volume, N, and their volume fraction, V,

$$N = \frac{3V}{4\pi R^3} \quad ....(6)$$

which when substituted in Eq. (8) results in:

$$\sigma = \left[\frac{\pi\gamma E_o}{D(1-\nu^2)(1+s/R)}\right]^{1/2}\left[1 + \frac{24V\ (1-\nu^2)(1+s/R)^3}{\pi}\right]^{-1/2} \quad ....(10)$$

For extremely small annular flaws ($s \rightarrow 0$), Eq. (10) reduces to:

$$\sigma = \left[\frac{\pi\gamma E_o}{D(1-\nu^2)}\right]^{1/2}\left[1 + \frac{24V\ (1-\nu^2)}{\pi}\right]^{-1/2} \quad ....(11)$$

However, it should be pointed out that for the limiting case $s \rightarrow 0$, Eq. (8) implies no presence of cracks and the strength of such a solid is governed by the initiation rather than extension of pre-existing flaws. Normally at this extreme (s=0) the linear elastic fracture mechanics concept is no longer applicable.

Figure 3 illustrates the variation of normalized strength (from Eq. (10)) with pore volume fraction and pore size.

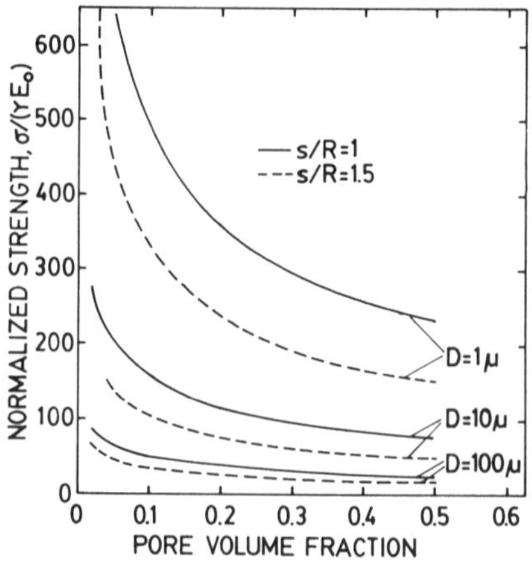

Fig. 3. Predicted variation of strength (from Eq. (10)) with pore volume fraction and pore size (for $\nu = 0.2$).

Evidently, Eq. (10) was derived assuming no interaction between the crack tip stress field and the stress concentration due to the pore. This assumption may not be justified for the case of a pore possessing no radial/annular cracks, where the presence of a stress concentration greatly assists in the initiation of flaws from the pore. However, as soon as the cracks are formed on the surface of the pore, either during stressing or in the course of specimen fabrication, such a crack pore configuration may be considered to be akin to the well known centre crack loaded case, when the stress concentration due to the presence of pore vanishes.

DISCUSSION

It is clear from the present analysis that at small pore sizes both the pore size and the pore volume fraction have a strong effect on strength of brittle solids (see Fig. 3). However, for a constant s/R, the effect of pore volume fraction on material strength diminishes with increased pore size because the contribution of the second term in Eq. (10) is much smaller than the first term. This type of behaviour was observed experimentally in borosilicate glass by Hasselman and Fulrath, 1967. Figure 4 illustrates the change of strength with pore volume fraction in barasilicate glass containing 60μm spherical pores. Inspection of Fig. 4 shows that after an initial sharp drop in strength at small pore sizes, further decrease in strength with increasing pore volume fraction becomes less severe. According to the present model this behaviour may be explained entirely on the basis of the s/R ratio.

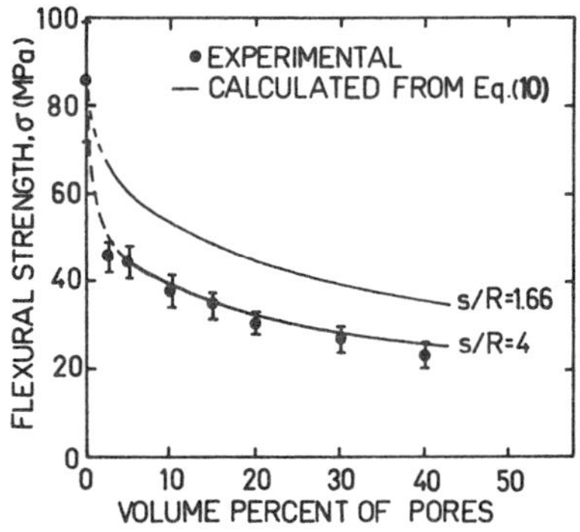

Fig. 4. Predicted and measured variation of strength with volume percent of spherical pores for a borosilicate glass (from Hasselman and Fulrath, 1967).

For a fully dense glass specimen, it is commonly found that the critical flaw size for fracture ranges from 50 to 100μm. If the glass has a fracture surface energy of $\gamma = 4J/m^2$, a Young's modulus of E = 60 GPa and a Poisson's ratio of $\nu = 0.2$, the strength of the fully dense glass (for V = 0 in Eq.(10)) is $\sim\sigma$ = 88.6 MPa. For an annular flaw size of 50μm (equivalent to one half inherent flaw size) and pore radius R = 30μm the s/R ratio becomes $\sim$1.66. When this value for s/R is inserted in Eq. (10), an overestimated value for strength of borosilicate glass is predicted (see Fig. 4). However, when s/R is increased from 1.66 to 4, excellent agreement is obtained between calculated and measured values for strength.

There are two factors which may influence this behaviour. First, the introduction of pores in a fully dense glass solid may also cause an increase in annular flaw size and a corresponding decrease in strength. Secondly, if the inherent flaw size is assumed to remain constant (s = 50μm) for all porosity levels, the introduction of spherical pores in a glass specimen may also lead to the development of stress concentration which then reinforces the annular crack tip stress field and ultimately leads to strength reduction.

Although it is often difficult to precisely measure the inherent flaw size in a porous solid, its fracture behaviour under stress requires a full knowledge of s/R for each porosity level and fabrication condition.

REFERENCES

Berry, J.P. (1960). J. Mech. Phys. Solids, 8, 194-206.
Bertolotti, R.L., and R.M. Fulrath (1967). J. Am. Ceram. Soc., 50, 558-562.
Bowie, O.L. (1956). J. Mat. and Physics, 35, 60-71.
Coble, R.L. and W.D. Kingery (1956). J. Am. Ceram. Soc., 39, 377-385.
Evans, A.G. and R. Tappin (1972). Proc. Brit. Ceram. Soc., 20, 275-297.
Hasselman, D.P.H. and R.M. Fulrath (1967). J. Am. Ceram. Soc., 50, 399-404.
Hasselman, D.P.H. (1969). J. Am. Ceram. Soc., 52, 600-604.
Krstic, V.D. and W.H. Erickson (1987). J. Mater. Sci., 22, 2881-2886.
Paris, P.C. and G.C. Sih (1965). Am. Soc. of Testing and Materials, STP 381, pp. 30.
Snedon, I.N. (1946). Proc. Roy. Soc. (London) A187, 229-260.
Sprigs, R.M. and T. Vasilos (1963). J. Am. Ceram. Soc., 46, 224-228.

MODELING CREEP IN MATERIALS WITH SOFT BOUNDARY PHASES

Mark Belchuk[1], Dan Watt[1],and John Dryden[2]

[1]Department of Engineering Materials, University of Windsor
Windsor, Ontario, Canada N9B 3P4
[2]Department of Materials Engineering, University of Western Ontario
London, Ontario, Canada N6A 5B9

ABSTRACT

A large number of ceramics, some metallic alloys, and a number of composite materials consist of hard, non-deforming crystals surrounded by a softer grain boundary phase. An important factor in the creep of these materials is the gradual redistribution of the boundary phase. A model material, consisting of Wood's metal (melting point 68°C) allows the experimental simulation of creep at essentially room temperature. A computer simulation of the Wood's metal flow has been done for comparison with experimental results.

KEYWORDS

Ceramics; creep; material transport; Grain boundary sliding; Wood's metal; fluid flow.

INTRODUCTION

As interest increases toward attaining higher operating temperatures and lighter weights in advanced engines so has the interest in ceramics as an engineering material for components in these engines. Liquid phase sintered ceramics are of particular interest due to their high strength and good formability. Under moderate temperatures the amorphous grain boundary phase softens, leading to creep dominated by grain boundary separation. The creep deformation in these ceramic materials is then generally governed by the properties of the second phase and its volume fraction.

There are, as well, a number of other types of material with essentially similar microstructures,suggesting that their creep behaviour may be governed by the same basic type of minimechanics. Tungsten carbide tools have a microstructure in which the hard carbides are surrounded by relatively soft cobalt metal (eg. Sigl and Exner (1987)) SiC particles are being used to reinforce aluminum to make a lightweight composite. Nickel based superalloys, appropriately heat treated, consist of cubic blocks of ordered γ' surrounded by a softer nickel base matrix (see Janowski, Harmon and Pletka (1987), Kear (1986) or White (1974)). The rheology of the earth's upper mantle may also be governed by similar creep behaviour. The common element in each of these systems is the material transport in a continuous minor phase. What is presently lacking is a mechanics framework, in which the appropriate constitutive flow equations can be entered, to provide a description of the creep behaviour of the total system.

This paper presents a theoretical approach to simulate the material transport in a model two-phase material containing a soft continuous grain boundary phase. The theory is an extension of some earlier work, particularly that by Drucker (1964), Lange (1974) and Dryden and Watt

(1981). It differs from these in that the present model does not assume that the free surfaces of the material are at infinity. It therefore holds promise in that, in principal, it can predict shape changes and frictional end effects during hot-pressing or during creep.

SLOW VISCOUS FLOW MODEL

Many multi-phase ceramics contain a phase which is comprised of a low melting point glass often found in thin layers along grain boundaries. When these materials are subject to stress at high temperatures the decreased viscosity of the glass phase may cause this phase to flow as a fluid. The equations to describe this viscous flow controlled creep follow.

Equation Development

The equations which describe fluid motion are the continuity equation, the momentum equations, and the energy equation. Enormous simplification can be achieved by introducing a two dimensional model of a two-phase material consisting of a matrix of regular, rigid rectangular grains embedded in a continuous glassy grain boundary phase. The grains measure a by b and with grain boundary thicknesses of h_x and h_z as depicted in Fig. 1. The material is subjected to a stress which causes the slow time dependent deformation (creep) of the viscous film.

The assumptions applied to the above model are as follows:

1) The body forces are neglected.
2) The pressure is constant throughout the thickness of the squeezed film.
3) The grain surface is large compared with the squeezed film thickness.
4) There is no slip at the boundaries.
5) The squeezed film behaves as a Newtonian fluid.
6) The flow is laminar.
7) Fluid inertia is neglected.
8) Fluid viscosity is taken as constant through the thickness of the film.
9) The fluid is incompressible.
10) The grains are infinite in the y-direction (plane strain).
11) The solution is steady state at an instant in time (quasi-steady state).
12) The temperature is constant.

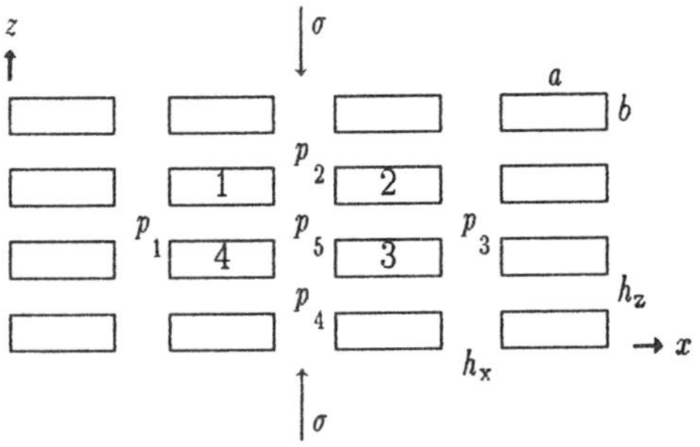

Fig. 1. Geometry of the creep model under compressive loading.

Application of these assumptions to the equations of fluid motion gives

$$\frac{\partial^2 u}{\partial z^2} = \frac{1}{\eta}\frac{\partial p}{\partial x} \tag{1}$$

for motion in the x-direction and

$$\frac{\partial^2 w}{\partial x^2} = \frac{1}{\eta}\frac{\partial p}{\partial z} \tag{2}$$

for fluid motion in the z–direction.

Through integration and the application of boundary conditions involving grains 1, 2, 3, and 4, equations (1) and (2) become

$$u = \frac{1}{2\eta}\frac{\partial p}{\partial x}(z^2 - zh_z) \tag{3}$$

and

$$w = \frac{1}{2\eta}\frac{\partial p}{\partial z}(x^2 - xh_x) \tag{4}$$

respectively. These equations represent the fluid velocity across a grain boundary film. Integrating these expressions across the film gives

$$q_x = -\frac{1}{\eta}\frac{\partial p}{\partial x}\frac{h_z^3}{12} \tag{5}$$

and

$$q_z = -\frac{1}{\eta}\frac{\partial p}{\partial z}\frac{h_x^3}{12} \tag{6}$$

which represent the flow rate of the fluid for unit depth. Assuming displacement rates (w_g, u_g) for the grains and then applying the continuity equation to (5) and (6) gives

$$\frac{\partial^2 p}{\partial x^2} = \frac{12}{h_x^3}\eta\, w_g \tag{7}$$

and

$$\frac{\partial^2 p}{\partial z^2} = \frac{12}{h_z^3}\eta\, u_g. \tag{8}$$

Integration yields

$$\frac{\partial p}{\partial x} = \frac{12}{h_z^3}\eta\, w_g\, x + C_1 \tag{9}$$

and

$$\frac{\partial p}{\partial z} = \frac{12}{h_x^3}\eta\, u_g\, z + C_2 \tag{10}$$

upon which further integration gives

$$p = \frac{6\eta}{h_z^3} w_g\, x^2 + C_1\, x + C_3 \tag{11}$$

and

$$p = \frac{6\eta}{h_x^3} u_g\, z^2 + C_2\, z + C_4. \tag{12}$$

Equations (9) and (10) represent the mass flux of fluid along any channel between grains and equations (11) and (12) give the fluid pressure along any fluid channel. The force on any grain face can be determined by integrating (11) and (12) along the length of a face to get

$$F_z = \frac{\eta}{h_z^3} w_g a^3 - \frac{p_5 + p_1}{2} a \tag{13}$$

for the faces between grains 1 and 4 and

$$F_x = \frac{\eta}{h_x^3} u_g b^3 - \frac{p_5 + p_4}{2} b \tag{14}$$

for the faces between grains 3 and 4. If the force in the x–direction is zero then the velocity of grain 3 in the x–direction is

$$u_g = \frac{h_x^3}{b^2 \eta} \frac{p_5 + p_4}{2} \tag{15}$$

The above equations are common in lubrication theory (Cameron, 1966) for squeezing a fluid film between two plates.

The equations presented above are only valid for the fluid film between grains and does not pertain to the fluid found at the nodes formed where the fluid film channels meet. At these nodes, pressure is assumed to be uniform throughout the node and according to continuity, mass flow is equal into and out of any node. Recall that equations (9) and (10) are the mass flux of fluid along any channel. Therefore a summation of these equations at each node must equal zero. For the node involving grains 1, 2, 3, and 4 in Figure 1, the summation yields

$$\frac{6a}{h_z^3} w_g a + \left[\frac{1}{a} + \frac{4}{b}\right] p_5 - \frac{p_1 + p_3}{2a} + \frac{p_2 + p_4}{b} = 0 \tag{16}$$

which for a generalized node is

$$\frac{6a}{h_z^3} w_g a + \left[\frac{1}{a} + \frac{4}{b}\right] p_{i,k} - \frac{p_{i-1,k} + p_{i+1,k}}{2a} + \frac{p_{i,k+1} + p_{i,k-1}}{b} = 0 \tag{17}$$

where only the pressures are unknown. For a matrix of grains as in Fig. 1 all of the nodes can be described using this equation leading to a system of simultaneous algebraic equations easily solved by computer.

APPLICATION OF THE MODEL

Most models spawn from a number of assumptions which often oversimplify the materials for which they are intended. The present model does not escape from this dilemma so a model material is introduced which conforms to the assumptions used in formulating the model.

Creep of a Model Material

The model material should have two phases, consisting of rigid blocks surrounded by a soft Newtonian fluid. Hardened steel was chosen to be the block material and Wood's metal, the soft phase. The Wood's metal is an alloy of tin, lead, cadmium, and bismuth whose melting point is 68°C. By conducting creep tests in compression on cylindrical samples of Wood's metal it was determined that it behaves as a Newtonian fluid during slow deformation near room temperature.

To date the model material has consisted of a stack of Wood's metal plates sandwiched between hardened steel blocks as seen in Fig. 2. The plates were then loaded at constant stress using a hydraulic press.

The variables of interest for creep of this model material are temperature, stress, and strain rate. All transducer measurements were monitored and recorded by a computer. In addition to temperature measurement by thermocouple, the sample temperature was also controlled to 35 ± 0.2°C. A load cell enabled the force to be recorded and a linear variable displacement transducer (LVDT) was used to measure the strain. All measurements were recorded with respect to the starting time of each experiment.

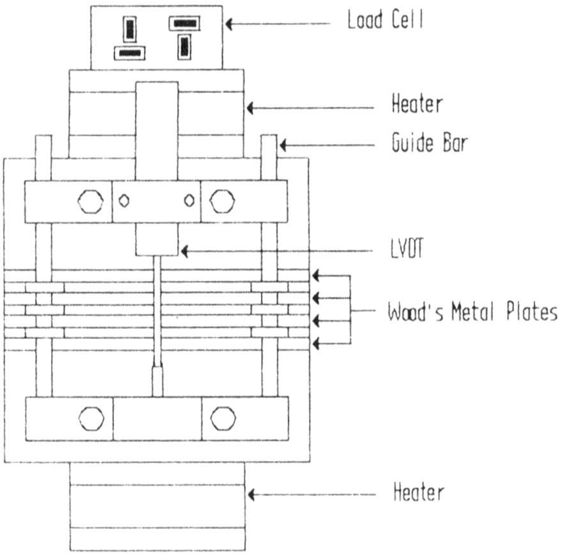

Fig. 2. Wood's metal model material setup.

The hydraulic system, appearing in Fig. 3, consisted of two pistons, one much larger than the other. The pressure, created by hanging weights on the small piston was transferred to the large piston resulting in a load 28 times that which was hung. It was expected that a constant load would be maintained on the samples, however it was found that static and dynamic frictional effects within the hydraulic system proved this to be unreliable. An effort is being made to determine the characteristics of this loading system in order to obtain meaningful creep data.

Fig. 3. The hydraulic loading system.

RESULTS AND DISCUSSION

Using the mathematical model creep simulation was conducted on a material containing 16 square blocks or grains, 4 blocks high by 4 blocks wide. Between the blocks is a material of known viscosity and thickness. A displacement rate is imposed on the top row of blocks causing a pressure distribution between the blocks according to equations (11) and (12). All the nodal pressures are calculated using equation (17) then the displacement rates and forces are determined from equations (13) and (14). The new displacement rates are used to recalculate the pressures and the iterations continue until an acceptable error level is achieved.

Typical output for a system consisting of blocks measuring 10 meters square surrounded by a fluid whose viscosity is 1.0 Pascal-seconds appears in Fig. 4 below. Initially the fluid thickness is 1.0 meters along all the boundaries. The output contains the nodal pressures in Pascals, the z-directed force in Newtons, the horizontal and vertical boundary thickening rate in meters per second, and the horizontal and vertical boundary thicknesses after one time step in meters. As expected the nodal pressures are symmetric. In addition, the total forces are equal for every cut taken across the width of the simulated sample. The boundary thicknesses calculated after one time step (1.0 seconds) show the shape changes (barrelling) that typically occur in compression tests due to frictional end effects.

Nodal Pressures

0	0	0	0	0
0	11.24	12.16	11.24	0
0	9.88	10.44	9.88	0
0	11.24	12.16	11.24	0
0	0	0	0	0

z-Directed Force				Total Force
-156.2	-217.0	-217.0	-156.2	-746.4
-160.5	-212.7	-212.7	-160.5	-746.4
-156.2	-217.0	-217.0	-156.2	-746.4

Rate of Horizontal Boundary Thickening

-0.1	-0.1	-0.1	-0.1
-0.1068	-0.1154	-0.1154	-0.1068
-0.1	-0.1	-0.1	-0.1

Rate of Vertical Boundary Thickening

0.0562	0.0608	0.0562
0.1056	0.1130	0.1056
0.1056	0.1130	0.1056
0.0562	0.0608	0.0562

Horizontal Thickness After 1 Time Step

0.9	0.9	0.9	0.9
0.8932	0.8846	0.8846	0.8932
0.9	0.9	0.9	0.9

Vertical Thickness After 1 Time Step

1.0562	1.0608	1.0562
1.1056	1.1130	1.1056
1.1056	1.1130	1.1056
1.0562	1.0608	1.0562

Fig. 4. Computer output of simulated creep.

CONCLUSIONS

There exist a number of multiphase materials whose separate phases have vastly different properties. For these materials there is a need for a mathematical model that will link the overall force-displacement behaviour to the material transport of the constituent phases.

Compressive creep experiments on cylinders of Wood's metal alloy have shown that it behaves as a Newtonian fluid and is therefore a candidate for use in a model material used to describe material transport in two-phase materials.

Computer simulation of a minimechanical model material demonstrates the redistribution of the boundary phase that may occur within a sample.

NOTATION

a	Grain length (L)
b	Grain height (L)
C_1	Constants of Integration
F_x	Grain boundary force in x-direction (MLt^{-2})
F_z	Grain boundary force in z-direction (MLt^{-2})
h_x	Grain boundary film thickness in x-direction (L)
h_z	Grain boundary film thickness in z-direction (L)
p	Pressure ($ML^{-1}t^{-2}$)
q_x	Volume flow rate in x-direction (L^3t^{-1})
q_z	Volume flow rate in z-direction (L^3t^{-1})
u	Fluid velocity in x-direction (Lt^{-1})
u_g	Grain velocity in x-direction (Lt^{-1})
w	Fluid velocity in z-direction (Lt^{-1})
w_g	Grain velocity in z-direction (Lt^{-1})
x	Rectangular coordinate (L)
y	Rectangular coordinate (L)
z	Rectangular coordinate (L)
η	Viscosity ($ML^{-1}t^{-1}$)

REFERENCES

Cameron, A. (1966). In A. Cameron (Ed.), *Principles of Lubrication*, John Wiley and Sons Inc., New York. 389-391.
Drucker, D. C. (1964). In V. F. Zackay (Ed.), *High Strength Materials*, John Wiley and Sons Inc., New York. 795-833.
Dryden, J. R. and D. F. Watt. (1981). In J. Pask and A. Evans (Ed.), *Surfaces and Interfaces in Ceramic and Ceramic–Metal Systems*, Plenum Publishing York. 261-270.
Hart, P. E., R. B. Atkin, and J. A. Pask. (1970). *J. Am. Ceram. Soc.*, 53, 83-86.
Janowski, G. M., B. S. Harmon, and B. J. Pletka. (1987). *Met. Trans.*, 18A, 1341-1351.
Kear, B. H. (1986). *Scient. Am.*, Oct., 159-167.
Lange, F. F. (1974). *Non-Elastic Deformation of Polycrystals with a Liquid Boundary Phase*, Westinghouse Electric Technical Report.
Sigl, L. S. and H. E. Exner. (1987). *Met. Trans.*, 18A, 1299-1308.
White, C. H. (1974) In W. Betteridge and J. Heslop (Ed.), *The Nimonic Alloys*, Crane, Russak and Co. New York. 83.

High Temperature Stability of Structural Ceramics

D. S. Wilkinson, M. Chadwick and A. G. Robertson

Department of Materials Science and Engineering
McMaster University
Hamilton, Ontario L8S 4L7, CANADA

ABSTRACT

Materials developed for high temperature applications are often microstructurally unstable under service conditions. The nature of these microstructural changes needs to be understood and their effect on properties determined. In this paper we discuss microstructural instabilities in two structural ceramic materials. Sintered silicon nitride contains glass which tends to crystallize during exposure at the expected service temperatures for certain heat engine components. This has been studied using analytical electron microscopy. In addition, the effects on creep strength and lifetime have been measured. We find that only partial devitrification occurs, with a stable glass phase developing. This reduces the lifetime of the material considerably, while increasing its strength marginally. In hot-pressed alumina, exposure at elevated temperature leads to swelling, even in the absence of applied stress. This is due to the presence of trapped gas under pressure in the residual porosity. Swelling is accompanied by extensive grain growth. A model for this process has been developed which correctly predicts the behaviour.

KEYWORDS

Silicon nitride, alumina, glass, devitrification, electron microscopy, swelling, grain growth, microstructure, instability.

INTRODUCTION

The development of advanced ceramic materials now makes it possible to consider such materials for load-bearing applications at elevated temperatures. One of the main concerns in this regard relates to the microstructural stability of these materials during extended service. If the microstructure changes with time then so will a range of material properties including those on which the design is based. Thus design for elevated temperature applications must not be based simply on initial or short term material properties (fracture toughness, for example), but on long term properties (e.g. creep rupture strength and lifetime) combined with an understanding of the processes by which microstructural degradation occurs.

Microstructural degradation can take a variety of forms depending on the material involved. For example, ceramics produced by hot pressing or HIP processing often contain trapped gas under pressure within the residual porosity. Even in sintered materials, gas pressure can develop due to internal reactions(1). When these materials are later subjected to elevated temperatures, diffusion is possible, enabling the pores to expand under pressure (2,3,4,5). This desintering process can weaken the materials, especially if it is allowed to proceed to the point where pore extensive pore coalescence leads to the development of microcracks.

Many ceramic materials are thermodynamically unstable at service temperatures. For example, significant amounts of glass are often present, as a result of additives used to promote liquid phase sintering. Upon cooling this liquid tends to form a glass which may be unstable at the expected service temperatures. Both debased aluminas and sintered silicon nitride ceramics exhibit this kind of behaviour. When subjected to elevated temperature the glass crystallizes, although not always fully. For example, Wiederhorn and Fuller (6) have shown that in a vitreous bonded alumina, annealing prior to creep leads to extensive crystallization. The result is a dramatic increase in strength and an increase in lifetime.

In this paper we review two studies currently underway to better understand various types of microstructural instability. The first is concerned with devitrification of grain boundary glass in sintered silicon nitride. The emphasis in this study has been placed on the use of analytical electron microscopy to better understand the nature of the process. The second is concerned with swelling and grain growth in hot-pressed alumina. Here a detailed experimental study has been combined with extensive modelling.

DEVITRIFICATION IN SILICON NITRIDE

We have studied the devitrification of grain boundary glass in a commercial silicon nitride, Kyocera SN220. This material contains Al_2O_3 and Y_2O_3 as sintering additives. In addition, like all silicon nitrides, some SiO_2 is present. After firing, the material has a microstructure consisting primarily of β-Si_3N_4 grains, plus some glass in the Y_2O_3-Al_2O_3-SiO_2 system (see Fig. 1a).

Following annealing at 1200°C for periods in excess of 100 hours we find a rather different microstructure, as shown in Fig. 1b, in which much of the glass has crystallized into several different phases. In terms of the mechanical properties of this and similar materials at elevated temperatures, it is of interest to ask what crystalline phases are formed, how much glass remains and is it stable, and how homogeneous is the microstructure? In terms of homogeneity, we are concerned to know whether or not the material differs between the surface and the interior of the samples, as well as the extent to which local inhomogeneities exist. Questions such as these can only be answered with the use of analytical electron microscopy techniques. In the current study we have combined three techniques- electron diffraction, energy-dispersive x-ray analysis (EDX), and electron energy loss spectroscopy (EELS) - to determine the structure and composition of SN220 following annealing. The results are briefly presented here. A more complete description of this work will be shortly available (7,8).

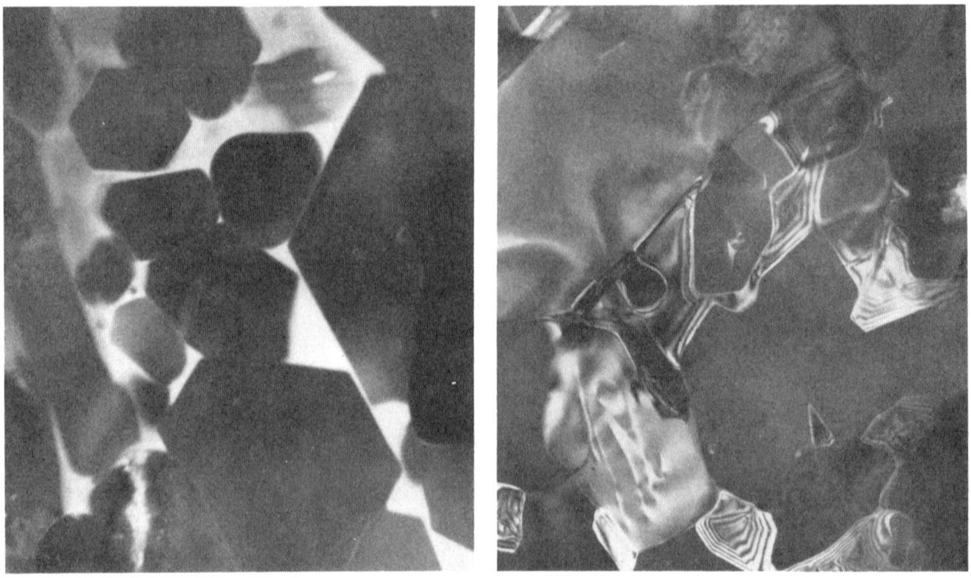

Fig. 1: a) A dark-field electron micrograph of as-received silicon nitride showing the glass phase (light) surrounding dark β-Si_3N_4 grains. b) Following annealing for 100 hours at 1150°C, this dark-field image shows that some of the glass has crystallized. These phases can be identified by electron diffraction.

The crystalline phases which form during annealing in an inert atmosphere are $Y_2Si_2O_7$ and $Y_3Al_5O_7$ (or YAG - yttria alumina garnet). When samples are annealed in air however, YAG is found only in the centre of the samples. Near the surface, a small amount of mullite ($Al_6Si_2O_{13}$) is found instead. The composition of the glass phase before and after annealing has been determined using both EELS and EDX. The results are illustrated on a ternary "behaviour" diagram in Fig. 2. The most interesting discovery from this analysis is that the composition of the glass following annealing in air is the same in the centre and near the surface of these samples, despite the different crystalline phases which form. This suggests that a stable glass composition exists for this system. Further evidence is provided by long time annealing experiments in which the composition and amount of glass in the material following 500 hours at 1250°C is the same as that found after 200 hours.

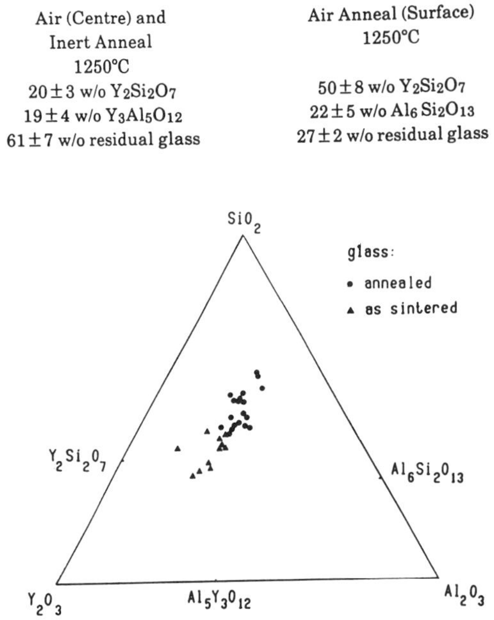

Fig. 2: A ternary "behaviour" diagram for the Y_2O_3-Al_2O_3-SiO_2 system, indicating the phases which are formed and the composition of the glass phase before and after annealing. Shown above the diagram are the weight percent of the original glass which is converted into each final product. The result is different near the surface of the air-annealed samples than in the interior and throughout the inert atmosphere-annealed samples.

Having established the composition of the amorphous and crystalline phases which form during annealing it is possible to use an elemental mass balance to estimate the amount of the original glass which actually crystallizes. The results of this analysis are shown at the top of Fig. 2. There is a considerable amount of glass still remaining throughout the material following annealing, amounting to about 22 weight percent of the original glass. The extent of devitrification is lower near the surface of air annealed samples than elsewhere because the diffusion of oxygen into the glass enhances the formation of mullite at the expense of Y_AG. This is consistent with microstructural observations which show the formation of glass films on the surface of air annealed samples.

The stability of the glass is likely to have important consequences for the creep properties of the material. It has often been suggested that good creep resistance could be achieved by devitrifying all (or nearly all) of the grain boundary glass in such materials. This is because the glass softens at elevated temperatures and enables deformation to occur, either by means of viscous flow in the glass (9,11), or through the dissolution-reprecipitation process (10). In either case, creep is enhanced by the presence of a thin film of glass along grain boundaries. Since most of the glass in silicon nitride and similar materials resides in pockets at grain boundary triple junctions, most of the glass needs to be devitrified before these films are removed.

In order to study this, we have performed a series of 4-point bend creep tests on material annealed for various lengths of time. The results are summarized in Fig. 3. They show that there is a small strengthening effect due to prior annealing at 25 or 100 hours. However, it is not very large. Of greater significance is the decrease in strain rate which is observed with increasing strain.

Over a wide range of applied stresses, and for all annealing conditions the strain rate decreases dramatically at a strain of about 0.25%. The cause of this behaviour is not clear. However, it can be explained assuming that creep is controlled by viscous flow of the glassy phase. This a transient process in which glass is squeezed out from grain boundary regions in compression and into large pockets or along boundaries in tension. As the thickness of the glass on the compressive boundaries decreases the material becomes stronger. A recent model for this process (11) suggests that the strain at which this strengthening should occur is about h/L, where h is the original glass layer thickness and L is the grain size. For a grain size of about 5 μm, this suggests a glass layer thickness of about 15 nm, which is not an unreasonable value for silicon nitride.

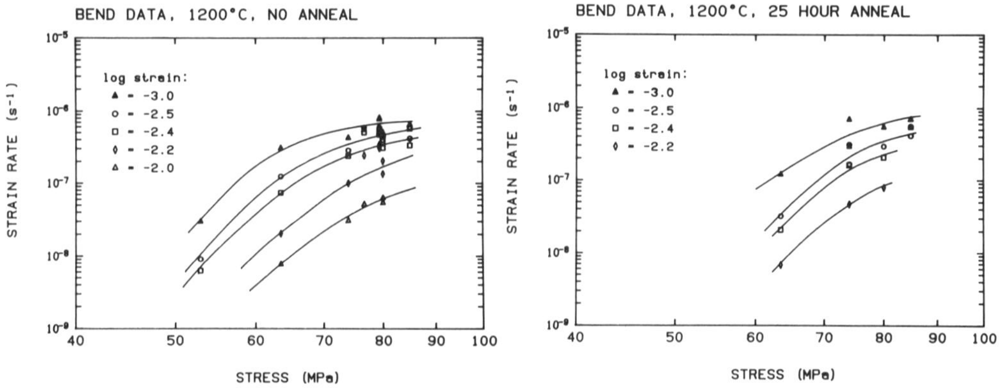

Fig. 3: a) The strain rate plotted as a function of nominal applied stress (actually the outer fibre elastic tensile stress in 4-point bending). These data are for material tested without prior annealing. The contours represent different outer fibre strains. A sharp drop in strain rate is observed at a strain of about 0.0025.
b) The same plot but for material annealed for 25 hours prior to testing. A small decrease in creep rate at any given strain is found. However the data is dominated by the sharp drop in strain rate at about the same strain as observed for the unannealed material.

Although the strengthening effect due to devitrification is modest, there is a noticeable, detrimental effect on creep rupture life. Thus annealing leads predominantly to embrittlement. There are several reasons for this. First of all when samples are annealed in air pits form on the surface due to oxidation. These can act as crack nucleation sites leading to creep rupture. However, even if the surfaces are repolished to remove these pits prior to testing, there remains a significant decrease in creep life. The most plausible explanation is that devitrification limits the ductility of the material. As noted already, devitrification affects primarily the pockets of glass at grain boundary triple junctions. These pockets do not greatly affect creep resistance (which is controlled by the thin glassy films along grain boundaries), but they do affect ductility.

As grain boundary sliding occurs during creep, stress concentrations build up which can be relieved by viscous flow in the glass pockets. Moreover microcracks running into these pockets can be blunted. If they are replaced by crystalline material, then the material is less accommodating and creep ductility decreases.

It is therefore clear that the high temperature mechanical properties of glass-containing ceramics are not always improved by annealing treatments which result in glass crystallization. The process involves a competition between increasing strength and decreasing ductility. When devitrification is incomplete loss of ductility dominates and creep life is reduced.

SWELLING IN HOT-PRESSED ALUMINA

We have studied the process of swelling in two different commercially available aluminas. Swelling involves the appearance, growth and coalescence of grain boundary voids, and is accompanied by simultaneous grain growth. The results for one of the materials studied are shown in Fig. 4a and 4b. They indicate a steady increase in the porous volume fraction (as measured by densitometry), and in grain size with increasing time and temperature. If the pore volume is plotted as a function of grain size, as in Fig. 4c, it becomes clear that the two processes are linked. An extensive investigation of the microstructure which develops during grain growth shows that pores remain at triple grain junctions throughout grain growth. They must therefore migrate with the grain boundaries. Thus, the number of pores observed per unit area on a polished plane rises initially, and then falls (see Fig. 4d). This is a direct result of pore coalescence during grain growth. The same general behaviour is seen in a different hot-pressed alumina, although the actual rate of swelling is less. The two materials differ primarily in their dopant and impurity levels. Thus differences in grain boundary and pore mobilities can be expected.

The process of simultaneous swelling and grain growth has been modelled, assuming that swelling is driven by gas trapped in pores (and not by gas generated internally during swelling), and that grain growth is controlled by a pore drag process. In addition we have considered a range of assumptions related to coalescence process. At one limit we assume that the number of pores per unit grain f_3, remains constant throughout. As a second case we assume the number of pores per unit of grain boundary area N, remains constant. These two assumptions bracket the range of the experimental observations. With these assumptions it is possible to develop a model (12) which accounts for all of the observed features. For example the predicted profiles for number of pores per unit area on a polished plane are plotted in Fig. 5 for both the constant f_3 and the constant N case. They are clearly consistent with the experimental results (Fig. 4d).

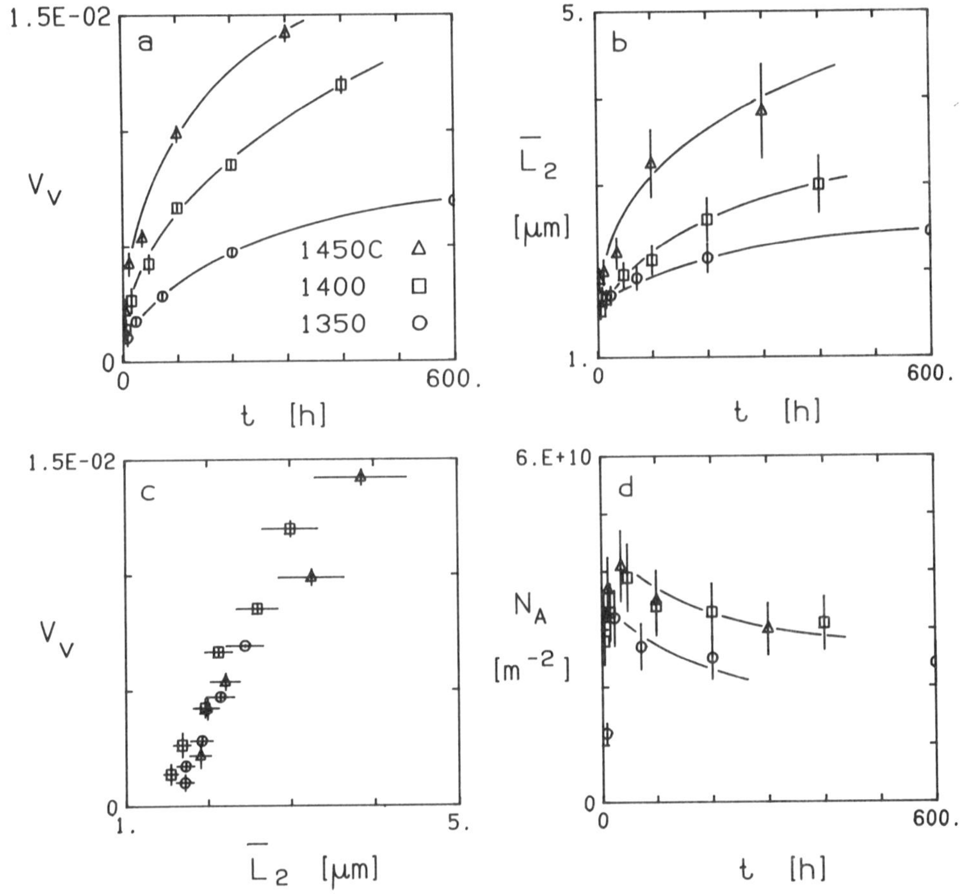

Fig. 4: a) The volume fraction of pores and b) the grain size, as a function of annealing time and temperature. These data are cross-plotted in (c) to indicate that the correlation between grain growth and swelling is independent of temperature. Finally, the number of pores per unit area on a polished plane is plotted in (d). The effect of pore coalescence during grain growth is clearly evident.

SUMMARY

Structural ceramics exhibit various types of microstructural instability at elevated temperatures. All of these affect in some way, the mechanical properties of the material after extended periods of time at temperature. Most often the effect is deleterious. In silicon nitride materials one of the main forms of degradation is glass crystallization. If the glass is fully crystallized then increased strength and lifetime are possible. However if only partial crystallization occurs then a severe loss of ductility may occur, with very little strength increase to compensate. In hot pressed materials swelling is often found. This creates porosity which may accelerate the development of damage leading to fracture. While this process has been studied in detail only for alumina, it may also occur in more complex materials, such as whisker reinforced aluminas.

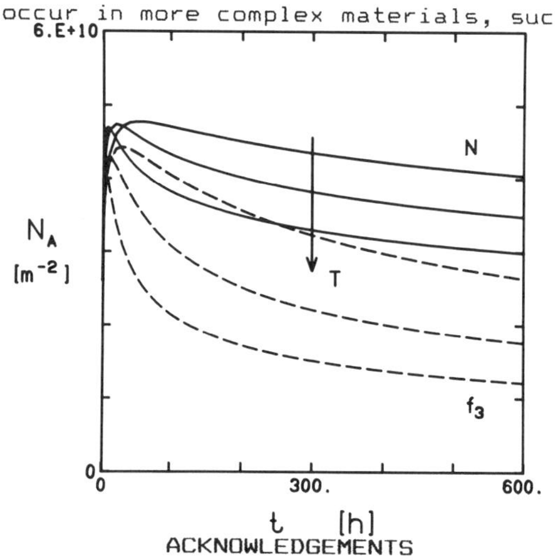

Fig. 5: The predicted pore number density according to the model. This should be compared with Fig. 4d. The same trends for N_A as a function of both time and temperature are observed.

ACKNOWLEDGEMENTS

Much of the electron microscopy of silicon nitride was performed in collaboration with Dr. T. Malis of the Metals Technology Laboratories, CANMET, Ottawa. We are also pleased to acknowledge the financial assistance of Alcan, Intl., URIF-Ontario, and NSERC-Canada.

REFERENCES

1. N. A. Haroun and M. A. A. El-Masry, Ceram. Intl. 7 (1981), 38.
2. A. A. Solomon and F. Hsu, J. Amer. Ceram. Soc., **63** (1980), 467.
3. S. P. Howlett and R. J. Brook, Proc. 2nd Irish Durability and Fatigue Conf. (1984), 320.
4. S. J. Bennison and M. P Harmer, Adv. in Ceram., **6** (1983, Amer. Ceram. Soc., Columbus, OH).
5. A. G. Robertson and D. S. Wilkinson, J. de Physique, Colloque C1, suppl. au no. 2, Tome 47, C1.661.
6. S. M. Wiederhorn and E. R. Fuller, Jr., Mater. Sci. Engin., **71** (1985), 169.
7. M. Chadwick and T. Malis, submitted for publication (1988).
8. M. Chadwick, D. S. Wilkinson and T. Malis, to be published (1988).
9. D. C. Drucker, in High Strength Materials (V. F. Zachary, ed.; Wiley, 1964), 795.
10. R. Raj, J. Geophys. Res., **87** (1982), 4371.
11. J. R. Dryden, D. Kucerovsky, D. F. Watt and D. S. Wilkinson, "Creep Deformation due to a Viscous Boundary Phase", submitted.
12. A. G. Robertson and D. S. Wilkinson, to be published.

FABRICATION OF Si_3N_4 VALVE DISCS

M.D. Pugh and R.A.L. Drew

Department of Mining and Metallurgical Engineering
McGill University
Canada

J.R. Wright and G.D. Webster
National Research Council

KEY WORDS:

Silicon nitride, engine components, processing ceramics, mechanical properties.

ABSTRACT

Si_3N_4 *was chosen as the candidate material to form part of the valve for a medium speed diesel engine. The fabrication procedure was by isopressing, green machining, low pressure sintering and final machining. The mechanical properties of the material are discussed in light of the critical defects causing fracture. The influence of processing and machining flaws on the strength is highlighted. The performance of the valve disc during engine trials is also discussed.*

INTRODUCTION

Non-oxide ceramics are receiving a great deal of attention as potential materials to replace certain components in heat engines, in particular those parts which are exposed to high temperatures, wear and corrosion (Bennett, 1986). Many of the envisaged applications are in gas turbine engines where the push is to increase the operating temperature of the engine beyond the capabilities of superalloys, i.e. $>1100°C$.

However, there are numerous applications where ceramic materials offer solutions to other problems in heat engines, that is, where high temperatures are not necessarily the major factor, or where raising the temperature by a small amount may lead to improvements in operating efficiency (Kabat, 1988). One example of such applications is in diesel engines where the operating temperatures are well below 1000°C. In such cases, the combination of temperature and corrosive combustion products leads to a deterioration of the most corrosion and high temperature resistant metallic alloys. Hence, the only solution is to introduce other materials, such as monolithic ceramics, which should offer greater resistance to this aggressive environment as a result of their superior properties (Drew, 1988).

A new design of Bombardier engine is presently undergoing trials in the Division of Mechanical Engineering (Fuels and Lubricants Laboratory) at NRC in Ottawa. The intention is to operate this engine with "residual" fuels. However, on combustion, this type of fuel produces a variety of corrosive gases, notably vanadium oxides. These gases attack the metal alloy valves and seats, leading to pitting and loss of compression in the combustion chamber. Since structural ceramics, in particular dense silicon nitride, should exhibit superior resistance to this environment than traditional metal alloys such as Stellite and other high temperature alloys, this ceramic was chosen to replace the lower section of the valve. The monolithic ceramic is in the form of a tapered disc (see Figure 1), which is attached mechanically to the remainder of the valve stem.

This paper outlines the fabrication of silicon nitride monolithic valve discs, and includes powder preparation through sintering of the machined discs. In addition, property determination and microstructural examination of the disc material will be presented. The performance of the ceramic valve component in the engine environment will also be discussed.

FABRICATION PROCEDURE

Powder Preparation

Silicon nitride is highly covalent in nature and hence the self-diffusivity is extremely low, even at the decomposition temperature of 1900°C. Therefore, densification can only be brought about by the use of sintering aids or additives which lead to the formation of a liquid phase and promotes densification (Terwilliger and Lange, 1975). The silicon nitride and sintering aids which were used for fabricating the valve discs are listed in Table I along with the manufacturers' specifications. All materials are >99.5% purity.

Table I: Specifications of Starting Materials.

Material and Grade	Supplier	Mean particle size (microns)	Surface area (m^2/g)	Other specifications
Si_3N_4 LC12	Starck	0.5	20	>90% α
Y_2O_3 5603	Molycorp	1.8	39	-
Al_2O_3 A16SG	Alcoa	0.5	9	α-Al_2O_3
AlN Grade C	Starck	1.0	6	

The total amount of additive used for sintering is typically 5-15%. The composition of the starting powders used to fabricate the discs was determined under a contract with Almax Industries and thus remains confidential (Drew, 1987). The silicon nitride and additive powders were attrition milled in 500g batches for 30 minutes using reagent grade isopropyl alcohol as a milling fluid to minimize hydrolysis of the powder (Zarnon, 1988). The milled slurry was separated from the media and microwave dried to remove the alcohol. This was followed by granulation through a 212 micron sieve.

In order to ensure complete removal of volatiles, the granulated powder was then dried at 500°C for 1 hour in an oven. The batches of powder were mixed together and packaged for shipment to Atlantic Research Laboratory (ARL).

Pressing and Green Machining

The procedure for pressing was developed through a number of iterations as result of discussion between McGill and ARL. The original method was to isostatically press a large block approximately 4" diameter and 4" thick in a rubber mould capped at one end with an aluminum plate. Cracking resulted due to uneven pressure gradients at the metal/powder interface. A solution to this problem was found by placing a bag of powder between the plate and the material to be pressed. The pressures used for isopressing were either 5000 psi or 25,000 psi. Various sizes of "rough-pressed" blocks were produced in this fashion.

In order to develop sufficient strength for conventional machining, presintering was carried out on the blocks at 1100°C for 1 hour, employing a slow heat and cool ramp to avoid thermal stresses. No shrinkage occurred during this stage.

Machining of the discs was carried out using conventional lath techniques and tooling. Details of the procedures used will be discussed in a separate report. In order to take account of the dimensional changes between the compacted and sintered state, linear shrinkages were established on blank specimens under the same sintering conditions described below. The values were used in predicting the green dimensions required to achieve the final net shape on sintering. A small allowance was made for final grinding. Examples of machined components are shown in Figure 1. A number of other test pieces were machined for use in materials characterization.

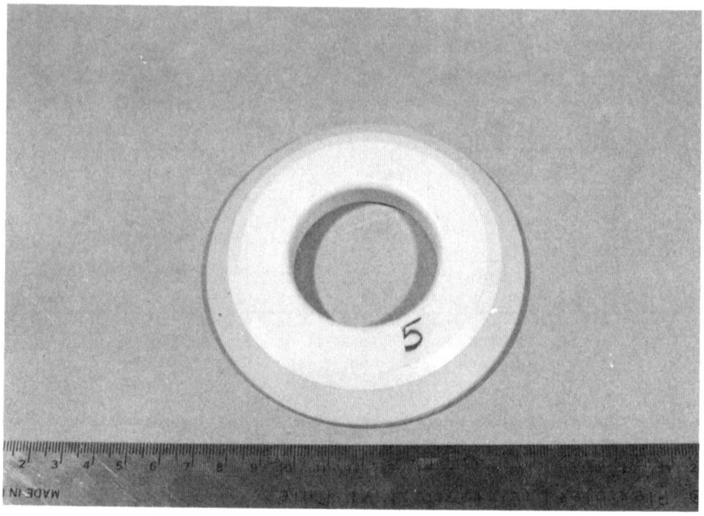

Figure 1: Green machined valve disc.

Sintering

The machined components were placed in a 5" diameter, 99% alumina crucible and were surrounded with a bed of 50% BN and 50% Si_3N_4 powder. The powder bed separated the components and suppressed the decomposition of the Si_3N_4 in the compacts during sintering, thus reducing weight losses. The samples were carefully weighed prior to loading them into the crucible.

The firing procedure was a heating and cooling rate of approximately 10°C/minute with a 1 hour soak at the sintering temperature of 1800°C. The density data and linear shrinkages for each pressing condition for the various pieces were measured by the Archimedian method.

Mechanical Properties

Strength assessment was made by 4-point modulus of rupture (MOR) with inner and outer knife edge dimensions of 15 mm and 30 mm, respectively. The bars were either (a) cut from sintered blocks using a diamond saw, or (b) originated from individually pressed bars. In either case the samples were diamond ground to a 200-mesh finish on a surface grinding machine; this ensured that the compressive and tensile faces of the pieces were flat and parallel. Typical final dimensions were 41mm x 7mm x 5mm. The edges were then bevelled to eliminate stress concentrations. Strength testing was carried out at room temperature on an Instron 1360 system employing a cross-head speed of 0.5 mm/minute.

RESULTS AND DISCUSSION

Sintering Behaviour

Table II gives the green and fired densities as well as the linear shrinkage values for the various pieces. The samples marked * were individual strength bars processed under controlled conditions at McGill. Weight losses were <1% in all cases, indicating that little decomposition had occurred during sintering. X-ray diffraction analysis indicated that the samples were composed entirely of β-Si_3N_4, as would be expected using the above sintering conditions.

Table II: Pressing and Sintering Data.

Description	Isopressing pressure (psi)	Green density (Mg/m^3)	Sintered density (Mg/m^3)	Linear Shrinkage (%)
Valve Disc	25,000	1.93	3.12	14.6
MOR samples (bulk)	25,000	1.93	3.06	13.8
MOR samples* (individual)	6,000	1.76	3.07	18.3
MOR samples* (individual)	25,000	1.92	3.07	14.1

Figure 2 indicates the effect of isopressing pressure on the green and fired densities of the material. As expected, there is a significant variation of green density with pressing pressure; however, only minor differences in sintered density are apparent.This is to be expected since densification is through a liquid phase which forms at temperatures above 1400°C and acts as the major driving force for densification. However, a much larger linear shrinkage occurs for lower isopressing pressures (viz. 18.3% at 6,000 psi) which makes close dimensional control on firing more difficult to attain.

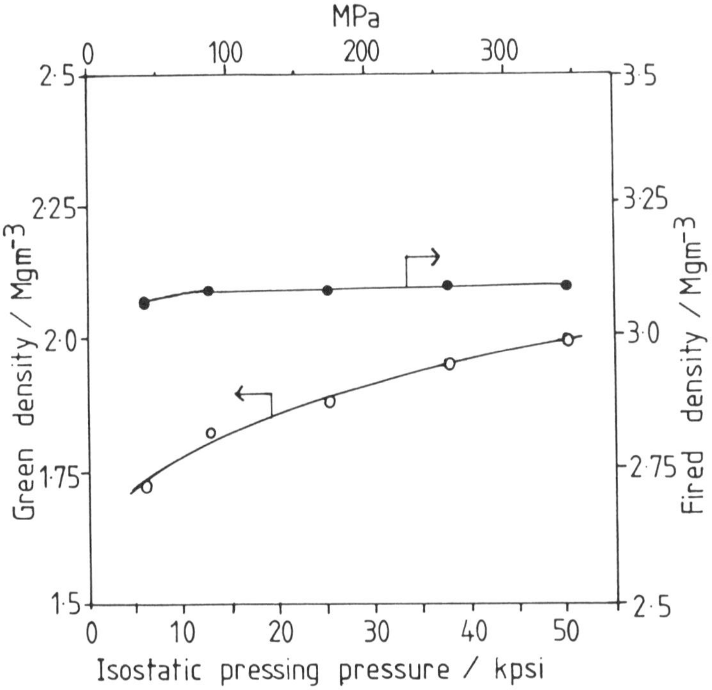

Figure 2: Green and fired densities as a function of isopressing pressure.

Mechanical Properties

Table III summarizes the results of strength measurements both for bars cut from a blank disc and for individual strength bars.

A comparison of Weibull plots for the strength data is given in Figure 3. The Weibull moduli (m) in Table III were calculated by linear regression analysis on these curves and shows the large scatter in the strength values. However, closer observation of the curves indicates a bimodal or even trimodal (bulk samples) distribution of strengths. This implies that there are a number of strength values which are relatively high and in a narrow range , but some of the data points occur at very low values of strength with a much smaller slope.

Table III: Summary of Mechanical Properties.

Sample description	Pressing pressure (psi)	Mean strength (MPa)	$\sigma_{0.632}$ (MPa)	Weibull modulus (m)	Number of samples
Bars cut from bulk (NRC)	25,000	360	405	3	22
Individual bars (McGill)	25,000	535	565	7	10
Individual bars (McGill)	6,000	535	570	7	11

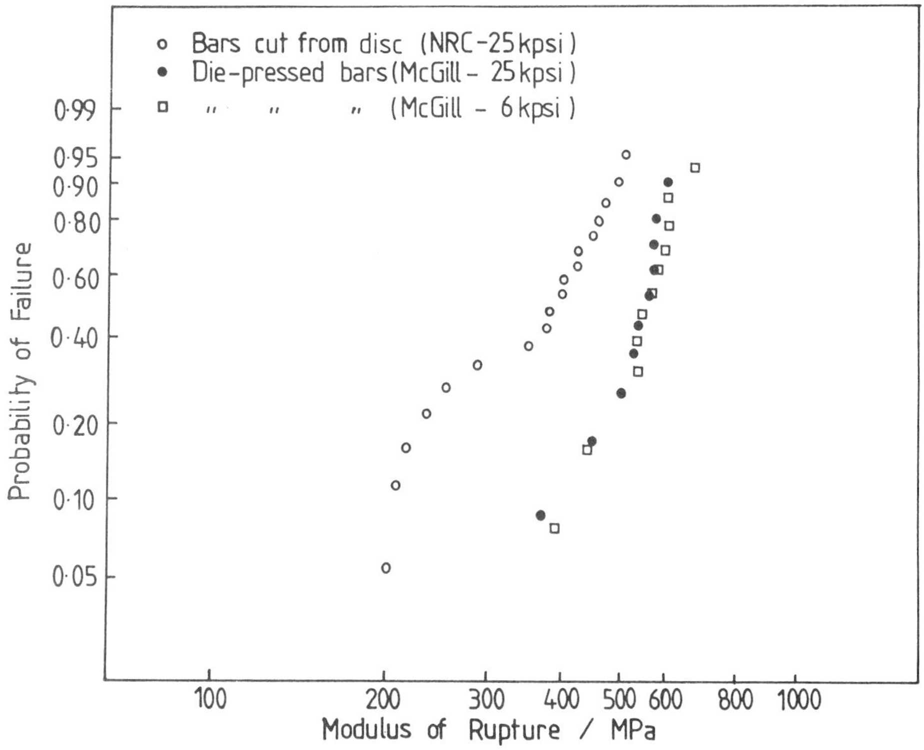

Figure 3: Weibull plots for bulk and individual MOR bars.

Microstructural examination of fracture surfaces was performed on the scanning electron microscope to identify the origin of failure for material pressed at 25,000 psi. Table IV summarizes the type of initiating flaws and their approximate dimensions for selected samples taken from the bulk disc and the individual fabricated bars after fracture. Examples of typical defects which caused fracture are shown in Figures 4-5.

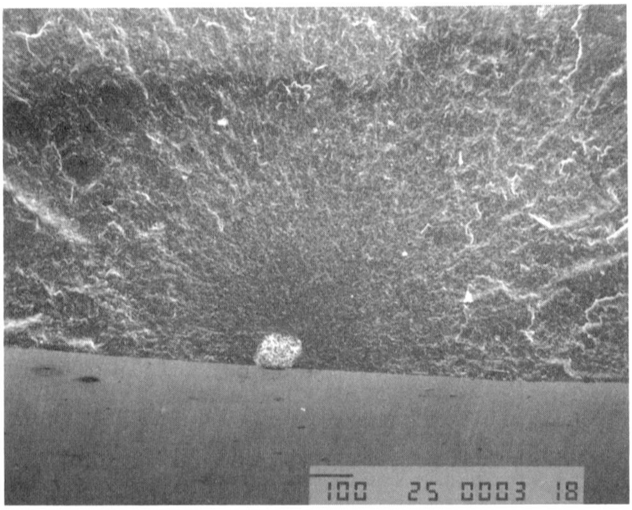

Figure 4: Surface pore revealed by grinding.

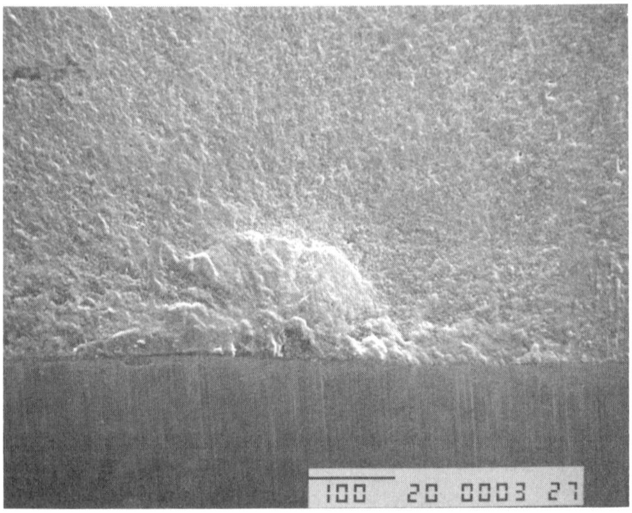

Figure 5: Half-penny crack left by machining.

Table IV: Flaw Origin and Dimensions (isopressed 25,000 psi).

#	Bulk Disc		Individual bars	
	Flaw origin	Flaw size (microns)	Flaw origin	Flaw size (microns)
1	Pore	300	Machining	50
2	Machining	100	Pore	75
3	Pore	50	Machining	35
4	Pore	100	Machining	25
5	Pore	150	Internal pore	100
6	Internal pore	150	Internal pore	80
7	Pore	50		
8	Internal pore	200		
9	Machining	150		

In general, a predominance of large pores was found in the bulk disc samples with defect sizes ranging from 50-300 microns. On the other hand, in the case of the individual bars, the defect sizes were in a narrower size range of 25-100 microns. These observations would explain both the low mean strength values of the bulk disc and the small Weibull modulus. One likely explanation for this is inadequate control over processing of the powder since most of the pores seem to originate from either poor sintering or burn-out of material (probably of organic origin). These inclusions must have been introduced between drying and pressing.

However, fracture initiation from machining flaws was common, particularly for the individual test bars. These are typified by the half-penny crack shown in Figure 5, and are a result of cutting and, more particularly, surface grinding of the bars prior to testing. The fact that these kinds of flaws were more common in the individual bars indicates that the population of large sintering flaws was much smaller when more strict powder processing was used.

Engine Performance

A fired valve disc was diamond-ground to final dimensions for assembly with the engine. A special design was required to take into account the differences in physical properties of the iron-based alloys and the Si_3N_4 ceramic. The complete valve with the ceramic disc in position can be seen in Figure 6.

Standard valve lapping procedures were used to obtain a seal between the ceramic disc and the standard Stellite seat in the cylinder head; a line contact occurred between the ceramic and the seat and was according to standard procedure.

The ceramic/metal valve assembly was then run in the engine under various conditions from idle upto full speed and the operating performance monitored throughout. The cylinder head was removed after 5 hours and the condition of the valve checked (Figure 7).

No damage was encountered and so the valve was run continuously, mainly at full power, for a further 69 hours at which point the ceramic disc failed.

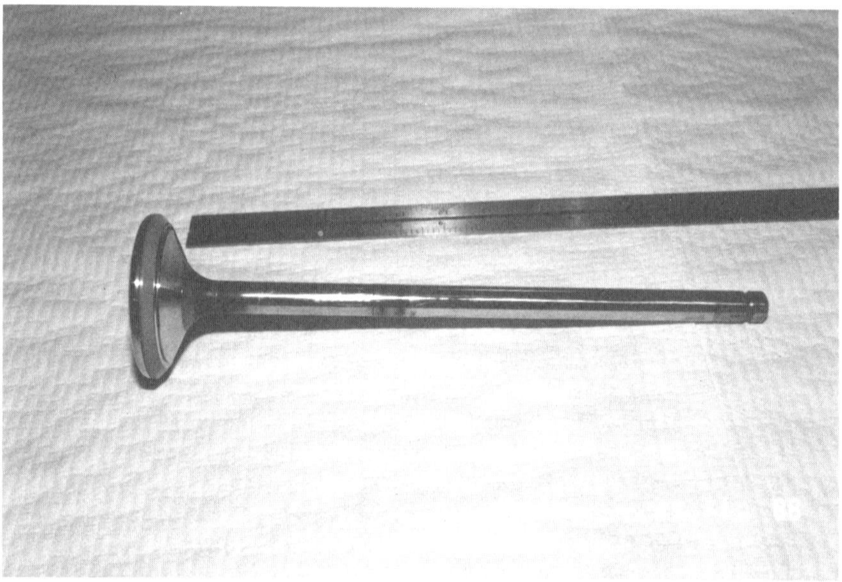

Figure 6: Assembled valve prior to operation.

A post-mortem failure analysis was performed and it was concluded that some design modifications were required. The minimization of point contacts and sharp radii were identified as important design considerations when employing ceramic components, especially in such a complex, composite structure. In addition, since the material properties of the disc would be similar to the destructively tested disc (viz. $\bar{\sigma}$ = 360 MPa), further improvements in processing these larger components is required in order to bring the mechanical properties upto those of the individual laboratory processed bars.

CONCLUSIONS

1. The ability to fabricate Si_3N_4 components using standard isopressing and green machining techniques has been shown.
2. The processing of large components to obtain good sintering and mechanical properties requires careful control to avoid the introduction of large critical defects.
3. The feasibility of integrating a Si_3N_4 disc into a standard metal engine valve assembly was shown. This enabled the engine to be operated effectively for upto 74 hours.

ACKNOWLEDGEMENTS

The authors would like to thank all the contributors to this work both at McGill and NRC. The financial support of both Almax and NSERC in the development work is also appreciated.

Figure 7: Valve assembly after 5 hours operation.

REFERENCES

Bennett, A. (1986) Requirement for Engineering Ceramics in Gas Turbine Engines. Materials Sci. and Tech., 2, p. 285.

Drew, R.A.L. (1987) Development of Sialon Ceramics. Final report submitted to Almax Industries.

Drew, R.A.L. (1988) Fabrication of Silicon Nitride Valve Discs. Report submitted to ARL-NRC.

Kabat, D.M., Garwin, I.J., and Hartsock, D.L. (1988) Ceramic Valve Analysis, Reliability and Test Results. Proc. Int. Congr. and Expo. of SAE, paper 880670.

Terwilliger, G.R., and Lange, F.F. (1975) Pressureless Sintering of Si_3N_4. J. Mater. Sci., 10, p. 1169.

Zarnon, L.C., and Drew, R.A.L. (1988). Dispersion Methods of Y_2O_3 in Si_3N_4 by Comminution. Paper submitted to J. Amer. Ceram. Soc.

Silicon Nitride and Silicon Carbide Ceramic Whiskers Synthesis and Phase Stability

Harue Wada
Department of Materials Science and Engineering,
University of Michigan, Ann Arbor, Michigan 48109.

ABSTRACT

Stability of phases in equilibrium is calculated for the Si-C-N-O system in order to apply to ceramic whiskers synthesis. Equilibria among SiC, Si_3N_4, Si_2N_2O and SiO_2 are evaluated at different carbon activities, nitrogen pressures and temperatures. Phase stability diagrams are constructed as a function of p_{O2} and p_{N2} for known a_C and temperature. Silicon nitride becomes a stable phase with increasing p_{N2} or decreasing temperature and a_C, while SiC becomes a dominant phase at lower p_{N2} or higher temperature when $a_C = 1$.

Si_3N_4, β'-sialon and SiC whiskers have been synthesized by carbothermal reduction of SiO_2. The process is based on the stability of phases in the Si–C–N–O system and involves an auxiliary halide bath. The synthesized whiskers are single crystal with growth direction <210> and <110> for Si_3N_4 and sialon, and SiC, respectively. The sialon whiskers, $Si_{(6-Z)}Al_ZO_ZN_{(8-Z)}$, is determined as Z = 0.8 - 1.1 from the aluminum content.

INTRODUCTION

Ceramic whiskers are in demand for application as reinforcing fillers in both metal– and ceramic–matrix composite materials. At present, silicon carbide whiskers are most widely used for this purpose due to their high strength. However, the application of other ceramic whiskers such as Si_3N_4 is expected to increase with increasing and diversifying utilization of composite materials.

β'–sialon whisker is another prospective material for the high temperature composite filler. β'–sialon is a solid solution of β–Si_3N_4 in which Si and N are substituted by Al and O [1]. The solid solution is thermodynamically more stable than the solvent; higher decomposition temperature due to lower vapor pressure. Calculation of the stability of phases in the Si–C–N–O system showed that both silicon carbide and nitride can be synthesized from SiO_2 by controlling the activities of carbon and nitrogen [2]. SiC and Si_3N_4 formations by the carbothermal reduction of SiO_2 are principally based on the SiO(g) formation

$$SiO_2\,(s) + C\,(s) \longrightarrow SiO\,(g) + CO\,(g) \qquad (1)$$

Ceramic whisker growth has been studied mainly for silicon carbide: SiC whiskers have been produced by various methods and these are well reviewed by Milewski et al. [3]. On the other hand, there are few studies on Si_3N_4 whiskers formation, especially by the reduction of silica [4,5].

In this investigation, single crystal whiskers of β–SiC, β–Si_3N_4 and β'–sialon have been synthesized at 1623 K on a laboratory scale under well–controlled experimental conditions. An auxiliary molten halide bath was applied to accelerate the SiO formation. The reaction mechanism in both the nitride and carbide syntheses is discussed in conjunction with the stability of phases and the molten halide bath.

PHASE STABILITY IN THE SI–C–N–O SYSTEM

Since the main purpose of this study is separate synthesis of nitride and carbide whiskers by a carbothermal reduction of silica by the same apparatus, the stability of phases is calculated for the Si–C–N–O system as a guideline. The equilibrium phases in the system are Si_3N_4, Si_2N_2O, SiC, SiO_2 (crystobalite), Si, C, and a gas phase. The main components of the gas phase would be N_2, O_2, SiO and CO/CO_2. Two carbon activity levels, $a_C = 1$ and 10^{-4} which are referred to the solid graphite, are used for the calculation. Temperature range is from 1573 to 2073 K.

Three-phase equilibria

Equilibrium between two solid phases and a gas phase is calculated first. For instance, equilibrium between Si_3N_4 and SiC is expressed as follows for the present temperature range:

$$Si_3N_4\,(\beta) + 3C\,(s) = 3SiC\,(\beta) + 2N_2\,(g) \qquad (2)$$

The ratio of the activities of Si_3N_4 and SiC, which are referred to solid Si_3N_4 and SiC, respectively, is written as

$$\log\,(a_{SiC}^3 / a_{Si3N4}) = \log K_2 + 3 \log a_C - 2 \log p_{N2} \qquad (2\text{–}2)$$

where K_2 is the equilibrium constant of Eq. (2). Since Si_3N_4 and SiC form almost no solid solution, their activities are either unity or less than unity. There are three possibilities for Eq. (2–2), i.e.,

if $\log\,(a_{SiC}^3 / a_{Si3N4}) = 0$, then $a_{SiC} = a_{Si3N4} = 1$

$\log\,(a_{SiC}^3 / a_{Si3N4}) > 0$, then $a_{SiC} = 1$ and $a_{Si3N4} < 1$

$\log\,(a_{SiC}^3 / a_{Si3N4}) < 0$, then $a_{SiC} < 1$ and $a_{Si3N4} = 1$.

In the first case both Si_3N_4 and SiC phases coexist, while SiC is the stable phase in the second case and Si_3N_4 is the stable phase in the third case. Equation (2–2) is calculated as a function of temperature for $a_C = 1$ and $p_{N2} = 1$ and 10 atm. At $p_{N2} = 1$ atm, Si_3N_4 and SiC are in equilibrium at 1647 K. SiC is stable at higher temperatures, and Si_3N_4 is the stable phase in the lower temperature range, while the temperature for equilibrium between the two phases moves upward to 1809 K when nitrogen pressure is increased to 10 atm.

Other equilibria such as Si_3N_4/Si_2N_2O, Si_3N_4/SiO_2, Si_2N_2O/SiO_2, Si_2N_2O/SiC, SiC/SiO_2, Si/Si_3N_4, Si/Si_2N_2O, Si/SiO_2 are calculated by the same principle. These equilibrium reactions in the Si–C–N–O system in the temperature range concerned in this study are summarized in Table I. The standard free energies of formation were obtained mainly from JANAF Tables [6] except of β–Si_3N_4 and Si_2N_2O, which were not available in JANAF Table. These two phases are important phases in the Si–C–N–O system and the free energies should be consistent with each other. Two values were accepted from Hendry's[7]. The standard free energies of formation of solid phases are summarized in Table II.

Table I. Equilibrium Reactions in Si-C-N-O System

Number	Reaction
(1)	$Si_3N_4(\beta) + 3C(s) = 3SiC(\beta) + 2N_2(g)$
(2)	$4\ Si_3N_4(\beta) + 3O_2(g) = 6\ Si_2N_2O(s) + 2N_2(g)$
(3)	$Si_3N_4(\beta) + 3O_2(g) = 3SiO_2(cr.) + 2N_2(g)$
(4)	$2\ Si_2N_2O(s) + 4C(s) = 4SiC(\beta) + O_2(g) + 2N_2(g)$
(5)	$2\ Si_2N_2O(s) + 3O_2(g) = 4SiO_2(cr.) + 2N_2(g)$
(6)	$SiO_2(cr.) + C(s) = SiC(\beta) + O_2(g)$
(7)	$3Si(l)$ or $(s) + 2N_2(g) = Si_3N_4(\beta)$
(8)	$4Si(l)$ or $(s) + 2N_2(g) + O_2(g) = 2\ Si_2N_2O(s)$
(9)	$Si(l)$ or $(s) + O_2(g) = SiO_2(cr.)$

Table II. Standard Free Energies of Formation of Condensed Phases

Species	$\Delta_f G°$, KJ/mole*	Reference
β-Si_3N_4	$-925.2 + 0.450T$	(7)
Si_2N_2O	$-658.3 + 0.131T$	(7)
β-SiC	$-72.832 + 0.007T$	(6)
SiO_2(cr.)	$-900.384 + 0.171T$	(6)
Si(l)	$-42.825 + 0.0647T - 0.24 \times 10^{-4} T^2$	(6)

* Standard states are Si(s), C(s), $p_{N2} = 1$ atm and $p_{O2} = 1$ atm.

Four-phase equilibria

When three solids and a gas phase are in equilibrium, the degrees of freedom are two from the phase rule: a_C and temperature. Combining the equations shown in Table I, phase stability is calculated at $a_C = 1$ and 10^{-4} and at temperatures of 1673 and 1973 K. For instance, the $Si_3N_4/Si_2N_2O/SiC$/gas equilibrium is calculated as follows: From the reactions Si_3N_4/SiC and Si_3N_4/Si_2N_2O in Table I,

$$\log K_2 + 3 \log a_C - 2 \log p_{N2} = -36909.385/T + 22.406 + 3 \log a_C - 2 \log p_{N2} = 0 \qquad (2\text{–}3)$$

$$\log K_3 - 2 \log p_{N2} + 3 \log p_{O2} = 13004.648/T + 52.959 - 2 \log p_{N2} + 3 \log p_{O2} = 0 \qquad (3)$$

At known a_C and temperature, the oxygen and nitrogen pressures are calculated for this invariant point. Other invariant points are calculated by the same method. Figure 1 shows the stability diagram for $a_C = 1$ and T=1673 and 1973 K, which are shown with solid and dashed lines, respectively. The most significant feature of the diagram is the effect of nitrogen pressure and temperature on the phase stability, especially at two invariant points where the effects can be seen simultaneously. A large increase in the nitrogen pressure is required to make Si_3N_4 the stable phase at 1973 K, and avoiding Si_2N_2O phase formation requires very low oxygen partial pressure. The effect of temperature is shown more dramatically when the carbon activity is reduced to $a_C = 10^{-4}$. These stability diagrams were used as a guide in experimental work.

MOLTEM SALT BATH

To enhance SiO formation, an auxiliary halide molten bath was applied. The chemical composition of bath was selected to meet the following requirements:
(1) A low melting point and a low viscosity, (2) No undesirable side reaction with SiO_2, carbon, and (3) No undesirable effect on ceramic whiskers.
By examining phase diagrams of halide systems, the NaF-AlF_3 system [8] was chosen as the auxiliary bath. Thermodynamic properties of he NaF-AlF_3 system have been well established[9]. Although the phase relationship in the NaF-AlF_3- SiO_2 system is not well known at present, a preliminary melting test showed that the system forms a Na-Al–fluorosilicate with lower melting point and viscosity.

EXPERIMENTAL WORK

Experimental condition was determined based on the phase stability as,

Reaction temperature—1523 to 1673 K; Total pressure—1 atm; Flowing gases—N_2 or $N_2 + 5\%$ H_2 mixed gas for Si_3N_4 whiskers, and Ar or Ar + 5% H_2 mixed gas for SiC whiskers.

A vertical, one end closed cylindrical graphite reaction chamber was used. The chamber was partitioned into upper and lower parts. The upper chamber contained several graphite plates and the top of the chamber was closed with a graphite cap. The cap and all the graphite plates had a 12 mm dia. hole in the center through which a graphite gas inlet tube was inserted from the cap downward to the lower chamber. The optimum condition of the process was determined based on preliminary tests. High purity flowing gases were further purified through a $Mg(ClO_4)_2$ column, then introduced directly into the lower reaction chamber with a flow rate 250 cc/min. Two halides, AlF_3 NaF and AlF_3 3NaF, were applied as the halide bath. The ratio of SiO_2/C/halide was kept constant as atomic ratio Si/C/Na = 1/3/1.

20 gr. of the well mixed starting materials, SiO_2 + carbon black + halide, were charged into the lower reaction chamber, and the reaction chamber was placed in a mullite reaction tube. After being purged overnight by N_2 or Ar, depending on the whiskers to be synthesized, the mixture was gradually heated to the reaction temperature. The reaction was continued for approximately 9 to 10 hours at that temperature, then cooled. After residual carbon black was oxidized in air at 700°C, whiskers were analyzed by X–ray diffraction, SEM and TEM.

EXPERIMENTAL RESULTS AND DISCUSSION

Whiskers

Whiskers grew in two regions: Region 1, inside the lower graphite chamber where the starting mixture was charged, and Region 2, inside the upper graphite chamber. About 60-74% of Si in the starting SiO_2 was yielded as whiskers after 10 hours. Figure 2 shows typical whiskers grown in Region 2. These whiskers were identified as a mixture of β–Si_3N_4 and β'–sialon by X–ray diffraction and none showed droplet on the tips. Whiskers have a uniform size, about 0.7 mm in diameter and about 10 mm maximum length. On the other hand, whiskers grown in Region 1 are less uniform in diameter as shown in Fig. 3. Figure 4 shows a SEM micrograph and X–ray element maps of Si and Al of β'–sialon whiskers. Some whiskers grown under N_2 gas flow were much thinner and longer. These thin whiskers were identified as silicon oxynitride, Si_2N_2O. As shown in Fig. 5, some of the Si_2N_2O whiskers grown in Region 1 have globules around them. The globule is identified as $AlNaSiO_4$ by X–ray diffraction and its Si content is lower but Al and Na contents are higher than in the whiskers as seen in Fig. 5. These globules were found only around thin Si_2N_2O whiskers grown in Region 1. Since the formation of Si_2N_2O whiskers is associated with higher oxygen partial pressure as shown in Fig. 1, oxidation of molten Al-Na fluorosilicate globules could occur upon vaporization of fluorine. The resulting $AlNaSiO_4$ phase solidifies at 1799 K [10]. Figure 6 shows typical β-SiC whiskers formed in Region 1. These β-SiC whiskers are thinner than Si_3N_4 whiskers formed in Region 1. Whiskers were analyzed by JOEL 2000FX at 200KV. Figure 7 shows a typical TEM micrograph and its electron diffraction pattern of a β'–sialon whisker. The whisker is a single crystal of smooth surface with <210> growth direction. All nitride whiskers show the same morphology whether grown in Region I or II, which may indicate that the growth mechanism of the nitride whiskers is the same in these two regions. A general equation of β'–sialon is $Si_{6-z}Al_zO_zN_{8-z}$, where the value of z reflects the atom fraction of Al and O substituted Si and N, respectively. The values of z determined by microanalysis of the At% ratio Si/Al are in a range of 0.8 to 1.1. The growth direction of β–SiC(3C) whiskers is determined as <110> by electron diffraction analysis. There are other studies [11,12] which reported <110>, but <111> has been a growth direction reported by the majority [13 -

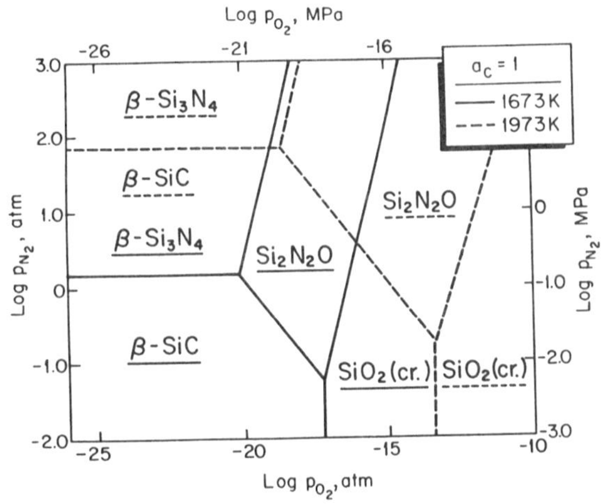

Figure 1. Phase stability diagram as a function of p_{N_2} and p_{O_2} at $a_C = 1$, T = 1673 and 1973 K.

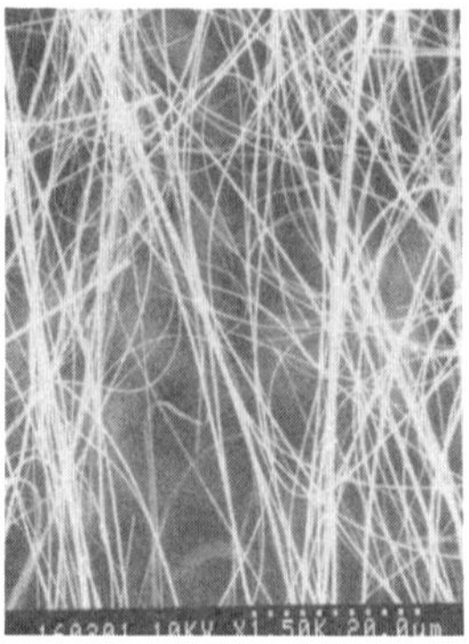

Figure 2. Typical nitride whiskers grown in Region 2 (Si_3N_4 + sialon)

Figure 3. Typical nitride whiskers grown in Region 1 (Si_3N_4)

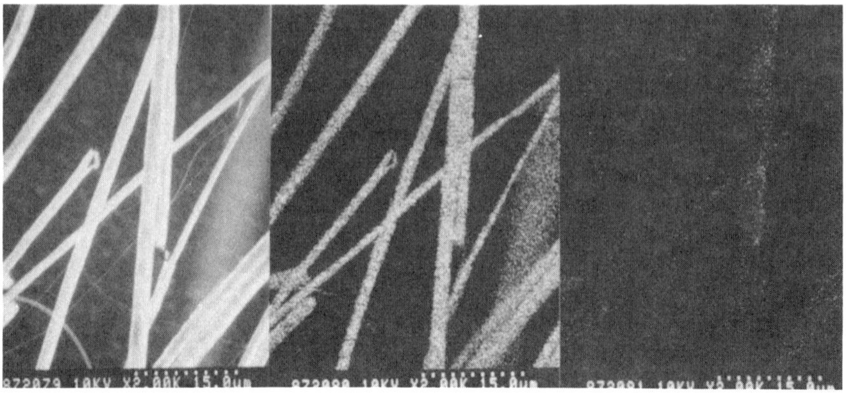

(a) micrograph (b) Si (c) Al

Figure 4. SEM micrograph and x-ray maps of sialon whiskers

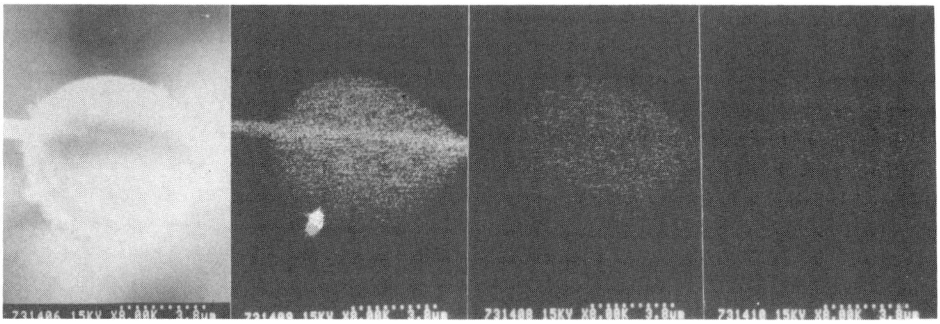

(a) micrograph (b) Si (c) Al (d) Na

Figure 5. Distribution of Si, Al and Na between Si_2N_2O whisker and a droplet

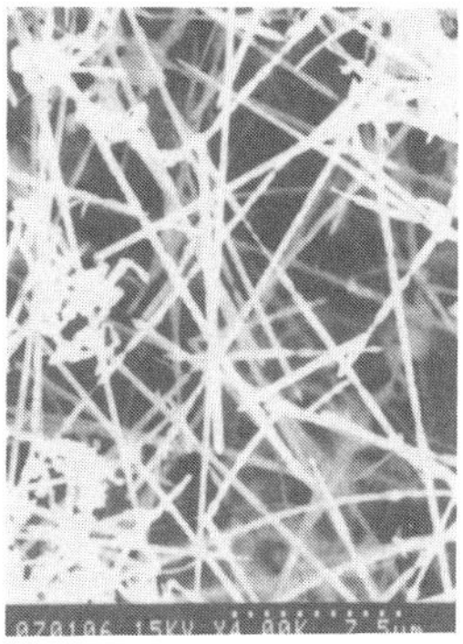

Figure 6. Typical SiC whiskers grown in Region 1

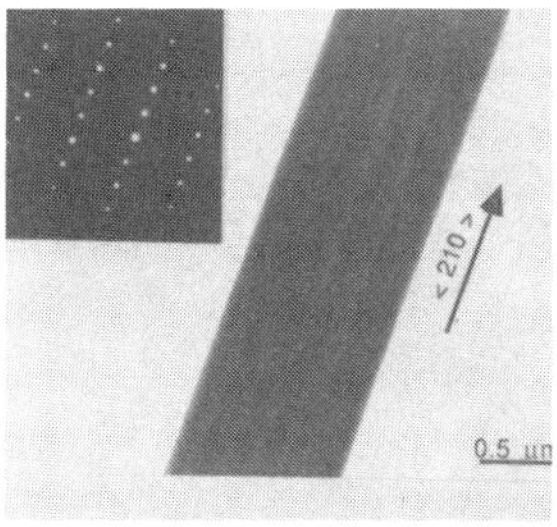

Figure 7. Electron diffraction pattern and growth direction of β'-sialon whisker

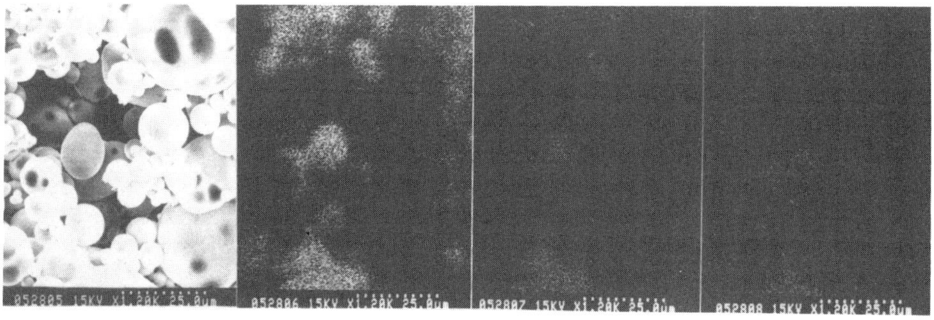

(a) micrograph (b) Si (c) Al (d) Na

Figure 8. Distribution of Si, Al and Na in $3NaF\ AlF_3$-SiO_2 bath

15] for β–SiC(3C) whisker.

Halide Bath

Figure 8 shows a SEM observation and X-ray element maps of an auxiliary [3NaF AlF_3] bath. The [3NaF AlF_3] was melted in a graphite crucible and kept 1100°C for 2 hr, and then quenched. This was identified as a mixture of [NaF] and [3NaF AlF_3] by X–ray diffraction. Figure 8 indicates that the SiO_2 in the starting material dissolved into the fluoride bath, forming an Al–Na–fluorosilicate. Figure 8 also indicates that the bath has a high surface tension. This high surface tension would increase contact between the fluorosilicate and carbon, and could also be a reason for the globules found around the oxynitride whiskers. The larger balls have crater–like openings, from which gases might be evolved. The condensed phase of vaporized gases from the bath was identified as a mixture of [NaF AlF_3] and [3NaF AlF_3].

The effect of fluoride addition on the viscosity of molten oxides has been well known. Kato and Minowa [16] measured changes in the viscosity of Al_2O_3-CaO–SiO_2 melt with oxides or fluorides addition. The viscosity was decreased from h = 4 to h = 0.7 poise by increasing NaF from 10 to 30 mol% at 1673 K. The effectiveness of fluoride in reducing the viscosity of oxide melts was observed as CaF_2>MgF_2>AlF_3>NaF. They concluded that fluorides reduce the unit size of molten silicate and hence reduce the viscosity, while increasing the surface tension of silicate melts. The conclusion seems to agree with the present observation in the Na-Al–fluorosilicate melt.

Reaction mechanism

The reaction mechanism of whisker formation in Region 2 is the VS mechanism with graphite as a substrate. The standard free energy changes of silicon carbide and nitride formation from SiO(g) are calculated as follows:

β–SiC formation :

$$SiO(g) + 3CO(g) = SiC(\beta) + 2CO_2(g), \qquad \Delta G_4° = -402{,}377 + 339\ T \quad \text{J/mole SiC} \tag{4}$$

β–Si_3N_4 formation :

$$3SiO(g) + 2N_2(g) + 3CO\ (g) = Si_3N_4(\beta) + 3CO_2(g),$$
$$\Delta G_5° = -1{,}333{,}017 + 876\ T \quad \text{J/mole } Si_3N_4 \tag{5}$$

Si_2N_2O formation :

$$2SiO(g) + N_2(g) + CO(g) = Si_2N_2O(s) + CO_2(g),$$
$$\Delta G_6° = -650{,}934 + 330T \quad \text{J/mole } Si_2N_2O \tag{6}$$

β'–sialon formation:

$$(6\text{-}Z)\ SiO(g) + (4\text{–}Z/2)N_2(g) + (6\text{–}2Z)CO(g) + Z\ Al(g)$$
$$= Si_{(6-Z)}\ Al_ZO_Z\ N_{(8-Z)} + (6\text{–}2Z)CO_2(g) \tag{7}$$

Equilibrium mole ratio SiO/N_2 is 1.5 and 2 for Si_3N_4 and Si_2N_2O, respectively. Therefore, to control the supply of nitrogen gas to the reaction sites is one of the important factors which affect the effective synthesis of Si_3N_4 whiskers particularly when an enhanced SiO formation is involved like in the present process. The CO/CO_2 reaction has been investigated by Wada and Wang [17] for β'–sialon formation. When $a_C = 1$, the CO/CO_2 reaction is in equilibrium. To form the sialon phase, Al and oxygen have to substitute for Si and nitrogen, respectively. The amounts of gas species necessary to form 1 mole of β'–sialon by Eq.(7) are shown in Fig. 9 for $a_C = 1$. No additional oxygen is required other than that contained in SiO at lower values of Z (Z<3), while balanced amounts of oxygen and nitrogen are required at higher values of Z (the $O_2/N_2 = 1/2$ at Z = 4).

The present process produces β-type nitride whiskers in the normal conditions, the α-type being found very seldomly. Although the research is still continuing, it should be noted that the major portion of the present process has been carried out under a very low oxygen potential as verified experimentally. The effect of Al on the α–β transformation of Si_3N_4 may involve the oxygen potential in the reaction system. Since formation of β'–sialon requires both Al and oxygen, the presence of Al stabilizes β type by 1) forming β'–sialon that is β type, and 2) oxygen in the system dissolves into β'–sialon and hence reduces p_{O2} in the ystem. The β–type is known to be stable at lower p_{O2} than the α–type. Therefore, the formation of α–Si_3N_4 whiskers reported may be related to higher oxygen potential in their reaction chamber.

Figure 9. Composition of gas phase in equilibrium with β'-sialon

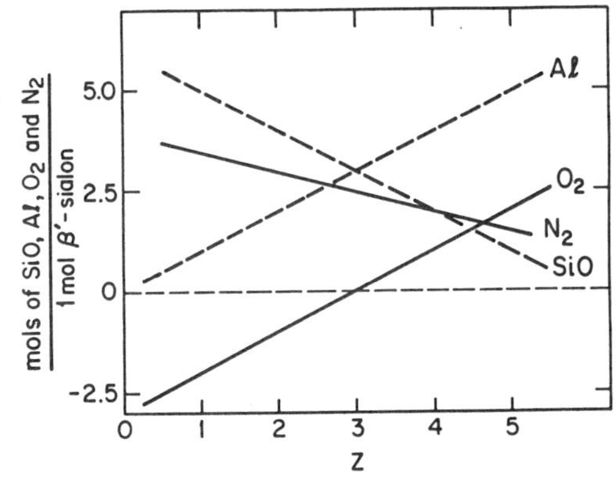

Acknowledgments

Supported by the NSF. Assistance in laboratory works by Mr.M.J.Wang is appreciated.

References

1) K.H.Jack, NATO Advanced Study Institute Series, 1983, p.52.
2) H.Wada, M.J.Wang and T.Y.Tien, To be published in J. Amer. Cer. Soc.
3) J.V.Milewski, F.D.Gac, J.J.Petrovic and S.R.Skaggs, J.Mat.Science, 20(85), p.1160.
4) H.Saito,T.Hayashi and K.Miura, J.Chem.Soc.Japan, 1981, p.1371; 1982, p.401.
5) T.Hayashi, S.Kawabe and H.Saito, J. Cer. Soc. Japan, 95(86), p.19.
6) JANAF Thermodynamical Table, 3rd Edition, 1986.
7) A.Hendry,"Nitrogen Ceramics",(F.L.Riley, Ed), Noordhoff, 1977, p.183.
8) "Phase Diagrams for Ceramists", Figure 501, Amer. Cer. Soc.
9) E.W.Dewing, Met.Trans., 1(70), p.1691; p.2211; ibid., 3(72), p.495; p.2699; Metall. Trans., 18B(87), p.409.
10)"Phase Diagrams for Ceramists", p. 474; p.475, Amer. Cer. Soc.
11)N.Setaka and K. Ejiri, J. Amer. Cer. Soc., 52(69), p.60.
12)G.A.Bootsma, W.F. Knippenberg and G.Verspui, J. Cryst. Growth, 11(71), p.297.
13) H.Saito and I.Yamai, J. Cer. Soc. Japan, 88(1980), p. 331 ; p. 265.
14) R.De Jong and R.A. McCauley, J. Amer. Cer. Soc., 70(87), C–338.
15)L.F.Allard, P.Pendleton and J.S.Brinen, Proceedings 44 Ann. Meet. Electr. Microscopy Soc, 1986, p.472.
16) M. Kato and S. Minowa, Tetsu–to–Hagane, 55(69), p. 260.
17)H.Wada and M.J.Wang, Proc. Int. Conf. Whisker and Fiber-Toughened Ceramics, ASM Int. 1988.

ADVANCED CERAMIC MATERIALS FOR METAL CUTTING

T.B. Troczynski, D. Ghosh, S. Das Gupta
and J.K. Jacobs

The Electrofuel Mfg. Co
Toronto, Canada

ABSTRACT

Properties of alumina and silicon nitride based ceramic cutting tools are compared with those of cemented carbides. Effect of temperature, cutting speed and tool material on machining productivity is discussed. It is shown that advanced ceramic cutting tools outperform carbides in machining of cast iron and nickel alloys. Microstructure and properties of ceramic/ceramic composite tool materials developed by The Electrofuel Mfg. Co. are presented.

KEYWORDS

Cutting tools; wear; ceramic composites; ceramic cutting tools.

INTRODUCTION

Each year over $100 billion is spent in North America on the machining of metal parts. These costs can be lowered by increasing the cutting speed and, in some instances, feed or depth of cut. The search for new materials to allow increased rates of metal removal resulted in the implementation, in the last 60 years, of two major classes of tools: cemented carbides and ceramics. Development of ceramic cutting tools (CCT) started in Germany and Great Britain at the beginning of the 20th century, more than 20 years before the introduction of carbides. Early CCT were made of alumina and gained wider industrial recognition only after World War II. Today, the majority of CCT are still alumina-based.

In this paper we review the principles of metal cutting operations and compare characteristic features of carbide and ceramic cutting tools. Superiority of CCT over carbides is shown in terms of increased chemical stability, wear resistance, hot strength and hardness. As a result, dramatic improvement in metal removal rates and productivity is achieved. Composite ceramic cutting tools manufactured in Canada by The Electrofuel Mfg. Co. will be characterised and compared with market materials.

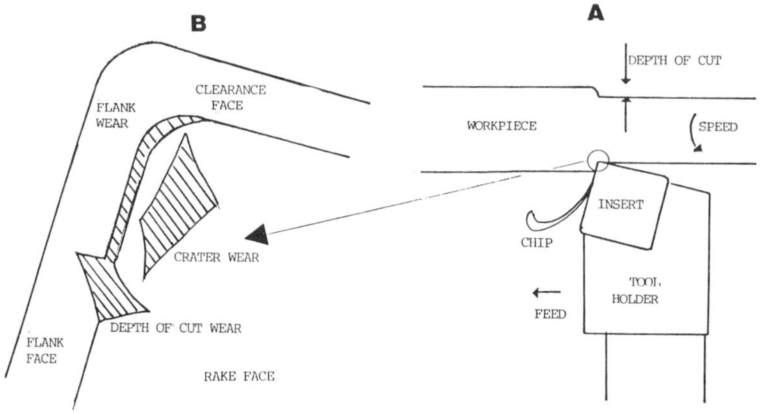

Fig. 1. Typical metal turning arrangement (A) and schematic regions of tool wear (B).

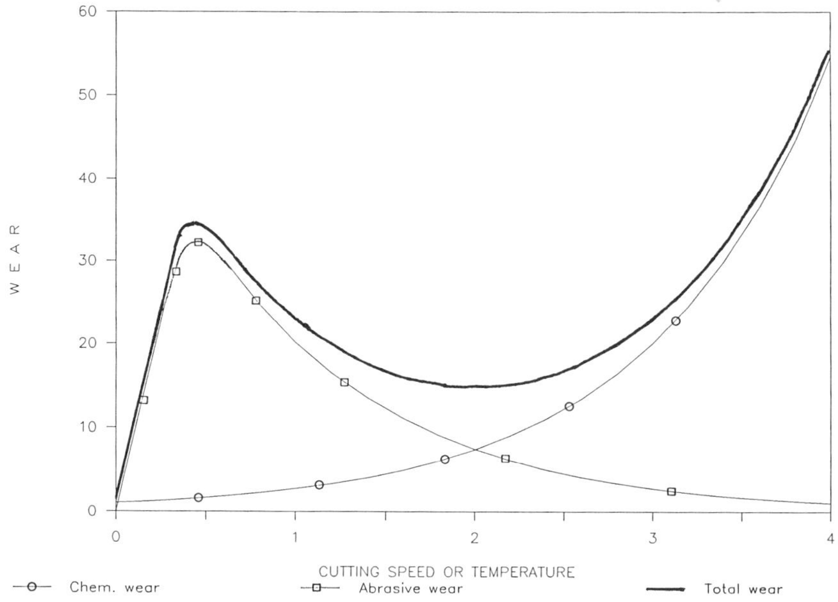

Fig. 2. Variation of chemical wear and abrasive wear of the tool vs. temperature or cutting speed.

METAL - CUTTING OPERATIONS

A typical metal turning operation is shown in Fig. 1A. The basic parameters of turning (depth of cut D, feed f and speed s) are defined in the figure. The insert vs. workpiece orientation is characterised by definition of flank, face and rake surfaces and respective angles. These

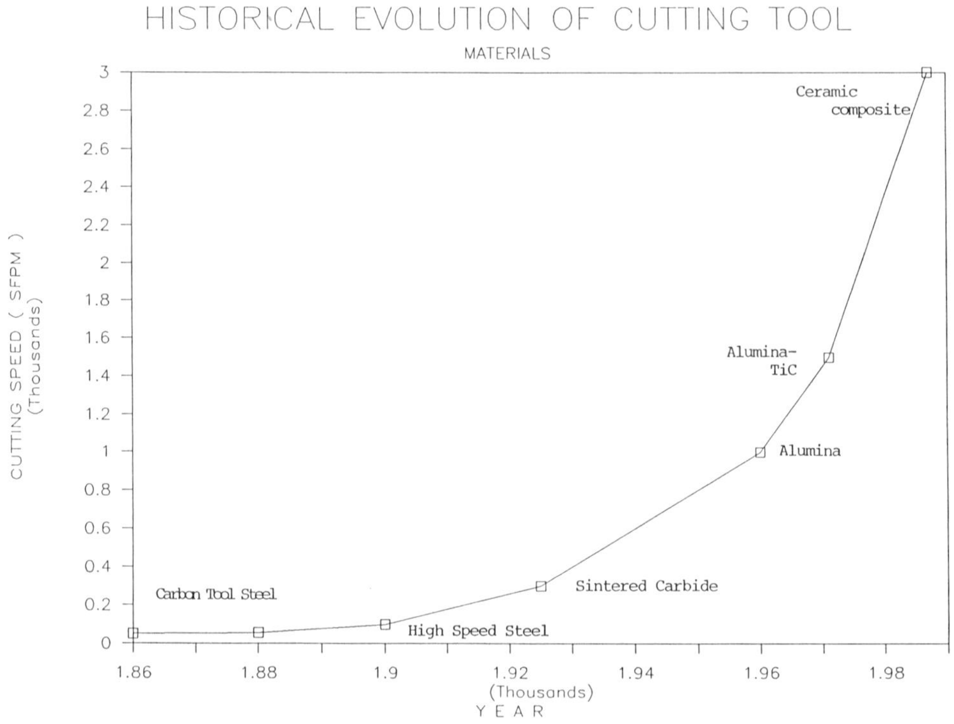

Fig. 3. Historical evolution of cutting tool materials.

surfaces carry the load and experience wear patterns shown schematically in Fig. 1B. It is essential that tools wear as uniformly as possible , e.g. without notching, cratering, microcracking etc. In this respect, flank wear area is of least danger as far as the catastrophic failure of the tool is concerned. The process parameters should therefore be optimized to avoid the depth-of-cut notch and trailing edge notch wear, Fig. 1B. The hottest area of the tool is the flank surface where high temperatures are generated by shear in the workpiece and friction with the tool. This area erodes due to material reaction and association at the contact surface. A characteristic crater forms and effective stress concentration heralds unexpected catastrophic failure of the tool.

Rate of metal removal is proportional to depth of cut, feed and speed. Consequently, the process efficiency can be improved by increasing any of the above parameters. For years this has been achieved by increasing D and f with high strength carbide in slow-speed turning. This approach is changing towards substantial cutting speed increase with implementation of ceramic cutting tools. Generally, increasing D and f results in the increase of mechanical load on the insert. Speed increase results in greater thermal and chemical loads on the insert. Therefore the thermal shock resistance of inserts is as important as mechanical and chemical resistance in high speed metal cutting. Materials of high thermal conductivity and fracture toughness and low thermal expansion coefficient are therefore favourable.

Heat generated in metal cutting dramatically accelerates the chemical interaction of the insert with the workpiece and the environment. Chemical compatibility between cutting tool and workpiece must be fulfilled. This condition favours materials of high free energy of formation, i.e. alumina or zirconia against carbides (WC, TiC, SiC). The nitrides hold intermediate position between oxides and carbides.

When the cutting speed is increased, the maximum local temperature in the cratering area, Fig. 1B, aproaches the workpiece melting point. Under this condition the tool wear is principally a combination of exponentially temperature dependent chemical wear WR, WR $\sim \exp(-a/T)$, and abrasive wear WA, WA $\sim \exp(-bT)$. The abrasive wear component decreases with temperature due to more rapid decrease of workpiece hardness compared to that of the tool. The curves schematically representing WR and WA are shown vs. temperature or cutting speed in Fig. 2, in arbitrary units. Considering total wear=WR+WA, an optimum temperature or speed exists where minimum wear is observed for each particular tool, workpiece and environment system.

CERAMIC CUTTING TOOLS

Refinement of the microstructure vs. property relationship for ceramic materials has resulted in significant improvement of CCT perfomance over the last 20 years. This is illustrated in Fig. 3. With the introduction of advanced CCTs we achieve at least an order of magnitude in speed increase over early carbon tool steels.

The biggest family of presently utilised CCTs is still based on aluminum oxide. Room temperature rupture strength of modern pure alumina tools

TABLE 1 Typical Room-Temperature Properties of Ceramic and Cemented Carbide Cutting Tool Materials.

MATERIAL	S (GPa)	E (GPa)	H (GPa)	Kc (MPa·$\sqrt{m}$)	L (W/mK)	A (10^{-6} /K)	R (W/$\sqrt{m}$)	M (GPa$^{5/4}$ m$^{3/8}$)
Al2O3	0.7	450	17	3	17	7	16	9
TiC/Al2O3	1	420	20	4	20	7	27	13
ZrO2/Al2O3	1	400	17	6	14	7	30	16
SiCw/Al2O3	0.8	400	19	5-9	20	6	50	17
Si3N4	0.8	300	15	5-7	25	3	160	15
Carbide	1.5	550	17	10-13	100	7	220	26

reaches 700 MPa. Special powder processing techniques result in strength over 1 GPa. However, low fracture toughness and thermal conductivity of pure alumina excludes use of this material in high thermal shock environments. Advances in ceramic/ceramic composites research allowed for consecutive introduction of improved alumina based materials. Their bend

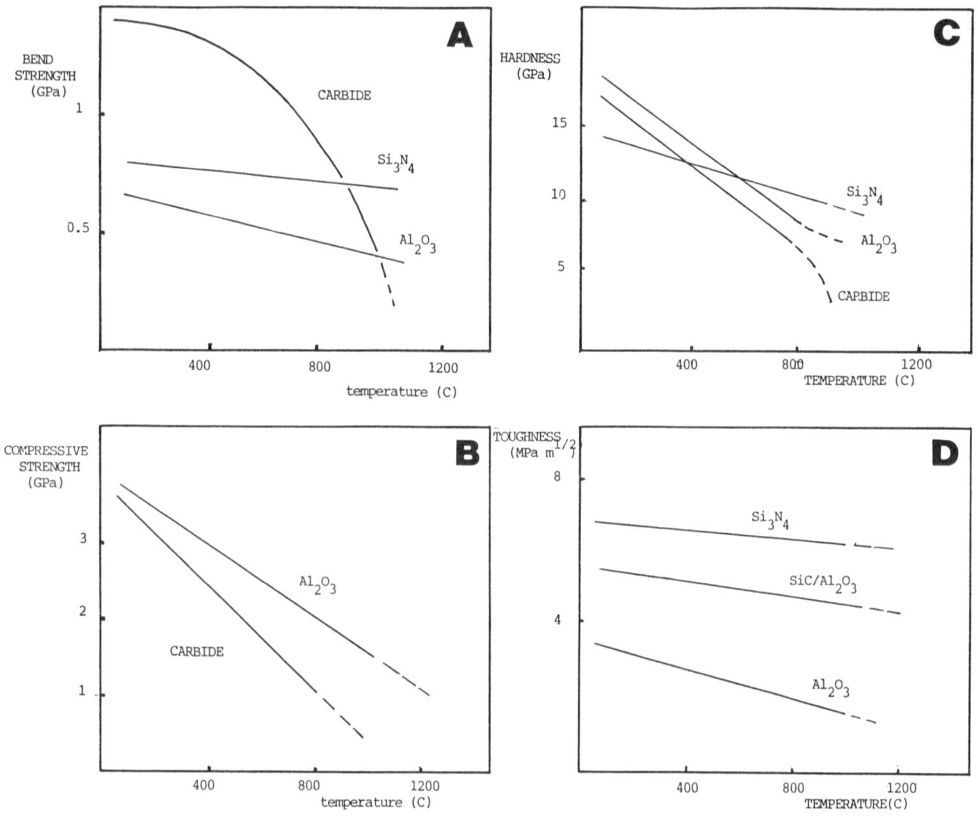

Fig. 4. Elevated temperature properties of cutting tool materials: (A) Bend strength, (B) Compressive strength, (C) Hardness (D) Fracture toughness.

strength (0.7 to 1 GPa) substantially exceeds 0.2 GPa, typical strength of traditional ceramics. The combination of strength and toughness suggests that the average processing flaw size in modern ceramic tool materials does not exceed 10 um. Thus the primary processing defects are no longer a source of catastrophic failure. In advanced ceramic tools, wear-generated stress concetrations set the critical conditions for fracture.

Overview of typical room temperature properties of ceramic and carbide tools appears in Table 1. The consecutive columns present strength S, Young's modulus E, hardness H, fracture toughness Kc, thermal conductivity L and thermal expansion coefficient A. The last two columns present two "figures of merit" : thermal shock resistance R and abrasive wear resistance M. The values of R and M are obtained from combinations of the basic parameters according to the following formulas:

$$R = (Kc \cdot L)/(E \cdot A)$$

$$M = \sqrt{H} \cdot Kc^{3/4}$$

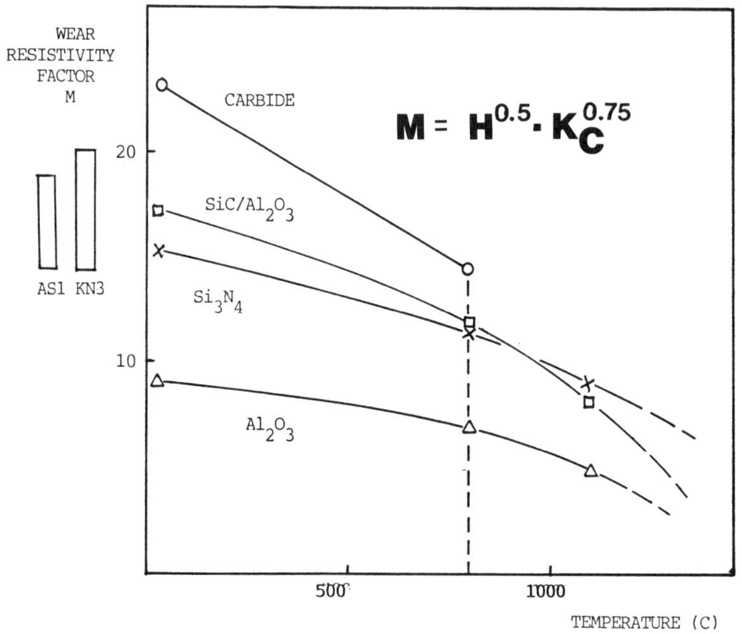

Fig. 5. Wear resistivity factor M for cutting tool materials at elevated temperatures.

Dispersion of 30 vol% of TiC particles in an alumina matrix, so called "black ceramic", allowed for a moderate increase in toughness and conductivity over pure alumina. The combined effect of these improved parameters resulted in material 70% more thermal shock resistant and 40% more abrasion resistant, Tab. 1. Further significant improvement of alumina tools resulted from dispersion of zirconia (transformation-toughened ceramics) and recently silicon carbide whiskers (SiCw). Although both modifications result in materials of comparable toughness, the later exhibits superior wear resistance thanks to higher hardness. Superior thermal shock resistance results from increased thermal conductivity and decreased thermal expansion coefficient. Recently, a new family of non-oxide CCTs based on silicon nitride was successfully introduced for machining cast iron and superalloys under severe conditions (high speeds, interrupted cutting, coolants, etc.). The advance was possible due to exceptional thermal shock resistance originating from high toughness at low thermal expansion coefficient of silicon nitride based materials.

Considering the cemented carbides characterised in the last row of Table 1, most properties seem superior to those of ceramics. This room temperature data are misleading, however, since the insert working temperature can reach 1200 degrees C. All the comparisons change when considered at elevated temperatures. Hot strength, hardness and toughness of the cutting tool materials appear in Fig. 4. The decrease of carbide properties with temperature is dramatic, with silicon nitride the winner over alumina as far as mechanical properties are concerned. The high temperature wear resistance factor M is plotted in Fig. 5. The curves support superiority of cemented carbide at low cutting speeds, i.e. temperatures <800 degrees C. However, chemical instability and plastic deformability of the WC-Co family of materials excludes their application at higher temperatures. This is schematically shown in Fig. 5 as a rapid decrease of wear resistivity factor M. Among ceramic tools, pure alumina shows the poorest performance; this is to blame for the relatively slow CCT introduction in practice over the last 30 years. Recently developed SiC-whisker/alumina and silicon nitride based compositions show wear resistance double that of alumina, with silicon nitride more suitable for very high speeds. Both types of advanced ceramic/ceramic composites are presently manufactured in Canada by The Electrofuel Mfg. Co.

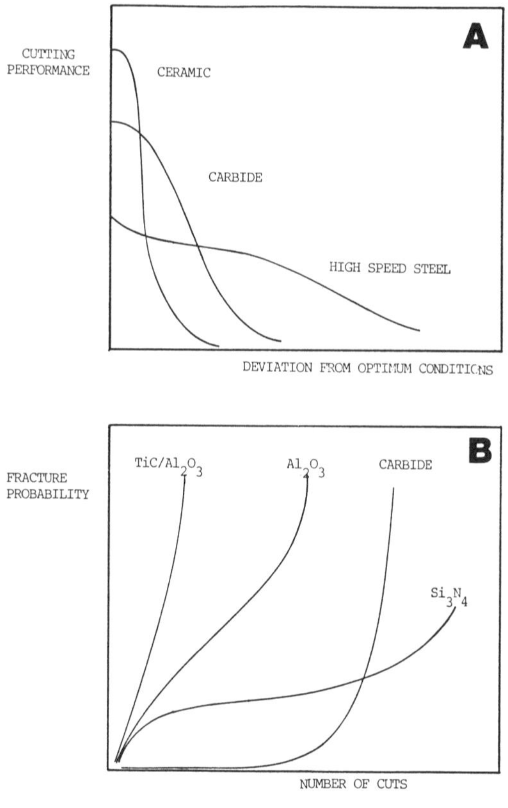

Fig. 6. Reliability of cutting tools: (A) Cutting performance vs. deviation from optimum conditions, (B) Failure probability vs. time or number of cuts.

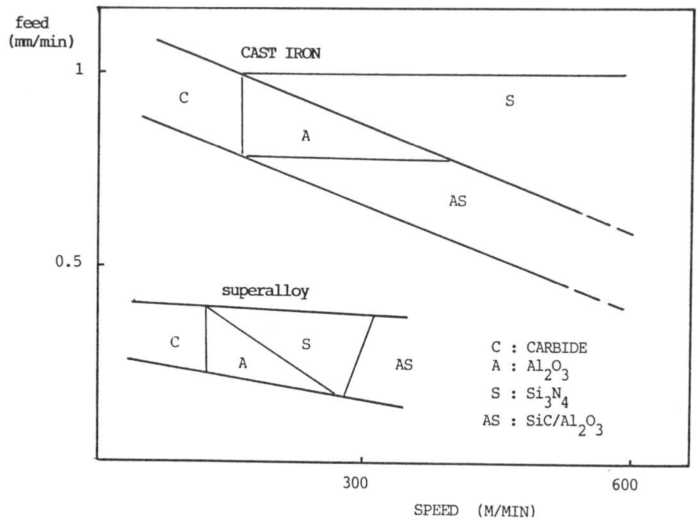

Fig. 7. Typical application map for ceramics and cemented carbides in machining of cast iron and superalloys.

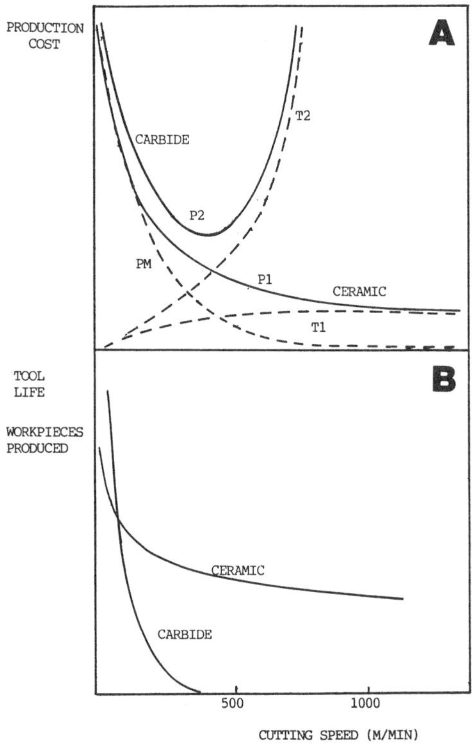

Fig. 8. Effect of cutting speed on productivity expressed as production costs (A) and tool life (B).

The biggest advantages of CCTs, their hot hardness, strength and chemical inertness, are partially offset by their reliability, especially low for poorly processed monolithic materials. Inspite of continuing improvements ceramic are presently stil the least forgiving among the cutting tool materials as far as accomodation to process variation is concerned. This is schematically presented in Fig. 6A, with cutting performance as an arbitrary dependent variable. The figure emphasizes the highest performance of CCT within precisely defined process conditions: high optimal speed; low vibration high stiffness machines and support; uninterrupted cutting. Determination of tool lifetime criteria is essential for cost effective metal cutting, especially in the machining of expensive superalloys. Low reliability of early ceramic tools is reflected in a gradual increase of catastrophic fracture occurance after certain work time or number of cuts. Instead, non-zero fracture probability extends to short times. This is illustrated in Fig. 6B for alumina, TiC/alumina, carbide and silicon nitride tools in cast iron machining. Significant improvement of reliability soá defined is expected for whisker-reinforced materials.

The cutting speed increase is limited even for ceramic tools and is practically determined by chemical composition and compatibility of tool and workpiece. Typically, alumina based tools are used for machining of carbon and alloy steels up to 4000 surface feet per minute (sfm) and cast iron up to 3000 sfm. SiCw/alumina inserts can not be utilised for steel machining due to the high solubility of silicon in steel.

TiC/alumina composites are presently the most widely utilised CCT. Their big advantage is their ability to perform hard part machining (55 to.65 Rc hardness) with ground-like finish. Since this kind of machining requires low speeds, typically 300 to 600 sfm, conventional lathes can be utilised with considerable savings on grinding equipment. High thermal shock and fracture resistance of Si_3N_4-based inserts allows cutting of cast iron and nickel alloys under severe conditions. SiCw/alumina inserts can be used under similar conditions, but at lower feed rate and higher speed, especially for nickel alloys. The application ranges (feed rate vs. speed) are compiled for high toughness CCT (Si_3N_4 and SiCw/alumina), alumina and carbide in Fig. 7. Clearly, the best forecast is obtained for composite ceramic, Si_3N_4-based for cast iron machining and alumina-based for nickel alloys.

The effect of cutting speed on productivity using ceramic tools (curve P1) and carbide tools (curve P2) is compared in Fig. 8A. The cost of a carbide tool (curve T2) increases rapidly above a certain speed, typically 300 m/min, due to accelerated wear and plastic deformation. The cost of a ceramic tool (curve T1) is almost flat vs. speed, as long as catastrophic fracture does not occur. The machine costs (PM) continuously decrease with the cutting speed, independent of tool type. The combination of curves PM+T2=P1 gives a sharp minimum while for PM+T2=P2 continuous cost decrease results. This attractive feature of CCT has been being fully exploited only recently, with wider implementation of ceramic/ceramic composite inserts in high-speed metal machining. Another aspect of productivity increase is illustrated by consideration of tool life or number of workpieces produced vs. cutting speed, Fig 8B. For carbide, tool life drops sharply to zero above 300 m/min while long term stability is expected for ceramics.

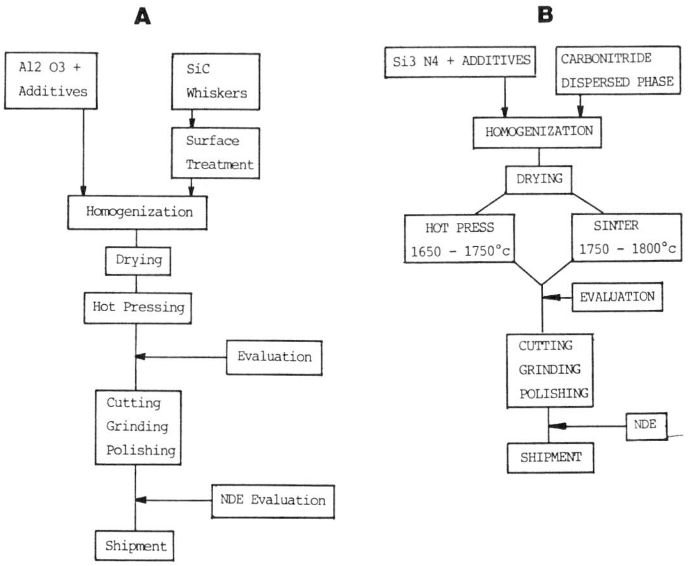

Fig. 9. Fabrication flowcharts for composite alumina (A) and silicon nitride (B) ceramic cutting tools.

TABLE 2 Room-Temperature Properties of Ceramic/Ceramic Composite Cutting Tools Produced by The Electrofuel Mfg Co. Ltd.

PROPERTY	Si_3N_4 COMPOSITE KN3	SiC_w/Al_2O_3 AS1
Color	Grey to Brown	Green
Density,%	99-100	99-99.8
Hardness Ra	91.5-93	94-95
Hardness HV,GPa	14-16	21-23
Youngs Modulus GPa	300-400	410
Fracture Toughness,$MPa\cdot\sqrt{m}$	6-9	6-8
Wear Resistivity Factor, $GPa^{5/4}m^{3/8}$	14-21	17-23

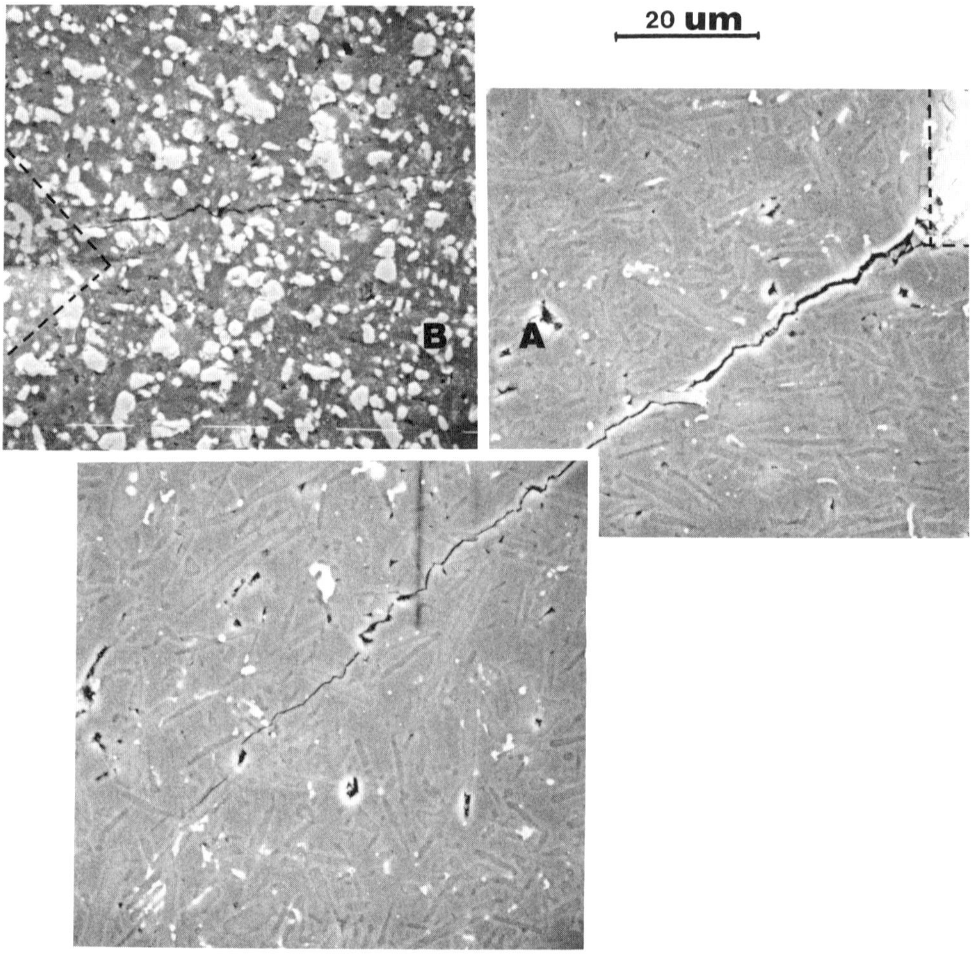

Fig. 10. Crack-particle interaction for SiCw/Al2O3 (A) and Si3N4-based (B) ceramic-ceramic composites.

COMPOSITE CERAMIC CUTTING TOOLS IN CANADA

Recently, Electrofuel has developed a number of composite ceramic cutting tools based on silicon nitride and alumina. The fabrication flowcharts for these materials are shown in Fig 9A and 9B respectively. SiC whisker reinforced alumina must be hot pressed, while Si3N4-based particulate composite can be sintered as well as hot pressed. Typical properties of these ceramics are shown in Table 2. The range of the wear resistance factor M is plotted in Fig. 5 for KN3 (trade name of a Si3N4 composite) and AS1 (trade name of a SiCw/alumina composite). These data compare favourably with typical properties of market materials. The composite ceramic cutting tools manufactured by Electrofuel are presently undergoing laboratory and field testing in high-speed machining of cast iron and superalloys.

The microstructures (polished section including 10 kg Vickers indentation crack) for SiCw/alumina and Si3N4 composite appear in Fig. 10A and 10B respectively. The whiskers interact with the propagating crack, causing its multiple deflection and arrest. If orientation of the whisker is favourable, i.e. parallel to the tensile crack opening stress, a pullout toughening mechanism is possible. This is not a frequent occurance however due to random orientation of whiskers in a plane normal to the hot pressing direction. All three toughening mechanisms can be observed on the polished section, Fig. 10A. Si3N4-based composite contains propriatory particulate dispersed phases of high hardness. It can also contain low volume fractions of SiC whiskers. Increased toughness of the composite seems to originate from multiple deflections of the crack, seeking a tensile stress field in the vicinity of the particle/matrix interface. Maximum matrix grain size is limited to the free path between inclusions ($\sim 5 \mu m$). A study of microstructure vs. performance for both composites is underway.

CLOSING REMARKS

It is characteristic that absorption of new technologies requires parallel development of "new thinking" from their users. We can recognize it clearly in a slow process of diffusion of ceramic materials into fields traditionally reserved for metals. Superiority of ceramics over metals when utilised in severe environments will be fully exploited only when people learn why and how to use them. It is hoped that the above presentation of characteristic features of advanced ceramic cutting tools will help in the teaching process.

It should be mentioned finally, that the predicted productivity increase is only an initial, relatively small benefit for society. Wide successful implementation of ceramic cutting tools will spin off into other application fields for wear resistive, high temperature structural ceramics: steelmaking, metal forming, aerospace and, in particular, heat engine materials.

THE JOINING OF REACTION-BONDED SILICON CARBIDE TO INCONEL 600 USING REFRACTORY METAL INTERLAYERS

J.R. McDermid, M.D. Pugh and R.A.L. Drew

Dept. of Mining and Metallurgical Engineering,
McGill University, Montréal, Canada.

ABSTRACT

In order to exploit the thermo-mechanical properties of engineering ceramics in advanced heat engine applications, it is necessary that they be joined to the existing metallic support structure. The purpose of the present work was to model a cylindrical butt joint, using the finite element method (FEM), to assess the effect of placing refractory metal interlayers (Nb, Mo and W) of various thicknesses on the residual stress distribution in Inconel 600 (IN600)-reaction-bonded silicon carbide (RBSC) joints resulting from joint cooling from the joining temperature. It was found that the most effective combination was an IN600-Nb-Mo-RBSC joint with the Nb and Mo thicknesses being 3mm, resulting in a reduction in the residual stress levels by a factor of 8. However, it was found that using refractory metal-RBSC joints reduced the stresses much more effectively while providing a simpler joint geometry. It was concluded that the simple refractory metal-RBSC geometry should be explored further in joining experiments.

KEYWORDS

Metal-ceramic joining; interlayers; reaction-bonded silicon carbide; finite element method; residual stresses.

INTRODUCTION

One of the areas for which the use of structural ceramics is envisaged is that of advanced heat engines. However, for reasons of fabrication costs and practicality, it is necessary that ceramic components be joined successfully to the existing metallic support structure. In the case of advanced heat engines, this may involve the joining of ceramics to nickel-based superalloys. Also, in order to exploit the high temperature properties of structural ceramics, metal-ceramic bonds must have both mechanical and chemical stability at temperatures approaching 1000°C. In particular, this investigation is concerned with the joining of reaction-bonded silicon carbide (RBSC) to Inconel 600 (IN600), a nickel-based superalloy.

There are two main areas of concern in metal-ceramic joining: (i) chemical incompatibility between the ceramic and metal, and (ii) coefficient of thermal expansion (CTE) mismatch. It has been shown by the authors (McDermid, 1988) and other workers (Mehan, 1976, 1979, 1981) that RBSC reacts extensively with nickle-based superalloys, forming low-melting nickel silicides and undesirable carbide phases, thus resulting in serious degradation of the thermo-mechanical properties of the starting materials. Also, the large CTE mismatch between the RBSC and the superalloy causes fracture in the ceramic during cooling of the joints from the bonding temperature.

One of the methods being proposed to alleviate the problem of high CTE mismatch stresses is the interlayer method (Suganuma, 1984, 1985). In this method, a metal of CTE intermediate to that of the metal and ceramic, usually a refractory metal (i.e. Nb, Mo or W), is placed between the two materials, thereby reducing the CTE mismatch stresses. However, the interlayer can also perform the additional function of being a chemical barrier between the metal, in this case a nickel-based superalloy, and the ceramic. Since the refractory metals mentioned above all form high-melting point silicides (~1350°C)(Hanson, 1958), this could reduce the probability of forming low-melting phases. However, most of the refractory metals are strong carbide formers, which may cause difficulties in terms of the chemical stability of the ceramic.

It is known that the mechanical properties (i.e. E, G and v), the CTE mismatch and joint geometry, especially the material thickness, have a strong influence on the magnitude of the residual stresses generated during cooling of the joint (Suganuma, 1984, 1985, 1987; Kirchener, 1987). Thus, selecting compatible joining materials and optimizing the joint geometry can be a difficult process to perform purely by experimental means.

One of the more useful techniques currently available for assessing the effect of changing the interlayer metal and joint geometry on the residual stress levels is the finite element method (FEM). The FEM assumes that a structure can be broken down into a finite number of discrete elements connected by nodes. Each element has an individual set of material properties and mechanical constraints defined for it; how each element interacts with its neighbors determines the stress distribution. FEM calculations can easily be carried out on a personal computer equipped with a math coprocessor.

There are, however, several limitations associated with the use of the FEM to obtain the residual stress distribution; the main one being that the FEM assumes elastic material responses only and cannot take into account the effect of plastic deformation on the stress distribution. Since plastic deformation of the metal will tend to relieve the stress within the ceramic, the use of the FEM to evaluate the magnitude of the residual stresses may, in fact, yield a higher value than that which would be observed experimentally. The second limitation of the FEM is that the effects of chemical reactions, occurring during the course of the joining process, on the CTE distribution and local mechanical properties cannot be taken into account. However, it has been shown by Suganuma and co-workers (1987) that trends in the residual stress distribution predicted by the FEM are followed by the experimental joints even though the actual values of the stresses may be much lower. Thus, the FEM is a useful technique for selecting mechanically compatible joint materials and optimizing joint design, although it must be emphasized that joining experiments must be performed to assess the role of chemical reactions on the joint quality.

The purpose of the present work is to investigate means of joining RBSC to IN600 using refractory metal interlayers (Nb, Mo and W) to compensate for CTE mismatch and to provide chemical barriers between the IN600 and the RBSC. In particular, this study will be concerned with the use of the FEM to determine the optimum joint geometry and the most mechanically compatible refractory metal for the interlayer(s).

EXPERIMENTAL PROCEDURE

Starting materials consisted of reaction-bonded silicon carbide, hot-rolled Inconel 600 rod and high-purity (>99.98%) refractory metals. X-ray diffraction showed that the RBSC consisted of α-SiC and cubic Si; image analysis of typical microstructures determined that the continuous Si phase comprised approximately 10 wt.% of the RBSC. The strength of the RBSC was determined using 4-point bend modulus of rupture (MOR) measurements on 40mm x 6.5mm x 2.5mm test bars cut from larger plates and surface ground flat prior to testing. It was found that the RBSC had a mean strength of 294MPa and a Weibull modulus of 7.4; the apparently low value of the Weibull modulus stems from the cutting and grinding of the test bars from the larger plates.

Finite element calculations were carried out using ALGOR SUPERSAP 7.1. FEM meshes modelled a cylindrical butt joint with one or two interlayers whose thicknesses were varied from 0 to 3.0mm. The thickness of the IN600 and RBSC was held constant at 6.0mm. Typical meshes used in the FEM calculations for single and double interlayer joint models are given in Figs. 1(a) and 1(b), respectively. All models were cooled through 1100K, i.e. ΔT = -1100K. The material properties used in the calculations, including yield stress values for the metals and MOR for the RBSC, are listed in Table 1 and are the mean

TABLE 1: Material Properties Used in FEM Calculations.

Material	E (GPa)	G (GPa)	ν	σ ($\times 10^{6}K^{-1}$)	σ_{YS} (MPa)
IN600	210	76	0.29	16.39	130
RBSC	380	156	0.22	5.04	294(MOR)
Nb	103	37.5	0.38	7.97	150
Mo	235	91	0.29	6.10	200
W	385	149	0.29	4.84	100

values of the property for temperatures between 298K and 1398K. All models are assumed to be in a zero stress state (i.e. unconstrained) at 1398K.

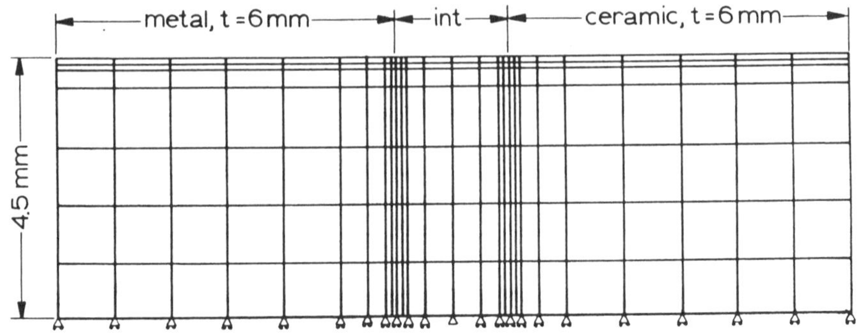

Fig 1 (a): Single interlayer FEM mesh.

RESULTS AND DISCUSSION

Finite element analyses were carried out in two sets; in the first, single refractory metal interlayers were placed between the IN600 and the RBSC; in the second, two refractory metal interlayers were used such that Nb was placed next to the IN600 (int.1 in Fig. 1(b))

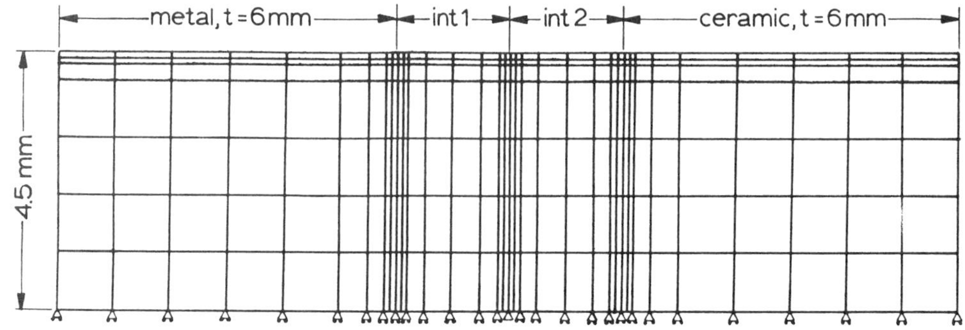

Fig. 1 (b): Double interlayer FEM mesh.

and Mo or W was used next to the ceramic (int.2 in Fig. 1(b)). Results for the single interlayer models, corresponding to the mesh in Fig. 1(a), are shown in Fig. 2. Results for the double interlayer models, which used a mesh like that given in Fig. 1(b), are given in Fig. 3, where Fig. 3(a) shows the results of models in which Mo was used as the second interlayer and Fig. 3(b) those in which W was used as the second interlayer. Results for Figs. 2 and 3 were plotted according to two criteria: the stress at the ceramic-metal interface and the peak stress (i.e. stress of highest magnitude) regardless of location of occurrence in the model.

From Fig. 2, it can be seen that the use of a single refractory metal interlayer greatly reduces the residual stresses at the ceramic-metal interface, Mo being the most effective at reducing the stresses. This is because the CTE and mechanical properties of Mo are the most intermediate of the refractory metals used in the models. However, it can also be seen that Nb is the most effective interlayer for reducing the overall peak stress. In the case of Mo and W, the peak stresses tended to occur at the IN600-interlayer interface. It can also be seen from Fig. 2 that the stresses tend to reduce in magnitude as the thickness of the interlayer increases; it was found that increasing the interlayer thickness beyond 3mm resulted in minimal changes in either the interfacial or peak stresses.

Results for the double interlayer models are given in Fig. 3; the results for Nb-Mo interlayers in Fig. 3(a) and those for Nb-W interlayers in Fig. 3(b). It can be seen that there is no significant difference in the residual stress levels at the ceramic-metal interface regardless of the interlayer combination used. The use of Nb-Mo interlayers, however, results in lower peak stress levels than those observed for the corresponding Nb-W interlayer. As with the single interlayer models, the magnitude of the residual stresses at the ceramic-metal interface tended to decrease with increasing interlayer thickness; the most effective interlayer being a Nb-Mo interlayer in which the thickness of each refractory metal is 3.0mm.

It should be noted that the thick interlayers needed to reduce the stress levels to magnitudes suitable for joining to RBSC would result in complicated joint geometries and joining procedures. A more practical joint geometry is shown in Fig. 4, in which the refractory metal is joined directly to the ceramic and the IN600 is removed from the model. In this model, the RBSC thickness is held constant at 6.0mm and the refractory metal thickness is varied from 0 to 6.0mm. Results from this series of models is shown in Fig. 5. It can be seen from this Fig. that W is the most compatible metal with RBSC, from the residual stress point of view, with residual stress levels of approximately 25MPa. These low residual stress levels result from the close matching of the CTE and mechanical property values (see Table I) of W and RBSC. Also, it can be seen that the residual stresses peak at a metal thickness of 3mm, becoming constant at thicknesses greater than 4 mm.

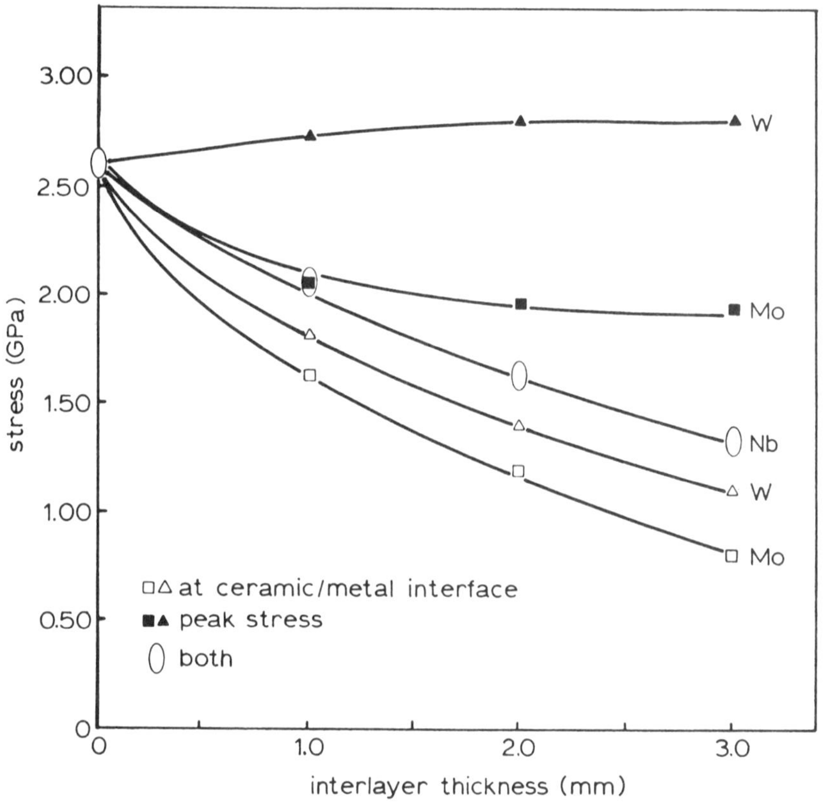

Fig. 2: Stress vs. interlayer thickness for single interlayer joints.

As discussed above, the FEM analyses do not take into account the effect of plastic deformation on the residual stress distribution. An examination of Figs. 2 through 4 will show that, except in the case of the simple W-RBSC joints (Fig. 4), stress levels are sufficiently high to cause plastic deformation in all of the metallic materials (see Table 1). This would serve to reduce the magnitude of the residual stresses in the joints. However, the simple refractory metal-ceramic joints give the lowest residual stress levels and the least complicated joint geometry.

A typical isostress contour map of a W-RBSC joint model is given in Fig. 6. As would be expected, the ceramic is largely under compression, the stresses being transformed from compressive to tensile across the ceramic-metal interface, with the metal being under tensile stress. It can be seen that the stress gradient across the interface is quite steep, emphasizing the importance of having a stress concentration-free braze layer and a smooth ceramic surface. However, since this is not the case in brazing, it can be said that there would be stress concentrations which would serve to increase local stresses. Also, the effect of chemical reactions cannot be accounted for in the FEM and their effect on the residual stress distribution is difficult to assess. Thus, it must be emphasized that, in order to obtain information about the effects of plastic deformation and chemical reactions on joint quality and reliability, the joining experiments must be performed.

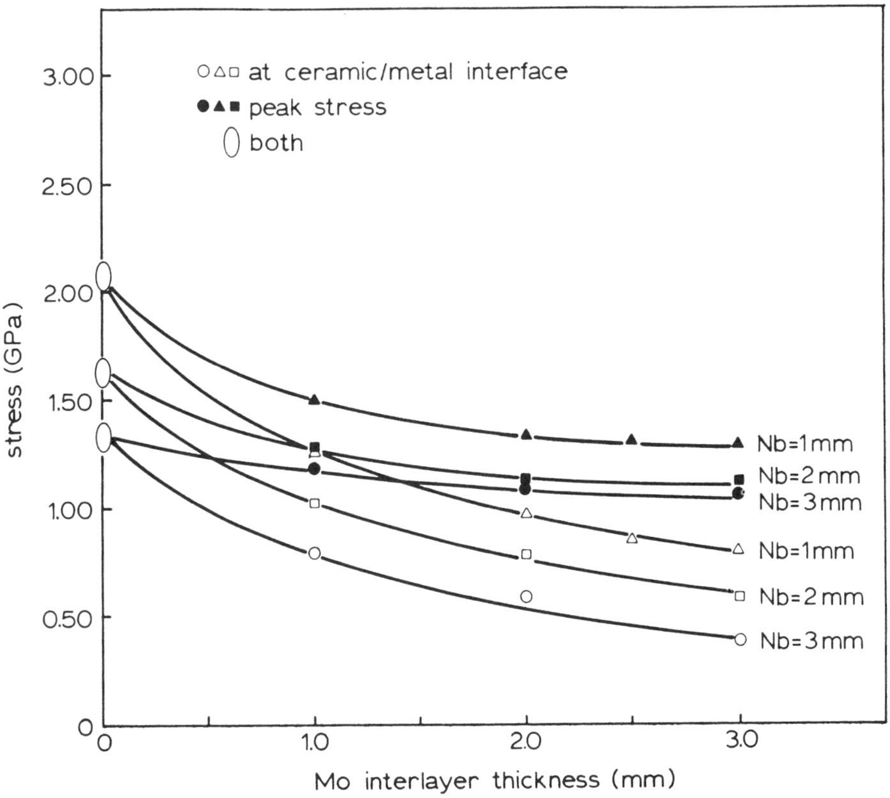

Fig 3 (a): Stress vs. second interlayer thickness for Mo.

CONCLUSIONS

1. The use of refractory metal interlayers between Inconel 600 (IN600) and reaction-bonded silicon carbide (RBSC) reduces the magnitude of the residual stresses from joint cool-down by as much as a factor of 4, as determined by the finite element method (FEM). The most effective interlayer geometry consisted of two 3mm thicknesses of Nb and Mo, the Nb being placed next to the IN600 and the Mo next to the RBSC.

2. A more practical joint geometry, with a greater reduction in the residual stress levels can be achieved with the use of refractory metal (i.e. Nb, Mo or W)-RBSC joints. The most effective refractory metal-RBSC combination is W-RBSC, with FEM stresses of 25MPa.

3. The FEM is a useful tool for assessing the effects of changing joining materials and joint geometry on the magnitude of the residual stresses and stress distribution. However, the effects of plastic deformation and chemical reactions on the residual stress distribution cannot be predicted using the FEM and, therefore, it is necessary that the joining experiments be performed to assess the effects of these factors on the joint quality.

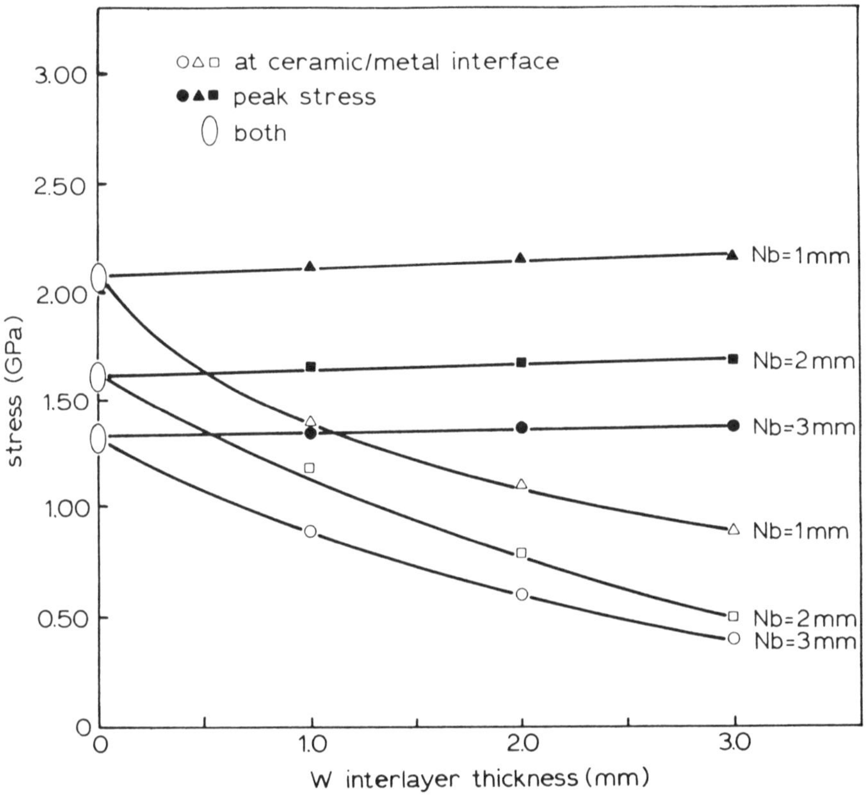

Fig 3 (b): Stress vs. second interlayer thickness for W.

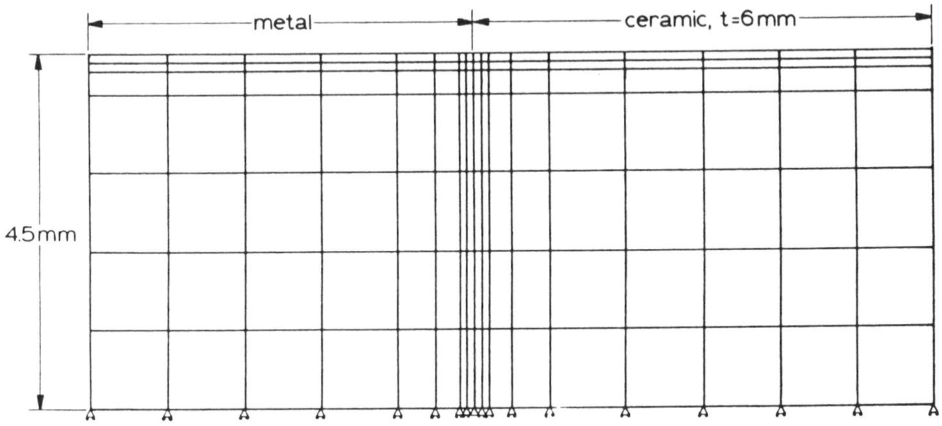

Fig. 4: FEM mesh used for simple refractory metal-RBSC joints.

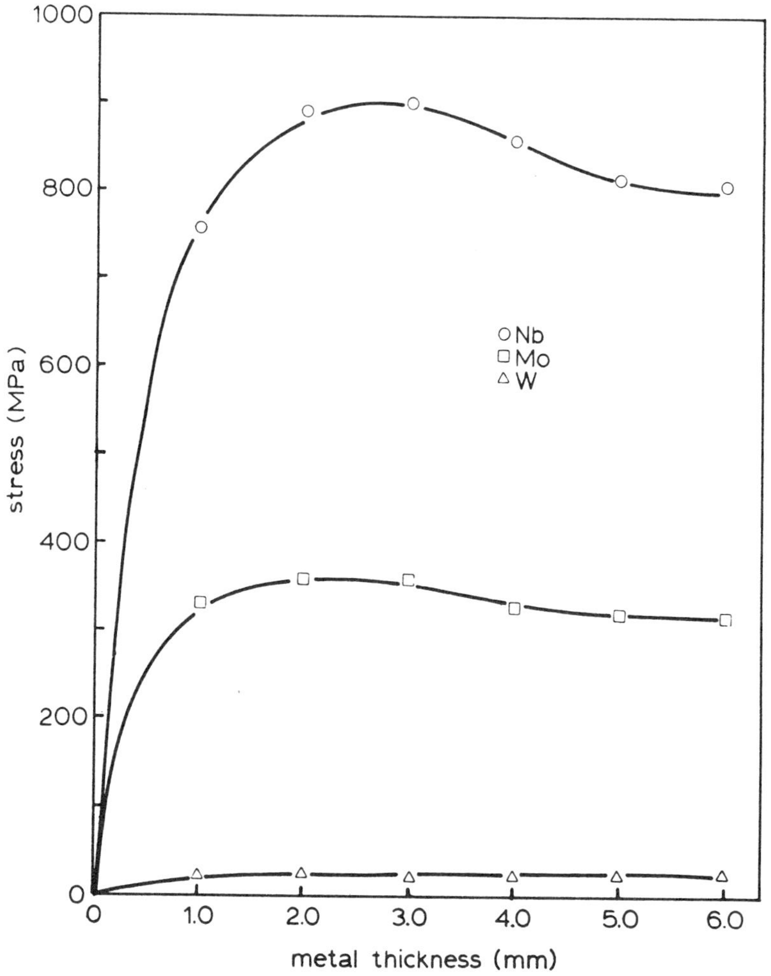

Fig. 5: Stress vs. refractory metal thickness for refractory metal-RBSC joints.

REFERENCES

Hansen, M. (1958). Constitution of Binary Alloys, 2nd. ed., Mc Graw-Hill, New York. p. 974, 1016, 1073.

Kirchener, H.P., J.C. Conway Jr. and A.E. Segall (1987). Effect of joint thickness and residual stresses on the properties of ceramic adhesive joints: I, finite element analysis of stresses in joints. J. Am. Ceram. Soc., 70 (2), 104-109.

McDermid, J.R., M.D. Pugh and R.A.L. Drew (1988). The joining of reaction-bonded silicon carbide to Inconel 600 using a nickel-based brazing alloy. Submitted to Met. Trans. A.

Mehan, R.L. and D.W. McKee (1976). Interaction of metals and alloys with silicon-based

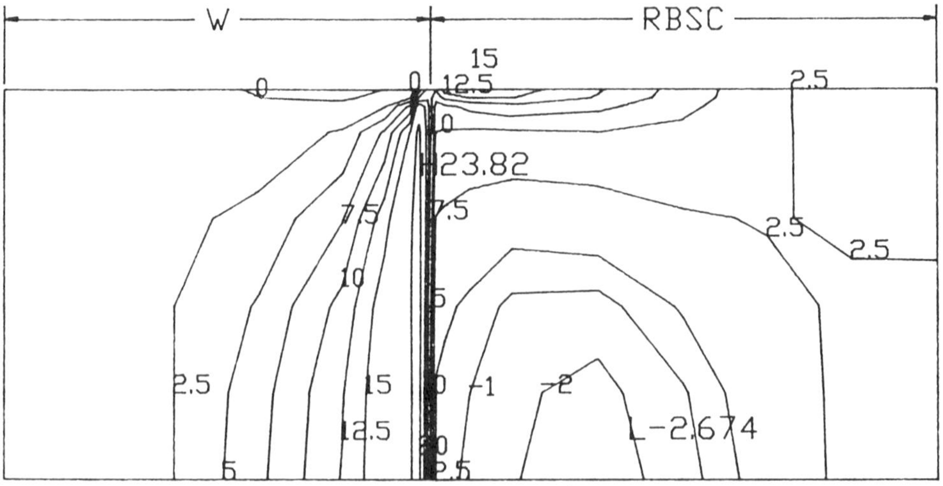

Fig. 6: Isostress contour plot (MPa) of a W-RBSC joint.

ceramics. J. Mat. Sci., 11, 1009-1018.
Mehan, R.L. and R.B. Bolon (1979). Interaction between silicon carbide and a nickel-based superally at elevated temperatures. J. Mat. Sci., 14, 2471-2481.
Mehan, R.L. and M.R. Jackson (1981). In J.A. Pask and A. Evans (Eds.)), Materials Science Research, Vo. 14. Plenum Press, New York. pp. 513-523.
Suganuma, K. and others (1984) Effect of interlayers in ceramic-metal joints with thermal expansion mismatches. J. Am. Ceram. Soc, 67 (12), C256-C257.
Suganuma, K. and others (1985). Method for preventing thermal expansion mismatch effect in ceramic-metal joining. J. Mat. Sci. Lett., 4, 648-650.
Suganuma, K. and others (1987). Influence of shape and size of residual stress in ceramic-metal joining. J. Mat. Sci., 22, 3561-3565.

A STUDY ON SiC WHISKER REINFORCED LITHIUM ALUMINOSILICATE COMPOSITES

G. McMahon*, S.S.B. Wang**, D.H.H. Quon**,
R.R. Sood*, R.T. Holt***, T.M. Maccagno***

*Ceramics Kingston Inc., Kingston, Ontario, Canada
**Ceramics Section, CANMET, Ottawa, Ontario, Canada
***National Aeronautical Establishment
National Research Council, Ottawa
Ontario, Canada

ABSTRACT

Lithium aluminosilicate glass ceramics reinforced with SiC whiskers have been fabricated using the hot pressing technique. Room temperature modulus of rupture values were evaluated through biaxial flexure tests. The modulus of rupture for the unreinforced material was 69 MPa. Minor strengthening effects were observed for composites containing 2.5 and 5.0 wt. % whiskers, while a decrease in the modulus of rupture occurred for the composite containing 20 wt. % whiskers. The increase in porosity with increasing whisker content is believed to be the major reason for this behaviour.

KEYWORDS

Lithium aluminosilicate, glass ceramics, SiC whiskers, biaxial flexure tests, porosity, modulus of rupture.

INTRODUCTION

Glass ceramics are polycrystalline solids in which fine, uniform crystals are grown by a controlled heat treatment of an oxide glass. The heat treating schedule often involves two steps. The first is performed at a temperature below the crystallization temperature to nucleate crystals throughout the glass. A nucleating agent, such as TiO_2, is frequently added to the glass to assist nucleation. As a result, more grains of finer size will be formed upon further heating above the crystallization temperature during the second step of the heat treatment procedure. The final structure obtained consists of fine crystals held together in a glassy matrix. A more comprehensive treatment regarding the formation and structure of glass ceramics is available in the work of McMillan (1979).

So far, the use of these materials has been hindered by their brittle nature and lack of toughness. However, by incorporating high strength, high modulus ceramic fibers, a material with improved mechanical properties may be obtained. This improvement is a result of the transfer of the applied load to the fibers after the matrix has fractured and the subsequent fracture and pullout of the fibers on continued load increase.

Current interest in fiber-reinforced glass ceramics for applications in gas turbine engines stems from several desirable attributes. These include the ability to control the viscosity and workability of glasses, permitting physical densification of composites without mechanical damage to the reinforcing material. They may also be made with a broad range of compositions, allowing control over both the fiber/matrix interaction and the coefficient of thermal expansion. Therefore, glass ceramic matrices may be developed with thermal expansion coefficients nearly matching those of the reinforcing material.

More extensive review articles on the development of fiber-reinforced glasses and glass ceramics have been presented by Prewo (1986) and Prewo, Brennan and Layden (1986), where the emphasis is placed on unidirectional reinforced glass ceramics incorporating non-stoichiometric continuously wound SiC filament as the reinforcing material. However, fine-scale SiC whiskers measuring 0.1 to 1.0 μm in diameter have become available recently as an alternative reinforcing medium. Since these whiskers are stoichiometric and have better mechanical properties than continuously wound filaments (Petrovic and others, 1985; Stacey, 1988) and because unidirectionally reinforced composites have transverse strengths less than 5% of the axial strength (Prewo, 1986), the present study is being performed to examine the potential of using SiC whiskers to multi-directionally reinforce a lithium aluminosilicate (LAS) glass ceramic. Such composites have potential applications in automotive gas turbine engines as a material for housings or piston crowns. The low coefficients of thermal expansion and thermal conductivity obtained in these materials make them candidates for such applications. LAS/SiC composites are also being considered for use as combustor liners in commercial aircraft engines.

EXPERIMENTAL

The composition of the glass material is given in Table 1 and is similar to Corning code 9608 described by Doherty, Lee and Davis (1967) except that no As_2O_3 was added; the Li_2O content was higher; and the SiO_2 content lower. This composition lies within the β-spodumene phase field of the ternary SiO_2-Al_2O_3-Li_2O equilibrium phase diagram.

TABLE 1 Composition of Glass Material

Component	Weight %
SiO_2	68.5
Al_2O_3	17.5
MgO	2.7
Li_2O	5.0
TiO_2	4.7
ZrO_2	0.2
ZnO	1.0
Na_2O	0.4

TiO_2 and ZrO_2 were added to act as nucleating agents which are required to promote bulk nucleation and subsequent crystal growth throughout the glass. MgO and ZnO were added as partial substitutes for Li_2O in order to improve the workability of the glass while also lowering the materials cost.

The raw materials used for glass preparation were analytical grade chemicals. Both lithium oxide and sodium oxide were obtained from their respective carbonates. Required amounts of raw materials were weighed and blended together using a roller mill. The mix was then calcined at 1000°C for one hour to decompose the carbonates and to pre-react the oxide constituents. The calcine product was thoroughly mixed in a ball mill for three hours, then loaded into a mullite crucible and melted in a globar furnace at 1350°C. The melt was allowed to reach equilibrium before being quenched in cold water. The resulting glass was crushed and ground to produce a powder with a particle size of less than 88 μm.

The composites were prepared with 0, 2.5, 5.0 and 20.0 wt. % whiskers. About 200 g of glass powder was dispersed in isopropyl alcohol using a magnetic stirrer. Silicon carbide whiskers were then added to the slurry and dispersed with the aid of an ultrasonic vibrator. After mixing, the alcohol was evaporated in a glass tray using a hot plate at 200°C. Cylinders approximately 3.9 cm in diameter and 5.0 cm in height were then cold pressed from the LAS/SiC whisker mix.

These cylinders were loaded into the graphite die of a hot press, and upon evacuating the chamber to 2.7 μPa (20 x 10^{-3} torr) the temperature of the furnace was slowly increased to 1280°C and held at this temperature for 15 minutes. The samples were hot pressed at an applied pressure of 13 MPa and maintained at that temperature and pressure for 5 minutes before being furnace cooled under vacuum.

From the cylinders, discs were prepared and these were then heat treated for two hours at 660°C, after which time the temperature was slowly raised to 900°C and held for three hours. Differential thermal analysis (DTA) measurements indicated that the crystallization temperature of the glass was 780°C, and so it was believed that the heat treating schedule described would provide proper nucleation and crystal growth.

Microstructural analysis was performed using x-ray diffraction (XRD), optical microscopy, scanning electron microscopy (SEM) and transmission electron microscopy (TEM). Specimens for microstructural analysis were prepared by grinding the samples to 600 grit using SiC paper, followed by polishing with a 0.05 μm alumina slurry. The polished samples were then etched in a 5% HF solution for approximately 5 minutes. Specimens for both optical microscopy and SEM were sputter coated with gold, a procedure which greatly enhanced grain boundary contrast in optical microscopy. TEM samples were prepared by ultrasonically drilling 3 mm discs, dimpling to a thickness of approximately 40μm, followed by ion milling to perforation at a gun voltage of 6 kV and an incident angle of 20 degrees. The samples were further thinned at a gun voltage of 4 kV and an incident angle of 10 degrees for three hours. Continuous cooling with liquid nitrogen during the thinning operation was used to limit microstructural artifacts which are frequently caused by the ion beam.

The biaxial modulus of rupture (MOR) of the disc samples was measured in accordance with ASTM standard F394-78. An MTS servo-hydraulic testing machine of 100 kN capacity was used for this purpose. The ram displacement rate was 0.05 mm/min and a linear variable displacement transducer (LVDT) with a full scale displacement of 2.5 mm was incorporated to measure deflection at the midpoint of the disc.

RESULTS AND DISCUSSION

The three XRD patterns presented in Fig. 1 show the structure of the LAS powder, the LAS/SiC composite after the first stage of the heat treatment, and the composite after the final stage of the heat treatment.

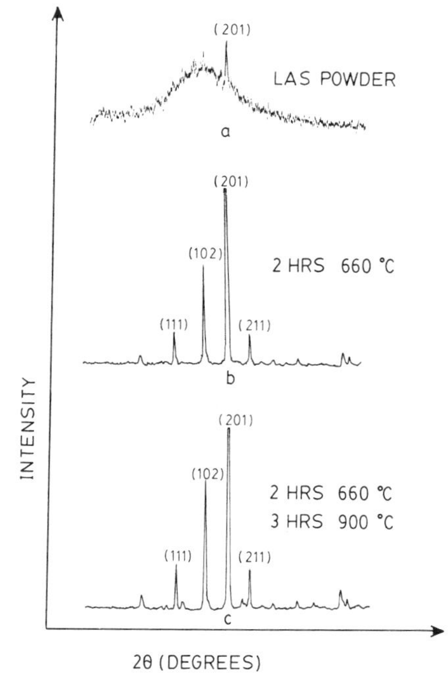

Fig. 1 X-ray diffraction patterns of (a) LAS powder (b) LAS after hot pressing and heat treatment at 660°C for 2 hours, and (c) after a further heat treatment at 900°C for 3 hours.

The broad peak observed in Fig. 1a is characteristic of a glassy material having only short range atomic order. The small crystalline peak on the right shoulder of the broad peak is due to the strong (201) reflection of tetragonal β-spodumene. The diffraction patterns in Fig. 1b and 1c are, respectively, from the material heat treated for 2 hours at 660°C, and the same material after it had been slowly heated to 900°C and held at that temperature for 3 hours. Both patterns indicate that the glass has crystallized into tetragonal β-spodumene, $LiAlSi_3O_8$. The SEM micrograph of an LAS/2.5% SiC whisker composite shown in Fig. 2 indicates a grain size of 5 to 10μ is achieved after the full heat treatment procedure.

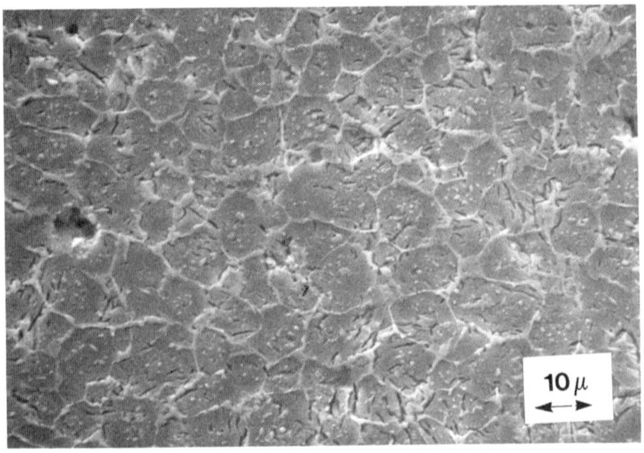

Fig. 2 SEM micrograph of an etched LAS/2.5% SiC whisker composite.

Results from the mechanical testing of the composites are presented in Fig. 3 and indicate that no significant improvement in the modulus of rupture was obtained with increasing whisker content.

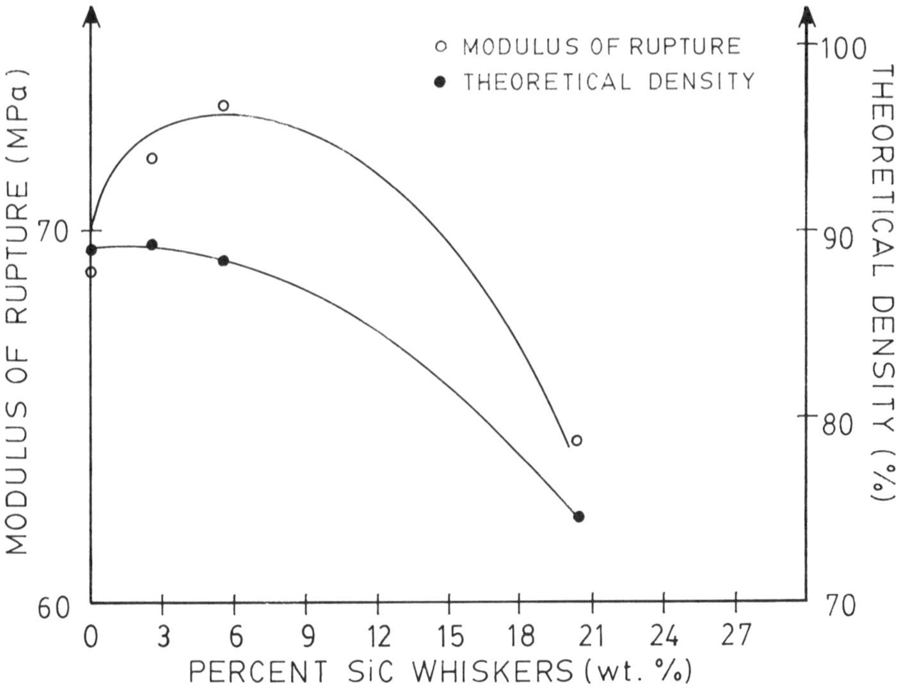

Fig. 3 Modulus of rupture and theoretical density as a function of SiC whisker content

The plot of sample density (as a fraction of theoretical density) versus whisker content shows that the likely reason for this behaviour is the increase in porosity which is observed with increasing whisker content. As pores may act as crack initiation sites, an increase in their number and size would be expected to decrease the modulus of rupture. Sensitivity to density variations was also observed by Gadkaree and Chyung (1986) on magnesium aluminosilicate composites containing 30, 40 and 50 wt. % whiskers. The porosity in the sample can be attributed to two separate causes. The first is improper degassing of the LAS/whisker mix during the hot pressing operation. The second and more important reason is porosity arising from poor distribution of whiskers. This behaviour is shown in Fig. 4 in a SEM micrograph of the fracture surface of an LAS/20% SiC whisker composite. The area of Fig. 4 indicated by an arrow is shown at higher magnification in Fig. 5. A high level of porosity is clearly associated with an agglomeration of whiskers.

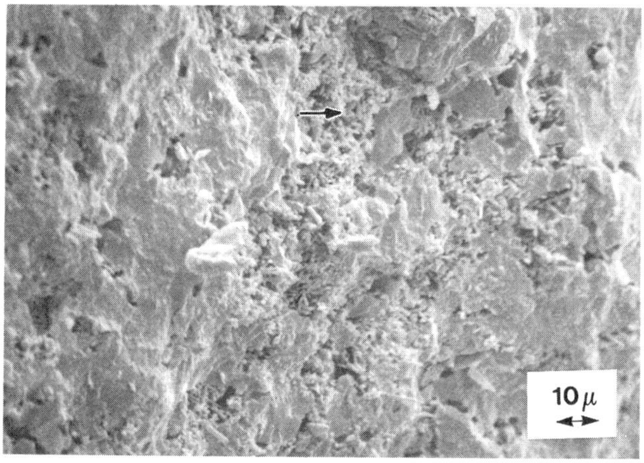

Fig. 4 SEM micrograph showing fracture surface of an LAS/20% SiC whisker composite.

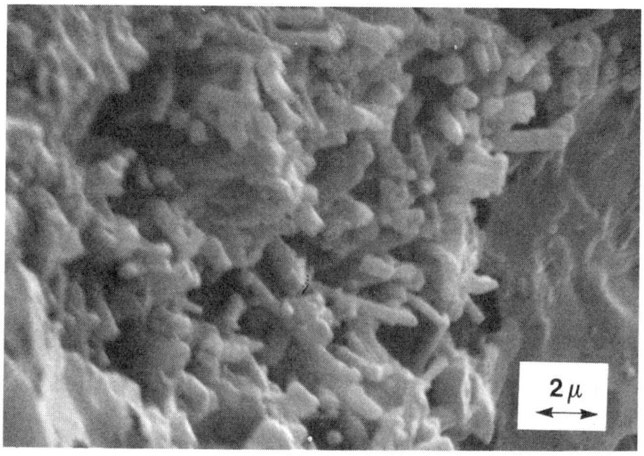

Fig. 5 Higher magnification SEM micrograph of an LAS/20% SiC whisker composite fracture surface.

To the best of our knowledge, there is no other published data on the biaxial modulus of rupture of β-spodumene LAS glass. The value of 69 MPa obtained in the present study was somewhat lower than the value of 83 MPa quoted by Grossman (1986) for the same material, but presumably obtained from a simple bend test. However, Godfrey and John (1986) have compared the MOR values for alumina and reaction-bonded silicon nitride determined from 3- and 4-point bend tests on bars and from ring loading tests on discs. Their Weibull modulus plots show that the discs failed at about 20% lower stress than the bars. Taking this into consideration, the results for the unmodified LAS glass in the present study compare favourably to the results quoted by Grossman.

Comparison of the modulus of rupture values for the composite materials is much more difficult because the majority of research involving LAS/SiC reinforced composites utilizes continuous Nicalon SiC fiber to produce unidirectionally reinforced or cross-plied composites. The axial modulus of rupture for a unidirectional reinforced LAS/SiC composite, containing 46 vol. % fibers, was quoted to be 690 MPa (Prewo, 1986), but the transverse modulus of rupture was only 25 MPa. However, cross-plied samples containing 50 vol. % fiber show modulus of rupture values measured in three-point bending to be 350 MPa (Brennan and Prewo, 1982).

It is felt that by modifying the whisker dispersion method and adjusting the hot pressing conditions, the porosity of the composites could be decreased, giving rise to mechanical properties which would approach those attained in the cross-plied composites. Also, the TEM bright field image in Fig. 6 shows that there is still a significant amount of residual glassy phase in the composite.

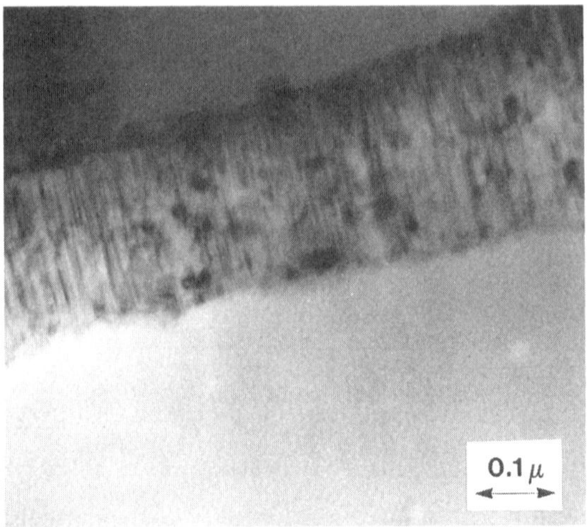

Fig. 6 TEM bright field image showing a whisker in a glassy matrix

The "granular" contrast observed in this region is likely a result of compositional fluctuations caused by phase separation (Vargin, 1965). We expect that a reduction in the amount of residual glassy phase by improved mixing techniques and by modification of the heat treating schedule should also lead to improvements in strength.

CONCLUSIONS

SiC whisker reinforced LAS composites have been produced by the hot pressing technique. No increase in strength was observed with increasing whisker content because the porosity of the composites also increased with increasing whisker content. The increase in porosity was associated with insufficient whisker dispersion within the composite matrix. Further studies regarding the dispersion of whiskers within the matrix and the optimization of the heat treating schedule to reduce the fraction of the residual glassy phase are being performed in an attempt to improve the mechanical properties of the composite material.

ACKNOWLEDGEMENTS

The authors would like to thank A. J. Hanson and G. McDonald for their assistance with the hot press, Dr. T.A. Wheat for helpful discussion and Dr. U. Erb and S.B. Sant for assistance with the TEM sample preparation and helpful discussion. Financial assistance through the National Research Council IRAP contribution number CA103-7-1090 is also acknowledged.

REFERENCES

Brennan, J. J., and K. M. Prewo (1982). Silicon carbide fibre reinforced glass-ceramic matrix composites exhibiting high strength and toughness. J. Mat. Sci., 17, 2371-2383.

Doherty, P. E., D. W. Lee and R. S. Davis (1967). Direct observation of the crystallization of Li_2O-Al_2O_3-SiO_2 glasses containing TiO_2. J. Am. Cer. Soc., 50, 77-81.

Gadkaree, K. P., and K. Chyung (1986). Silicon-carbide-whisker-reinforced glass and glass-ceramic composites. Ceram. Bull., 65, 370-376.

Godfrey, D. J., and S. John (1986). Disc flexure tests for the evaluation of ceramic strength. In W. Bunk and H. Hausner (Ed.), Ceramic Materials and Components for Engines, Verlag Deutsche Keramische Gesellschaft 657-665.

Grossman, D. G., (1986). Glass ceramics. In M. B. Bever (Ed.), Encyclopedia of Materials Science and Engineering, Vol. 3, Pergamon Press, Oxford, 1966-1972.

McMillan, P. W., (1979). Glass Ceramics. Academic Press, London.

Prewo, K. M., (1986). The development of fiber reinforced glasses and glass ceramics. In R. E. Tressler, G. L. Messing, C. G. Pantano and R. E. Newnham (Ed.), Tailoring Multiphase and Composite Ceramics, Plenum Press, New York, 529-547.

Prewo, K. M., J. J. Brennan and G. K. Layden (1986). Fiber reinforced glasses and glass-ceramics for high performance applications. Ceram. Bull., 65 305-322.

Stacey, M. H. (1988). Production and characterization of fibers for metal matrix composites. Mat. Sci. Tech., 4, 227-230.

Vargin, V. V., (1965). Catalyzed Controlled Crystallization of Glasses in the Lithium Aluminosilicate System. Consultants Bureau, New York.

IMPACT BEHAVIOR OF FIBER REINFORCED GLASS AND CERAMIC MATRIX COMPOSITES

D.F. Hasson* and S.G. Fishman**

*Dept. of Mech. Engrg., U.S. Naval Academy
Annapolis, Maryland 21402, U.S.A.
Office of Naval Research
Arlington, Virginia 22217, U.S.A.

ABSTRACT

Glass and ceramic matrix composites with continuous fibers were tested with instrumented impact apparatuses. The composite architecture was crossply ($0/90^{o}$). The glass matrix composite had a borosilicate matrix with untreated high modulus carbor fibers, while ceramic matrix composites had a ceramed modified lithium aluminosilicate matrix with silicon carbon fibers or prewoven silicon carbide 2–D fabric. All the $0/90^{o}$ components showed that the edge on orientation had a lower dynamic rupture work per unit area than the interlaminar orientation, with the exception of glass matrix composite materials cut at 45° to the final ply axis. For the ceramic matrix composites the continuous silicon carbide fiber material was tougher than the composite with the 2–D prewoven fabric. For all the composities the percentage of dynamic rupture work which is attributed to crack initiation is less than twenty, and thus most of the rupture work is attributed to fracture of the fibers and fiber bundles after fiber/matrix interfacial bond failure with the attendant frictional sliding mechanism toughness contribution.

KEYWORDS

glass matrix composites; ceramic matrix composites; borosilicate glass; ceramed lithium aluminosilicate glass; instrumented impact tests; high modulus uncoated (HMU) carbon fibers; silicon carbide fibers; dynamic rupture work; notch orientation; stress intensity factor.

INTRODUCTION

Twenty years have passed since Bowen (1968) reviewed and proposed possible ways of improving the strength and toughness of ceramics by the incorporation of fibers into the ceramic matrix. In the next year Crivelli–Visconti and Cooper (1969) reported on the improvement of toughness of silica by the incorporation of continuous carbon fibers. In recent years process developments have allowed the fabrication of glass or ceramic matrix composites (CMC's) with improved toughness and tensile strengths suitable for structural applications (Prewo, Brennan and Layden, 1986). The impact behavior of these CMC materials is illustrated in Fig. 1. The force versus time curve shows elastic behavior to point 'A'. Between 'A' and 'B' periodic matrix cracks form (Phillips, Sambell and Bowen, 1972), while the strong fibers remain intact which provides a post–cracking load capability during matrix microfracture. Pugh (1967) called this toughening mechanism "bridging". Phillips (1972) and later Davidge (1987) note that at 'B' fracture initiation of fiber reinforcement occurs, which is followed by crack propagation which consists of fiber and fiber bundle failure plus fiber pullout from the matrix from 'B' through 'C'.

Final fracture corresponds to where the force reaches zero.

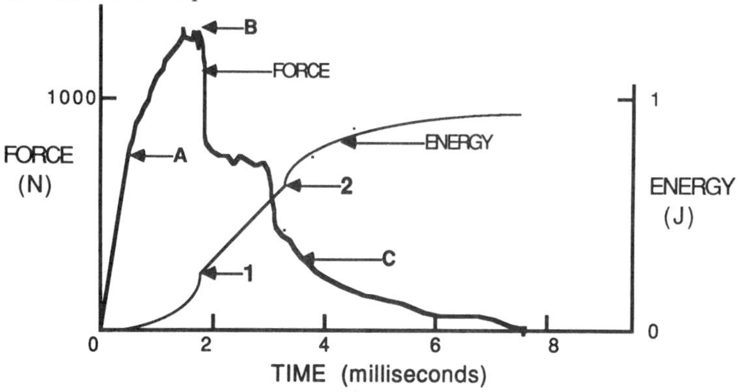

Fig. 1 Impact force and energy versus time for ceramic matrix composites.

During crack propagation, fiber pullout, which is controlled by the matrix/fiber interfacial bond, contributes to the composite toughness, especially if the interfacial bond is weak, because energy is absorbed by the frictional sliding of the fiber through the matrix (Marshall and Evans, 1985). In unidirectionally reinforced composites, the matrix cracking stress, the interfacial frictional sliding resistance and the composite ultimate stress are characteristic stresses and are not sensitive to matrix damage, because the fibers have sufficient strength to support the load in the presence of the matrix cracks. Properly processed composites, therefore, exhibit the damage tolerant behavior which is imperative if ceramic composites are to find application in structures subjected to tensile loading (Aveston, Cooper and Kelly, 1971; Evans and coworkers, 1985). Unfortunately, the low fiber–matrix bond strength important in these composites results in inferior transverse properties. As is the case in the reinforced epoxy composites, such considerable anisotropy can be overcome by the use of laminates. Recent investigations of the tensile and compressive response of symmetric ($0/90^{\circ}$) laminated SiC reinforced lithium aluminosilicate glass ceramics have shown that the stress–strain characteristics of these laminate layers in combination with knowledge of residual stresses present in the matrix (Lankford, 1985; Shaizero and Evans, 1986). Although the mechanical response of continuous fiber reinforced CMC's has been extensively studied under conditions of static loading, only limited information is available on the response to dynamic loading (Prewo, 1980; Brennan and Prewo, 1982; Hasson and Fishman, 1987). The objective of the present study is to increase the data base on the effect of dynamic load and notch orientation on the toughness and strength of various CMC materials.

EXPERIMENTAL DETAILS

The CMC materials were fabricated by the United Technologies Research Center (UTRC). Typical process details have been described (Prewo and Minford, 1985). The architecture was crossply ($0/90^{\circ}$). Each crossply layer was 200 microns thick. The CMC materials included one with a glass matrix and carbon fibers and another with a ceramed glass matrix and silicon carbide fibers.

The glass matrix composite had 42 volume percent of 8 micron high modulus untreated (HMU) in a Corning Code 7740 borosilicate glass. This material was used primarily for the effect of notch orientation on the toughness.

Two ceramed glass matrix composites were fabricated with a modified Corning lithium aluminosilicate (LAS III) matrix. One composite utilized 15 micron Nicalon silicon carbide fibers, and the other was fabricated from bolts of prewoven SiC 2–D fabric. Ceramic grade SiC yarn was woven into bolts of 29 x 29 12 harness satin weave fabric (HSW). The 29 x 29 refers to the number of warp ends per inch and fill picks per inch. The architecture of the 12 HSW composite is essentially the same as the 0/90° unitape ply couple, except for the staggered crossover points every 12 strands. The fiber volume percent was 44 for both composites. The prewoven SiC 2–D fabric 12 HSW represents an approach to produce a composite at lower cost and with the ability to be fabricated in a complex shape.

The span for the impact specimens was 40 mm. The specimen thickness, B and specimen width, W were kept as close to 10 mm as dictated by Charpy geometry. The notch depth, a, was machined so that the ratio (a/W) was 0.2 with a notch tip radius of 178 microns for all specimens. The specimen notch orientations are shown in Fig. 2 with the primary orientations being L–S (interlaminar) and L–T (edge on). For the notch orientation study in the borosilicate glass/HMU material specimen were cut at 45° to the top ply –0° axis which is the reference for the longitudinal, L–direction. The original square fabricated plates had thickness of 7.3 mm, 2.5 mm and 3.4 mm for the borosilicate/HMU carbon, LAS III/SiC and LAS III/HSW CMC's, respectively.

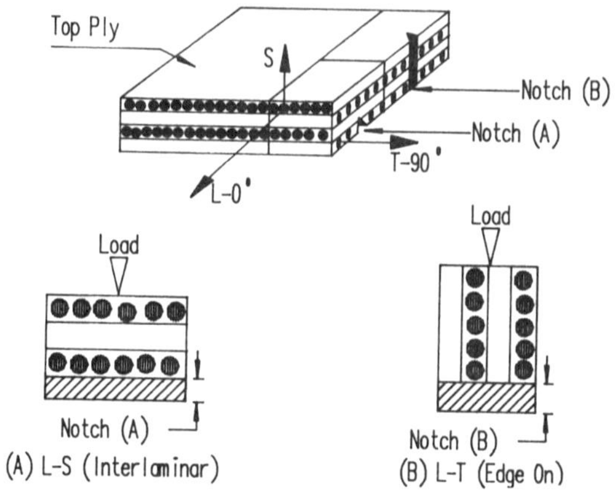

Fig. 2 Specimen notch orientations.

All the L–S tests and several of the L–T tests were performed with an instrumented drop tower at a test velocity of 1.2 m/s. Other tests, mostly the LAS III/SiC and LAS III/HSW L–T specimens, utilized an instrumented low blow Charpy pendulum with a test velocity of 0.5 m/s. The tests were performed at room temperature.

ANALYSIS METHODS

As compared to monolithic ceramics, the mechanical property and toughness evaluation of CMC materials is more complex (Phillips and Davidge, 1986). Impact damage is even harder to assess and fracture criteria require development (Davidge, 1987). It was felt appropriate, therefore, to provide explicitly the data reduction equations which were utilized herein.

Although multiaxial reinforced composites require complex stress analyses and there are stress concentrations in the areas of the loading points, the simple elastic beam theory equation was utilized to determine the maximum tensile stress at the notch tip, σ_{max}. This approach has been used previously by others (Prewo, 1980). The relation for σ_{max} is given as

$$\sigma_{max} = \frac{3}{2} \frac{F_{max} L}{B(W-a)^2} \quad (1)$$

where F_{max} is the maximum load from the force versus time curve at 'B' in Fig. 1, and the other terms are defined by the Charpy geometry. It is noted that in the Charpy three point bend test compression, shear and tensile stresses are all applied simultaneously and fracture can occur by one or a combination of these modes.

The conditional linear elastic fracture mechanics (LEFM) mode I stress intensity factor, Kq, as given by the American Society for Testing and Materials E399 specification (1981), is

$$K_q = \frac{F_{max} L}{BW^{3/2}} \cdot f\left[\frac{a}{W}\right] \quad (2)$$

where F_{max} is the maximum load as in equation (1) and f (a/W) is a multiterm function described in the reference specification.

The dynamic rupture work per unit area, U_d is given by

$$U_d = \frac{V \int_0^{t_f} F \cdot dt}{B(W-a)} \quad (3)$$

where the numerator is the integral of the force, F versus time, t curve, shown in Fig. 1, times an instantaneous velocity, V which is computed at various time intervals to account for decreases in the pendulum velocity from initial impact at t=o to final fracture at time, t_f; and the denominator is the fracture ligament area.

It is noted that of the two toughness parameters given by equations (2) and (3), the dynamic rupture work per unit area, U_d, is more meaningful as a comparator of laminated composite materials. Phillips and Davidge (1986) noted the nonlocalized complex fracture mechanism which involves a multiplicity of cracks at and away from the separation zone in the matrix. These cracks occur during matrix microcracking and prior to the maximum load. This multiplicity of cracks does not comply with the macroscopic scale of the LEFM analysis, that is, there is no singular crack length, a to use in equation (3). At best, then K_q is only useful for guidance in material development.

RESULTS AND DISCUSSION

The effect of notch orientation on the impact toughness properties is illustrated by the borosilicate glass/carbon fiber composite, and the data are given in Table 1.

TABLE 1 Effect of Notch Orientation on Impact Toughness

Orientation	σ_{max} MPA	K_q MPa(m)$^{1/2}$	U_d J/cm^2
L–S	281	11.8	6.0
L–S/45°	130	5.5	7.4
L–T	298	14.8	3.4
L–T/45°	295	10.0	11.0

The interlaminar (L–S) dynamic rupture work is greater than for the "edge on" orientation (L–T), because in the L–S orientation all the fiber pullout contribute to the failure and friction as the crack propagates. The resulting fracture, as shown in Fig. 3(a) includes compression failure where the impact occurs, interlaminar shear and tensile fracture at the notch. For the L–T orientation the fracture is essentially tensile, as shown in Fig. 3(b), and only half of the fiber layers contribute to the fiber pullout and frictional toughening mechanism. (See Fig. 3(c)).

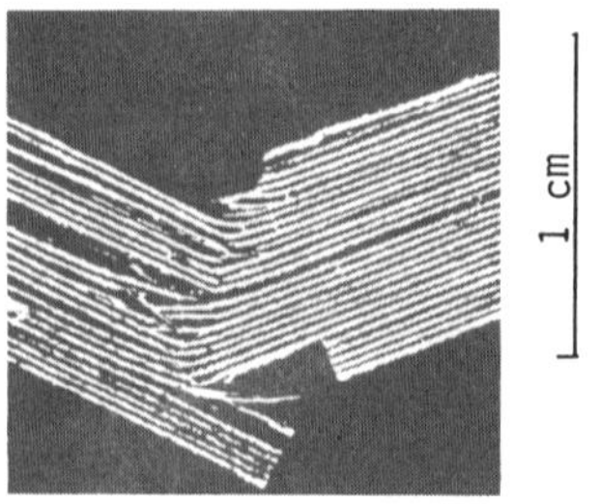

(a) L–S (Interlaminar)

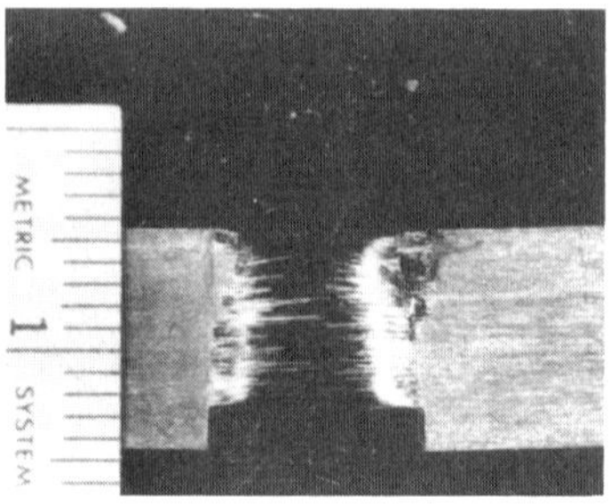

(b) L–T (Edge On)

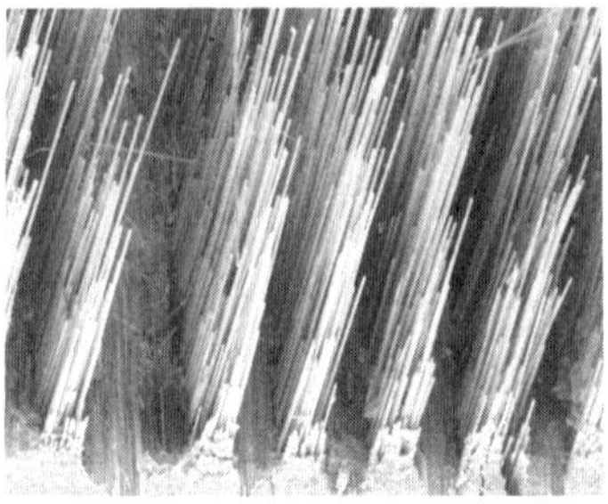

(c) L–T

200 μ

Fig. 3 Fractographs of borosilicate glass/carbon fiber composite.

The L–S and L–T specimens cut at 45° to the longitudinal axis showed an increase in U_d. The large increase for the L–T/45° is attributed to the fact that the fibers at 45° in the alternate layers could carry load and hence inhibit the rapid low energy fracture path shown for the L–T specimens in Fig. 3(c).

The toughness results for crossply LAS III/SiC CMC materials are given in Table 2. The values for the prepreg tape L–S orientation materials are comparable to those reported by Brennan and Prewo (1982) of σ_{max} = 380 MPa, K_q = 10 $MPa(m)^{1/2}$ and U_d = 1.3 J/cm^2 for LAS I/SiC material.

TABLE 2 Impact Toughness Results for LAS III/SiC CMC Materials

SiC Fiber	Orientation	σ_{max} MPa	K_q $MPa(m)^{1/2}$	U_d J/cm^2
Prepreg tape	L–S	479	11.8	9.2
	L–T	617	31.5	6.4
Prewoven 12 HSW	L–S	347	9.8	3.4
	L–T	484	——	4.9

The prepreg tape material has higher impact toughness than the prewoven 12 HSW material. The higher toughness is due to higher tensile strength and increased fiber pullout frictional resistance for the prepreg tape SiC fiber reinforced material. The load–time curve for the LAS III/12 HSW material in the L–T orientation showed the time to maximum load was much larger than the time for the onset of matrix microcracking, which indicates a large amount of deformation during matrix microcracking. Consequently, a stress intensity K_q toughness value was not computed.

The energy versus time data was examined to determine whether the total absorbed energy was due mostly to energy required for initiation or to energy required to propagation. The schematic energy versus time trace shown in Fig. 1 is representative of how the energy was partitioned. The initiation energy, which is denoted as point '1', is about 10 to 15 percent for the L–S orientation and 20 to 30 percent for the L–T orientation. At point '2', where the fiber pullout frictional sliding becomes dominant, it shows that this mechanism is the main contributor to toughness during crack propagation.

CONCLUSIONS

For all the CMC materials the edge on orientation had a lower dynamic rupture work per unit area than the interlaminar orientation, with the exception of glass matrix composite materials cut at 45° to the final ply axis. For the ceramic matrix composites the prepreg tape SiC fiber material was tougher than the 2–D prewoven fabric composite. For all the composites the percentage of dynamic rupture work which is attributed to crack initiation is less than thirty, and thus most of the rupture work is attributed to fracture of fiber and fiber bundles after fiber/matrix interfacial bond failure with the attendant frictional sliding mechanism toughness contribution.

ACKNOWLEDGEMENTS

This work was supported by the Office of Naval Research. The continued laboratory expertise of W. Umlandt is also noted. The careful manuscript preparation of Ruth Chittum is also appreciated.

REFERENCES

Aveston, J., G. A. Cooper and A. Kelly (1971). The properties of fiber composites (IPC Science and Technology Press Ltd., Surry, UK), 15–21.

Bowen, D. H. (1968). Fibre reinforced ceramics. Fibre Sci. Tech. 1, No. 2., 85–112.

Brennan, J. J., and K. M. Prewo (1982). Silicon carbide fibre reinforced glass–ceramic matrix composites exhibiting high strength and toughness. J. Mater. Sci. 17, 2371–2383.

Crivelli–Visconti, I. and G. A. Cooper (1969). Mechanical properties of a new carbon fibre material. Nature 221, 754–755.

Davidge, R. W. (1987). Fibre reinforced ceramics, Composites 18, No. 2, 92–98.

Evans, A. G., Thouless, M. D., Johnson–Walls, D. P., Luh, E. Y. and Marshall, D. B. (1985). Some structural properties of ceramic matrix fiber composites. In W. C. Harrigan, Jr., J. Strife and A. K. Dhingra (Eds.), International Conference on Composite Materials ICCM–V, The Metallurgical Society/AIME, Warrendale, 543–553.

Hasson, D. F. and S. G. Fishman (1987). Impact behavior of ceramic matrix composite materials. In F. L. Matthews, N. C. R. Buskell, J. M. Hodgkinson and J. Morton (Eds.), Sixth International Conference on Composite Materials, Vol. 2, Elsevier Applied Science, London, 2.40–2.47.

Lankford, J. (1985). Compressive strength and damage mechanisms in a SiC fiber reinforced glass–ceramic matrix composite. In W. C. Harrigan, Jr., J. Strife and A. K. Dhingra (Eds.), International Conference on Composite Materials ICCM–V, The Metallurgical Society/AIME, Warrendale, 587–602.

Marshall, D. B., and Evans, A. G. (1985). Failure mechanisms in ceramic–fiber/ceramic–matrix composites. J. Amer. Cer. Soc. 68(5), 225–231.

Phillips, D. C., and R. W. Davidge (1986). Test techniques for the mechanical properties of ceramic matrix fibre composites. Br. Ceram. Trans. J. 85, 123–130.

Phillips, D. C., R. A. J. Sambell and D. H. Bowen (1972). The mechanical properties of carbon fibre reinforced pyrex glass. J. Mater. Sci. 7, 1454–1464.

Prewo, K. M. (1980). The importance of fibres in achieving impact tolerant composites. Phil. Trans. R. Soc. Lond. A294, 551–558.

Prewo, K. M., J. J. Brennan and G. K. Layden (1986). Fiber reinforced glasses and glass–ceramics for high performance applications. Am. Ceram. Soc. Bull. 65(2), 305–313.

Prewo, K. M., and E. J. Minford (1985). Graphite fiber reinforced thermoplastic glass matrix composites for use at 1000°F. SAMPE J. 21–22, 26–33.

Pugh, S. F. (1967). The fracture of brittle materials. Brit. J. Appl. Phys. 18, 129–162.

Sbaizero, D. and Evans, A. G. (1986). Tensile and shear properties of laminated ceramic matrix composites. J. Amer. Cer. Soc. 69, 481.

INTERFACIAL STUDIES ON ALUMINA REINFORCED BY CARBON FIBERS OR SILICON CARBIDE WHISKERS*

S. Karunanithy
Atlantic Research Laboratory
National Research Council of Canada
Halifax, N.S., Canada

ABSTRACT

The properties of alumina reinforced by carbon fibers or SiC whiskers are affected by the interfacial reactions. In carbon fiber reinforced alumina, precoating with boron carbide appears to prevent the oxidation of the fibers at elevated temperatures. In this case, the oxide layer formed on the coating material functions as an oxygen barrier. In the whisker reinforced composites, however, the silica layer formed on SiC whiskers during sintering would prevent further oxidation of the whisker surface. The deterioration of the properties of these composites, in air at temperatures above 1000°C, can be attributed to the chemical changes at the interface.

KEYWORDS

Coatings on carbon fiber, silicon carbide whiskers, composites, interface.

INTRODUCTION

Ceramic matrix composites are being explored for their potential applications as structural materials. Reinforcement of the matrix with fibers or whiskers results in increased toughness and increased strain at fracture [1,2]. The use of carbon fibers to reinforce oxide ceramics, glasses and glass ceramics has been investigated [3,4]. These composites show extensive fiber pull-out during fracture, as required to dissipate the fracture energy. However, they also exhibit open porosities at the fiber/matrix interface [3], as it is difficult to induce atomic diffusion to or from the surface of the carbon fiber. Attempts to reduce these porosities by increasing the pressure in the hot pressing technique proved unsuccessful [5].

Recent studies [6,7] on carbon fiber reinforced-ceramic materials have been concentrated on developing a better interface. These efforts have been aimed at increasing the strength of the composite, and providing additional protection to the fibers in oxidizing environments. In addition, the flexural strength and fracture toughness of alumina-carbon fiber composites have been reported to increase when the fibers are coated with materials such as SiC and

* NRCC 29361

TiC [6]. These improved mechanical properties have been attributed to "good adhesion" at the interface. The coatings can also prevent the occurrence of strong chemical reactions between the fiber and the matrix. Attempts have been made to use boron nitride (BN) as a coating material for carbon fibers used to reinforce mullite because without a coating the fibers would fuse to the mullite matrix during the sintering process [7]. The first part of this paper deals with the use of boron carbide (B_4C) as a coating material for carbon fibers to reinforce an alumina matrix.

Another important reinforcing material is silicon carbide in whisker form. These whiskers normally have a thin layer of silica (SiO_2) on their surface, and this layer plays an important role in determining the interfacial properties of whisker reinforced-alumina. Studies on the fracture toughness and the strength of alumina-SiC whisker composites have demonstrated the importance of the chemical reactions on the whiskers, in high temperature applications of these composites. A substantial increase in the formation of amorphous silica, and mullite crystals at the whisker/matrix interface has been attributed to its increased toughness and drastic loss of strength in air at temperatures above 1100°C [8,9]. As alumina itself is stable to 1800°C in air [10], any failure in mechanical properties should be attributable to the chemical changes in the whiskers and the interface material. The second part of this paper deals with interfacial reactions of this composite.

EXPERIMENTAL

The materials used in this study are listed in Table 1.

TABLE 1. Properties of Materials

Material	Density (Mg/m^3)	Average Size of Constituents (μm)	Coefficient of Thermal Expansion x 10^{-6} K^{-1}
Alumina RC-HP-DBM	3.96	0.5	8.5
Carbon Fibers (PAN based 99.5% C)	1.80	8 (diameter)	radial = 8 axial = 0
Silicon Carbide Whiskers (Tateho 1S1)	3.18	0.5 - 1.5 (diameter) 5 - 20 (length)	3.1
Boron Carbide	2.50	1 - 2	5.2

Boron carbide (B_4C) was deposited on the carbon fibers from a slurry containing B_4C powder and a binder in methyl ethyl ketone (MEK). Long fibers (2-3 cm) were coated with the slurry, dried in air for a few hours and chopped into 1-2 mm lengths. The coating was checked with an optical microscope, and batches of fibers with uniform coating were selected for slipcasting. The average thickness of the coating was in the range of 2-5 μm.

Sample shapes of the composites were made by slipcasting. A slurry of alumina (total solids content 70% w) in water was prepared, blended with carbon fibers or

SiC whiskers and stabilized, at pH 5.2 for carbon fibers and 4.5 for SiC whiskers. The fiber or whisker content of the slip was in the range 10-30 vol %.

The slip was cast into conventional plaster molds to give cylindrical shapes 1 cm in diameter and 2 cm in height. The green castings were allowed to dry in air and then sintered inside the graphite susceptor of a vacuum furnace (Balzers VSG 02) powered by a 12 kW induction generator. In the case of alumina reinforced with B_4C coated carbon fibers, before vacuum sintering the samples were heat-treated at 400°C in air to burn out the binder.

The electron micrographs were obtained in a JEOL JXA-35 scanning electron microscope equipped with a Kevex energy-dispersive X-ray microanalysis system. FT-IR spectra were obtained in KBr pellets using a BOMEM model DA 3.02 spectrometer.

RESULTS AND DISCUSSION

1(a). Composites of alumina with carbon fibers

The successful slipcasting of alumina containing carbon fibers requires uniform mixing of the fibers in the alumina slurry. The fiber distribution in the slip was homogeneous in the range of 10-30% v. Fracture surfaces of sintered samples showed fibers mixed uniformly in the bulk of the sample, with some isolated areas of high fiber concentration. However, as seen in Figure 1, the fiber/matrix interface shows extensive open porosities. The strong covalent bonds in the fiber would make atomic diffusion from its surface difficult, therefore one should induce a weak chemical reaction at the interface to "wet" the fibers with the matrix.

Fig. 1. Interface of carbon fiber-alumina composite

The reaction of alumina with carbon has been studied in detail [11-13], and the following reactions have been proposed:

$$Al_2O_3 + 3\ C = Al_2OC + 2\ CO \quad (1)$$
$$2Al_2O_3 + 3\ C = Al_4O_4C + 2\ CO \quad (2)$$
$$Al_4O_4C + 6\ C = Al_4C_3 + 4\ CO \quad (3)$$

Thermodynamic data for the formation of these compounds have been calculated [11]. These studies indicate the possibility of an interfacial reaction between carbon fibers and the alumina matrix at temperatures above 1600°C in vacuum or reducing atmosphere. Formation of the oxycarbides such as Al_2OC and Al_4O_4C in these reactions before the formation of Al_4C_3 has been confirmed by differential thermal analysis [11] and electron microscopy [13]. The oxycarbide, Al_2OC, formed as a transient liquid phase in the reaction of alumina with carbon in an inert atmosphere is found to be a good sintering aid for SiC.

An attempt was made during this work to form an oxycarbide layer at the fiber/matrix interface to induce wetting of the fibers by the matrix and to examine the effect of this interfacial material on the fracture toughening of the composite. Two factors have to be considered in attempting to promote a reaction at the fiber/matrix interface for this purpose:

(1) The reaction should be controlled just enough to form an interfacial layer, with minimum damage to the surface of the fibers. A strong reaction will either fuse the fiber to the matrix or damage it permanently.

(2) If the interface formed is too brittle, it will result in the formation of microcracks around the fiber after the sintering.

Based on earlier work on carbothermal reduction reactions [11-13] and several attempts to sinter these materials, it was found that the minimum temperature for inducing a reaction at the interface was 1650°C. SEM analysis on the fracture surface showed fibers pulling out with some alumina grains still attached to the surface (Fig. 2). Energy dispersive X-ray analysis, (EDX) on the surface of the fibers in the fracture surface showed the presence of Al, O, Si, and C on the fiber surface. Qualitative interpretation of these results, based on the increase in oxygen content on the fiber-surface, suggests that the wetting of carbon fibers by the matrix could have been caused by one or more of the compounds SiO_2, Al_2SiO_3, Al_2OC, $Al_4O_4C_4$ and Al_4C_3 formed at the interface. These materials induce bonding which is weak enough to allow the fiber pull-out during the fracture. The brittle nature of these interfacial materials, however, would have a negative effect on the strength of the composite.

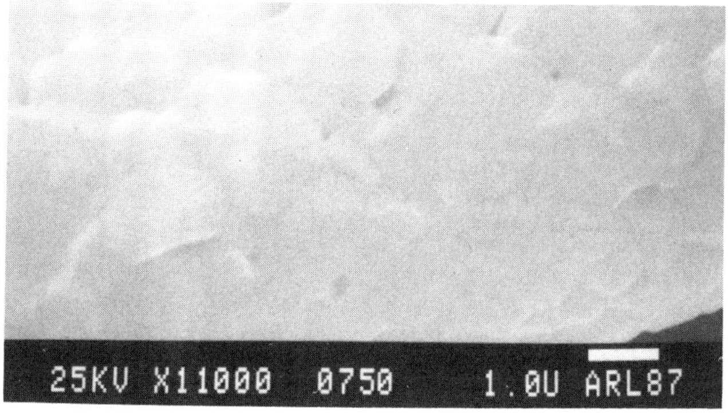

Fig. 2. Alumina grains bonded to the carbon fiber

I(b). Composite of alumina with carbon fibers coated with boron carbide

Boron carbide exhibits useful qualities such as high strength, and low density (14). In spite of its hardness, it could still be a useful coating material for carbon fibers due to its ability to form a protective oxide coating at temperatures above 600°C in air according to:

$$B_4C + 4O_2 = 2B_2O_3 + CO_2 \qquad (4)$$

Once the oxide layer is formed, further oxidation of the coating is retarded up to 1100°C (15). The boric oxide B_2O_3 formed in this reaction is known to be most suitable for wetting carbon than any other glassy material (16).

Boron carbide coatings were deposited from a slurry containing the powder and a binder. After drying, the coated fibers were slipcast with alumina. The binder was burned out in air at 400°C and the sample was sintered in vacuum. The fracture surface of this composite showed fiber pull-out with the coating material still remaining on the fibers (Fig. 3). Although the porosity at the interface appeared reduced, it could not be completely eliminated by this method. The reduction in the interfacial porosity, however, could be attributed to the formation of the glassy phase on the coating material which could heal some of the porosities at the interface. Attempts were made to examine the stability of the composite at elevated temperatures. Fracture surfaces of the composites heated in air at 1000°C for 1 hour showed coated fibers still remaining in the matrix. Fibers pulled out during the fracture were covered with the oxide and clearly indicated crack initiation at the interface. In addition, the difference in thermal expansion coefficients between the matrix and the glassy phase formed during oxidation would contribute to the loss of fracture strength of this composite during thermal cycling. In this respect, elevated temperature applications of this composite could only be possible where such thermal cycling effects are minimized.

II. Composite of alumina with SiC whiskers

The composites used in this study were reinforced with 'Tateho-SCW-1S' type SiC whiskers, which have no surface silica on them. Surface studies on these whiskers by Si(2p)-ESCA techniques have identified only the SiC in the whiskers, while the silica on the other types of SiC whiskers such as Silar-SC-9 was easily detectable by the same technique (17).

Fig. 3(a).

Fig. 3(b).

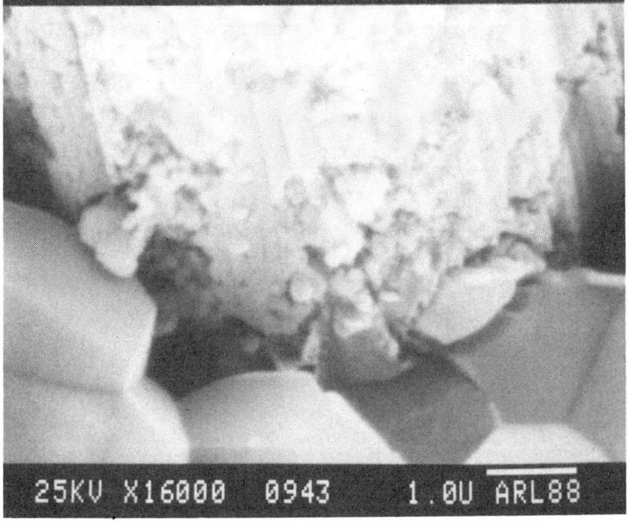

Fig. 3(a) and (b). Interface of alumina reinforced with B_4C coated carbon fiber

SEM analysis on the fracture surfaces of this composite showed that the whiskers were well-distributed in the final product. In addition, there was no evidence of any interfacial layer formed by chemical reaction between the matrix and the whiskers. Since the samples used in this study were sintered in a vacuum furnance, rather than in flowing argon, evidence for the evolution of gaseous material(s) from the sample at elevated temperatures was detectable by the increase in the chamber pressure. Since the alumina matrix is stable at the sintering temperature involved, the gaseous materials released from the composite probably are due to the oxidation of the whiskers according to:

$$SiC_{(s)} + 3/2\ O_{2(g)} \rightarrow SiO_{2(l)} + CO_{(g)} \quad \Delta G_{1873K} = -\ 797.7 \text{ kJ/mole} \qquad (4)$$

$$SiC_{(s)} + O_{2(g)} \rightarrow SiO_{(g)} + CO_{(g)} \quad \Delta G_{1873K} = -\ 473.1 \text{ kJ/mole} \qquad (5)$$

Higher values of the interfacial shear stress have been observed [17] with the composites containing Tateho type SiC whiskers, which is attributable to the roughening of the whisker surface due to the reactions shown above.

FT-IR spectrometry was used to follow the changes in the composite. Samples were ground and pellets were made with KBr for the analysis. Since grinding of the sample induces local heating of the material, all the samples were ground under identical conditions so that a comparison could be made. Figure (4) shows the FT-IR spectra obtained on the whiskers and the composite material. The spectrum of SiC whiskers shows a characteristic Si-C vibration (around 800 cm^{-1}) and a weak absorption due to Si-O vibration (around 1080 cm^{-1}). The latter is due to the small amount of silica formed on the whiskers due to local heating during the grinding process. However, the composite material analysed under the same conditions shows a large increase in the intensity of the Si-O vibration compared to the Si-C vibration. This observation confirms the whisker oxidation during the sintering process. The layer of silica formed on the whiskers due to this oxidation effect should protect the whiskers from further oxidation. A sample of the composite was heat-treated in air at 1000°C for 1 hour. This sample showed almost the same amount of silica (with respect to silicon carbide) as the original composite. The composite material, however, is known [7] to lose its strength rapidly around 1000°C. Therefore this strength-loss could be attributed to the changes occurring at the interfacial material (i.e. silica) rather than further oxidation of the whisker surface. Possible changes of the surface silica include crystallization to form crystoballite with a corresponding decrease in volume of the interface material. Detailed studies of the interfacial changes and their effects on the strength of these composites are presently being carried out in our laboratory.

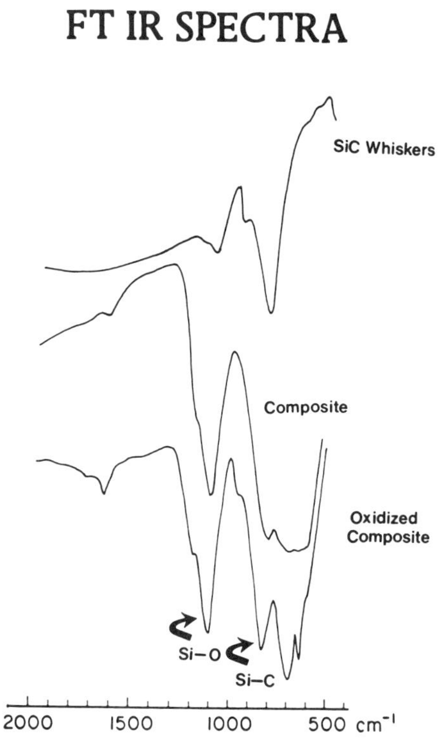

Fig. 4 FT-IR spectra of whiskers and composites

CONCLUSION

The chemical changes at the interface dominate the properties of both carbon fiber and SiC whisker reinforced-alumina. The carbon fibers in the composite, however, can be protected from oxidation by precoating them with B_4C. In the whisker reinforced composite, an increase in the SiO_2 content of the SiC whiskers during the sintering process contributes to the protection of the whiskers in air at elevated temperatures.

ACKNOWLEDGEMENTS

The author gratefully acknowledges Dr. M. Falk for the use of FT-IR spectrometer and stimulating discussions, and Dr. P. Odense and Mr. D. O'Neil for taking the electron micrographs.

REFERENCES

(1) Davidge, R.W. (1987). Fiber-reinforced ceramics. Composites, 18, 92-98.

(2) Hillig, W.B. (1987). Strength and toughness of ceramic matrix composites. Ann. Rev. Mater. Sci., 17, 341-83.

(3) Sambell, R.A.J., Phillips, D.C. and Bowen, D.H. (1972). Carbon fiber composites with ceramic and glass matrices, Part 1, Discontinuous fibers. J. Mater. Sci., 7, 663-665.

(4) Sambell, R.A.J., Briggs, A., Phillips, D.C. and Bowen D.H. (1972). Carbon fiber composites with ceramic and glass matrices. Part 2, Continuous fibers, J. Mater. Sci., 7, 676-681.

(5) Sambell, R.A.J., Phillips D.C. and Bowen, D.H. (1974). The technology of carbon fiber reinforced glasses and ceramics, in "Carbon fibers, their place in modern technology", Proc. of Plastics and Polymers Conf. London, 105-113.

(6) Fitzer, E. and Gadow, R. (1986). Fiber reinforced ceramics via sol/gel route, Mater. Sci. Res., 20 (Tailoring Multiphase Compos. Ceram.), 571-607.

(7) Singh R.J. and Brun, M.K. (1988). Effect of boron nitride coating on fiber-matrix interactions, Adv. Ceram Mater., 3 235-237.

(8) Marra, R.A. and Bray, D.J. (1986). Thermochemical characterization of SiC whiskers in Al_2O_3 matrices, Ceram. Eng. Sci. Proc., 7, 945-946.

(9) Becher, P.F. and Tiegs, T.N. (1988). Temperature dependence of strengthening by whisker-reinforced alumina in air Adv. Ceram. Mater., 3, 148-153.

(10) Barin, I. and Knacke, O. (1973). In Thermochemical properties of inorganic substances, Springer-Verlag Pub., Vol. 1 p.32.

(11) Cox, J.H. and Pidgeon, L.M. (1963). An investigation of the aluminum-oxygen-carbon system, Can. J. Chem., 41, 671-683.

(12) Worrell, W.L. (1965). Carbothermic reduction of alumina: a thermodynamic approach, Can. Met. Quart., 4, 87-95.

(13) Klug, F.J., Pasco, W.D. and Borom, M.P. (1982). Microstructure development of aluminum oxide: graphite mixture during carbothermic reduction, J. Am. Ceram. Soc., 65, 619-624.

(14) Encyclopedia of Materials Science and Engineering, (Pergamon Press), (1986). 1, 401.

(15) Gmelin Handbook of Inorganic Chemistry, (1981). Supplement on Boron, 2, 183.

(16) Strife, J.R. and Sheehan, J.E. (1988). Ceramic coatings for carbon-carbon composites, Am. Ceram. Soc. Bull., 67, 369-374.

(17) Homeny, J. and Vaughn, W.L. (1987). Whisker reinforced ceramic matrix composites. MRS Bull., 12, 66-71.

CERAMICS FOR THE CONTINUOUS CASTING OF STEEL

N.T. Cotterhill and C. Martin
Thor Ceramics Systems Inc.
Hamilton, Ontario
Dr. N.A. McPherson and G.E. Holling
Thor Ceramics Limited
Clydebank, Scotland

Ceramics have been used in the manufacture of liquid steel for as long as steel has been produced. However, the type and quality standards of the ceramics being used have had to change to meet the ever changing needs of steel producers. These needs resulted from demands for better steel quality and also the introduction of new steelmaking and casting processes. The growth of continuous casting has risen at a remarkable rate, as shown in Table 1. The benefits of this process over ingot casting and slabbing have been documented in a number of publications (Fitzgerald, 1982; Fitzgerald, 1980).

The ceramics used in the continuous casting process have therefore come in for extremely close scrutiny. For example, traditional fireclay lined steel ladles were subjected to steel being tapped into them at significantly higher temperatures. As a large proportion of the steels are deoxidised with aluminium in the ladle, residual silicon levels on low carbon strip steels were too high as a result of the reaction below, which was kinetically favoured by the higher ladle temperatures of the continuous casting process:-

$$3\ SiO_2 + 4\ Al \longrightarrow 2\ Al_2O_3 + Si$$

Aluminium levels of the steel were difficult to control to tight limits, such as 0.02-0.05%. This problem led to the adoption of ladle linings in alumina (Harrage and Patrick, 1980) and currently with the advent of secondary steelmaking processes, there has been an increase in use of dolomitic linings for ladles, with enhanced slag lines.

Areas of ceramic development in the continuous casting of steel can now be considered.

Ladle to Tundish Stream Protection

Fifteen years ago there were virtually no plants employing any form of ladle to tundish stream protection, and in hindsight, the reasons for many strand breakouts and high slab rectification rates were rather obvious. Extensive reoxidation of the steel was taking place between ladle and tundish, generating alumina in aluminium killed steels. Prior to the adoption of ceramic shrouding between the ladle and tundish, several inert gas shrouding systems were developed and effectively reduced reoxidation, but were latterly costly (Mercer and McPherson, 1979).

Ceramic shrouding was being introduced in the late 1970's and early 1980's with either fused silica or graphitised alumina materials. The latter is probably in more common use, and has the flexibility of being used when casting high and low manganese content steels. Ceramic shrouding produces a physical barrier between the flowing steel and the atmosphere. Figure 1 shows a typical design of a ceramic ladle shroud. Figure 2 illustrates that reoxidation between ladle and tundish can be reduced by 20% by the use of the ceramic ladle shroud. (Nitrogen pick-up is used as a measure of reoxidation.) For some applications such as low carbon aluminium killed slabs being rolled for exposed part autobody panels or D and I can, this level of reoxidation was still too high. The problem manifested itself as alumina lines on cold reduced coil surfaces or as splits which occured during pressing of the sheet. (Originating at internal alumina inclusions.) A similar problem was experienced with stainless steel containing titanium, where titanium carbonitride particules caused sheet surface defects (Lindenberg and co-workers, 1984).

Reoxidation was still occuring as air was being sucked between the contact faces of the ladle collector nozzle and the shroud throat. To counteract this source of reoxidation, a porous insert shown in Fig. 3 can be used to distribute an insert gas such as argon above the contact area.

The development of this system involves the use of a porous material capable of distributing the desired amount of argon to the critical area. The porous material has to be able to withstand oxygen lancing between ladle changes in a sequence. The material to satisfy this requirement is a zirconia mullite, which is a hard bonded fused oxide grain with excellent thermal shock and erosion resistance.

Ceramic ladle shrouds can be isostatically pressed in heights from 500 mm to 1700 mm. If required, the bottom end of the ladle shroud can have a corrosion resistant component such as graphitised zirconia. Normally a ladle shroud should have a minimum immersion depth in the steel in the tundish of 200 mm, otherwise vortexing will occur and the protective tundish covering will be entrained into the steel. A ceramic ladle shroud used in the correct manner is probably the most significant aspect in the continuous casting process for achieving consistently clean steel.

Tundish Flow Control Ceramics

According to an IISI study on "Refractory Materials for Steelmaking", the ceramic systems used to control the flow of steel from tundish to mould in continuous casting are:

Stopper
Slide Gate
Metering Nozzle
Stopper and Slide Gate
Metering Nozzle and Slide Gate

As can be seen from Fig. 4(a), (b), (c) and (d), the stopper is the dominant mode of flow control on slab and bloom casters. As a result a significant amount of work has been carried out on developing the designs required for ceramic stopper rods, particularly in Europe. There does appear to be a tendency to use slide gate systems in North America and Japan in preference to stopper rods.

The stopper rods currently in use are monolithic and are mainly of graphitised alumina ceramic. As with most aspects of continuous casting ceramics the role of the stopper rod has changed from that of a simple flow control device to that of a precise flow control device and a metallurgical tool (Amory and co-workers, 1982). However, the ceramic stopper has to be capable of casting a wide variety of steel types under and equally wide variety of conditions. Although the graphitised alumina stopper satisfied the majority of conditions it has been found necessary to develop stoppers with noses in enhanced ceramic materials. For example, long sequences can be accommodated by enhancing the stopper nose with graphitised zirconia as shown in Fig. 5. To cater for corrosive steels such as calcium treated, and low carbon free cutting steels, the nose can be enhanced with magnesia. Magnesia enhancement has also been found to be beneficial when casting some stainless steel types. The shape of the stopper rod is also important to achieve good control of the steel flow into the mould. For example, Fig. 6 shows that small amounts of opening can result in large increases in steel flow. The use of profiled stopper rod noses may increase, to satisfy the high standards being set for automatic level control variations in the mould.

As stated earlier, the stopper has also some metallurgical functions. For several years now it has served as a useful device for passing argon into the mould of the continuous caster. The argon serves two main functions:-

1. Reduces alumina build up in the bore of the submerged entry nozzle and therefore can increase sequence length and reduce premature stoppage of casting.
2. Enhances non metallic inclusion flotation (particularly alumina) in the mould and improves steel cleanness.

The mode of the operation of an argon stopper varies from plant to plant, but it is essential to eliminate possible points of air ingress into the argon flow path, either via refractory joints or supply pipe joints. By operating the system with a positive back pressure in the pipework, consistency of operation can be achieved.

An alternative flow control system has been developed - The Rocon Rotary Valve. A number of publications describe the concept and it is shown schematically in Fig. 7 (Purdie and McPherson, 1988;

McPherson and Cotterhill, 1987; Roberts and co-workers, 1986). It consists of a hemispherical dome with an outlet port which is held stationary in the tundish base. Above that there is a compressively loaded stem with a mating hemispherical face and inlet port. The stem can be rotated to allow steel flow between tundish and mould when the respective ports are aligned.

Although the system has been relatively successful when casting low dissolved oxygen steels, it has been found necessary to utilise nitride bond graphitised alumina in critical areas when casting higher oxygen steels (60 - 120 ppm). The nitride bond is basically a secondary skeletal structure which serves to further reduce the effective pore size, probably by formation of a proportion of fibers of Si_3N_4 in the pores. This nitride tends to be non-wetting.

Tundish to Mould Stream Protection

For the reasons already discussed under ladle to tundish stream protection, a graphitised alumina ceramic nozzle is used. This ceramic nozzle can have enhancement in the throat area with magnesia, and in the area of the mould level enhancement by graphitised zirconia can be used as shown in Fig. 8. Alternatives to graphitised zirconia have been advocated recently in the form of plasma sprayed coating (Hoffken and co-workers, 1988).

The main function of the mould level enhancement is to counteract the corrosive effects of continuous casting mould fluxes. Attack of the mould level area can occur due to the action of alkali oxides and flourides in the mould flux. Examples of typical continuous casting mould fluxes are shown in Table 2. The greater the resistance of the ceramic nozzle to the corrosive conditions at the mould level then the longer will the period of stable casting be.

Although the rate of wear of the ceramic nozzle at the mould level is a limiting factor in the effective life of the nozzle, a further factor is the rate of alumina deposition on the inner bore of the nozzle. The alumina is normally present to some greater or lesser extent within the steel entering the nozzle when casting aluminium killed steels. The mechanism of alumina deposition and build up is complex but it has been shown that producing a fine dispersion or argon bubbles on the surface of the inner bore of the nozzle can significantly retard the alumina build up rate. This has been achieved by the use of the natural permeability of graphitised alumina as shown in Fig. 9. The flow of gas away from the bore has been effectively stopped by the use of an impermeable membrane.

Another option is to coat the inner surface of the graphitised alumina nozzle with a Si Al ON compound. It has also been suggested that boron nitride could be used in this area.

With the ability to develop ceramics with good external wear resistance and an ability to reduce alumina build up, a suitable combination is emerging which could be used for nozzles used in the development of casting of thin slabs (50 mm), i.e. nozzles with 10 mm wall thicknesses. There is a great deal of evidence to show that the casting of thin slabs is gaining momentum. (Ehrenberg and co-workers, 1988; Iverson and Kappes, 1988) and will require a further major step in ceramic development to keep pace.

Horizontal Continuous Casting

A number of new horizontal continuous casting plants have been constructed recently mainly for the production of stainless steel blooms/billets. Within the mould assembly shown in Fig. 10 is a component identified as the break ring. This is the joint between mould and tundish and it acts in a way to keep the starting position of the solidifying front constant, similar to the meniscus in a conventional continuous casting mould. Boron nitride has been used extensively as the material for break rings when casting stainless steels. However, a recent innovation has been the development of Si Al ON materials, and it has also been reported (Toothill, 1988) that yttrium stabilised Si Al ON has shown dramatic improvements in erosion resistance. The use of 10% to 20% boron nitride in beta SiAlON has allowed the casting of a wide range of steels. Irrespective of which ceramic materials are finally chosen they must satisfy the following requirements:

1. Precise machinability
2. High thermal shock resistance
3. Low thermal expansion co-efficient
4. High temperature strength
5. Good erosion resistance

In association with these property requirements there has to be an acceptable cost/performance ratio.

Conclusions

The trend in improving and developing refractories for the continuous casting of steel has continued. New casting develpments such as thin slab casting and horizontal continuous casting have resulted in nitride bond refractory designs and plasma spraying having to be considered for routine use.

REFERENCES

Amory, D, J.C. Dhuyvetter, M. Jeanneau and J. Leclerc (1982). Continuous Casting of Slabs suitable for deep drawing-parameters influencing inclusion cleanliness. 4th International Iron and Steel Congress; Continuous Casting, London, England.

Ehrenberg, H.J., G. Diederich, L. Parschat, P. Pleschiutschigg and W. Rahnfeld (1988). M D H Thin Slab Technology, Casting of Thin Slabs at Mannesmannrohren - Werke AG, Huckingen Works. 4th International Conference, Continuous Casting, Brussels, Belgium.

Fitzgerald, F. (1980). Continuous Casting; Present and Future. The Continuous Casting of Steel Conference, Glasgow, Scotland.

Fitzgerald, F. (1982). Continuous Casting - Growth Development and Future Trends. 4th International Iron and Steel Congress; Continuous Casting, London, England.

Haggart, J.E. and G. Patrick (1980). Steelmaking Practice at Ravenscraigh BOS Plant. The Continuous Casting of Steel Conference, Glasgow, Scotland.

Hoffken, E., H. Lax and G. Pietzko (1988). Development of Improved Immersion Nozzles for Continuous Slab Casting. 4th International Conference, Continuous Casting, Brussels, Belgium.

Iverson, F.K. and P. Kappes (1988). Innovative Mini mill Concept for Flat Production at Nucor's new Crawfordsville, Indiana Plant. 4th International Conference, Continuous Casting, Brussels, Belgium.

Lindenberg, H. - U.; W. Knackstedt, H. - J. Kohler, W. Blesterfeld, K. - D. Unger and R. Thielmann (1984). Metallurgy of Stainless Steel Continuous Casting. Stahl und Eisen, Vol. 104 pp 51 - 58.

McPherson, N.A. and N.T. Cotterhill (1987). The Development of the Rotary Valve Tundish Flow Control System. Proceedings of the International Symposium on Advances in Refractories for the Metallurgical Industries, Winnipeg, Canada.

Mercer, R.E. and N.A. McPherson (1979). Experience of Slab Casting at BSC Ravenscraig Works. Proceedings of 62nd NOH and BOS Conference, Detroit, U.S.A.

Purdie, J. and N.A. McPherson (1988). The Develpment and Operational Experience of a Rotary Valve for Tundish Flow Control. 71st Steelmaking Conference, Toronto, Canada

Roberts, G.J., S.J. Lee, J. Monks and M. Howell (1986). The Rotary Valve, A New Concept in Tundish Flow Control. Refractories for Continuous Casting Conference, Aachan, Holland.

Toothill, D. (1988). Application of HORICAST for Continuous Casting of Stainless and High Alloy Steels. Ironmaking and Steelmaking, Vol. 15, pp 43 - 48.

TABLE 1 Growth of World Continuous Casting Ratio

YEAR	1970	1973	1977	1981	1983	1985	1987
% CC	4.4	9.3	18.4	31.5	40.0	45.8	53

TABLE 2 Chemical Analysis of a Range of Mould Fluxes Used For Continuously Casting Slabs

% CaO	34	32.5	34	34	20
% SiO_2	33	26	34	34	37
% Al_2O_3	6	7	2.5	7	6
% Na_2O	11	7	10.5	3.3	20
% F	6.5	6	7.5	4.25	-

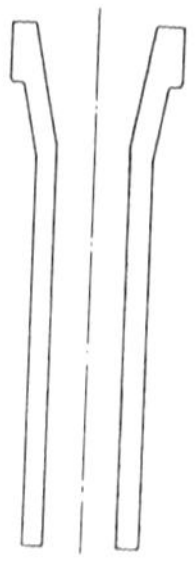

Fig. 1. Typical design of ceramic ladle shroud

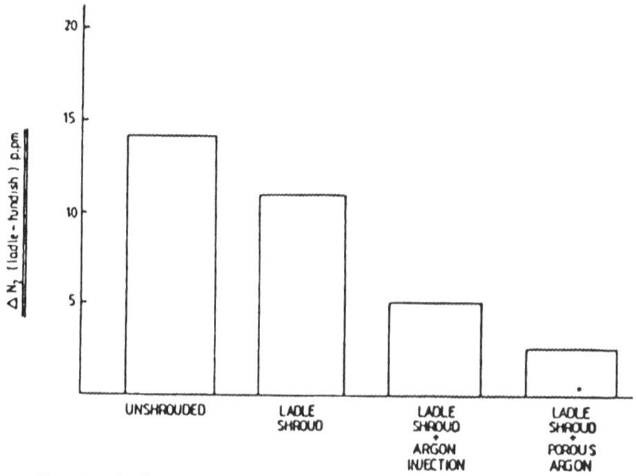

Fig. 2. Ladle to tundish nitrogen pick up for various shroud systems

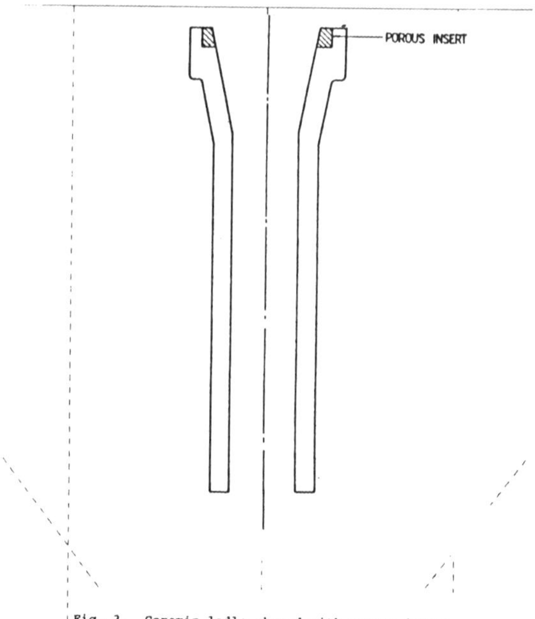

Fig. 3. Ceramic ladle shroud with porous insert

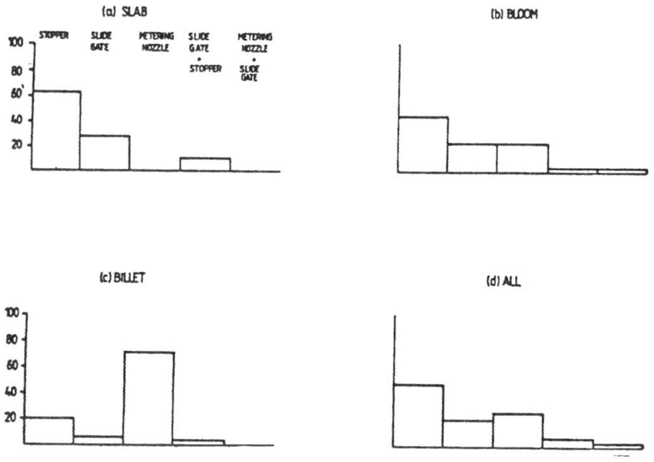

Fig. 4. Current tundish flow control systems

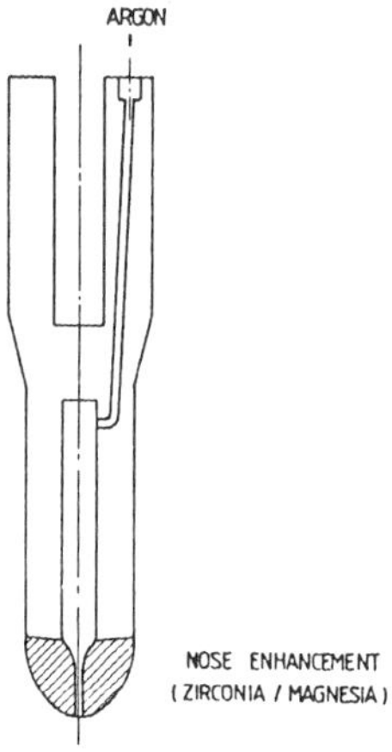

Fig. 5. Stopper rod showing argon by-pass and nose enhancement

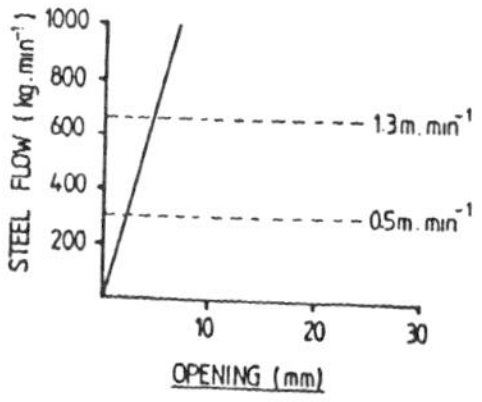

Fig. 6. Steel flow rate as a function of stopper rod opening

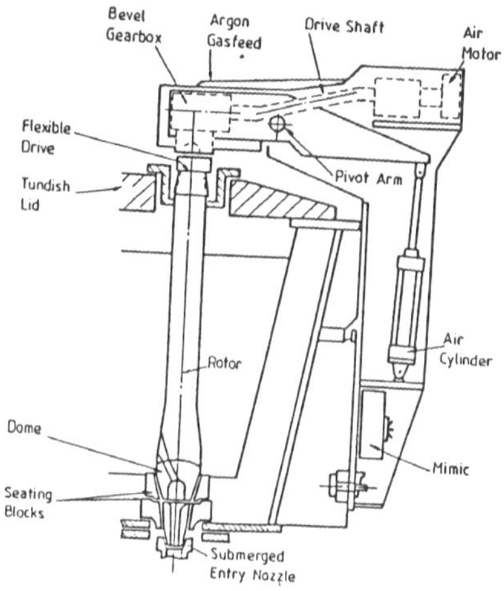

Fig. 7. Tundish rotary valve system

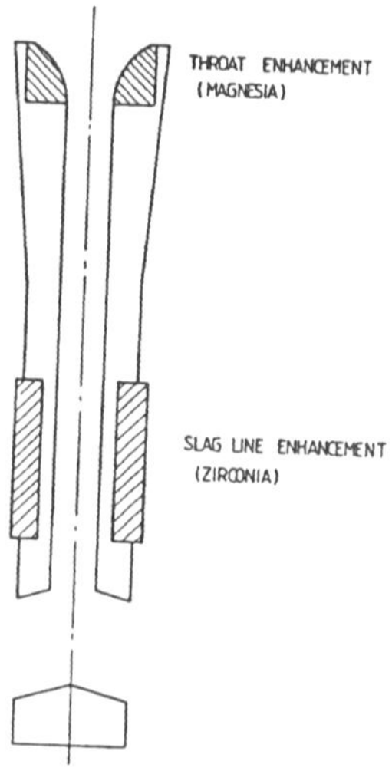

Fig. 8. Ceramic submerged entry nozzle showing throat and slag line enhancement areas

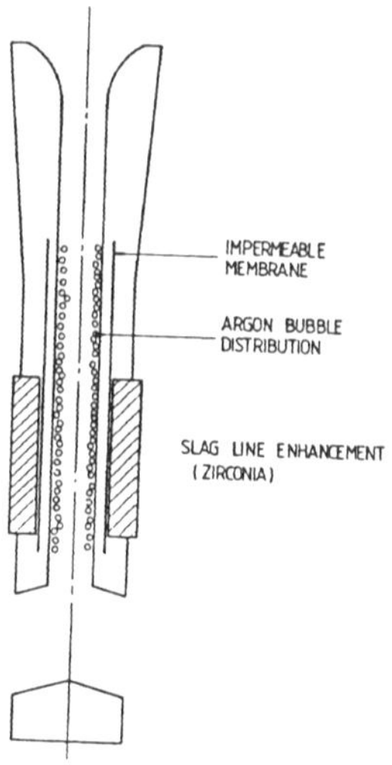

Fig. 9. Ceramic submerged entry nozzle showing effect of argon injection

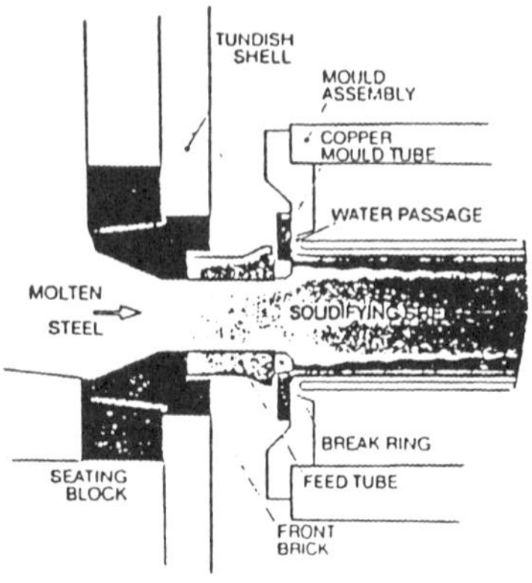

Fig. 10. Schematic representation of tundish - mould joint for horizontal continuous caster

Interface Control in Materials

K.T. Aust and G. Palumbo

Department of Metallurgy and Materials Science,
University of Toronto, Toronto, Canada M5S 1A4

ABSTRACT

In the past, interface control in polycrystalline solids has largely involved the control of crystal size in the μm range and the formation of low angle boundaries, e.g. by obtaining a strong preferred crystal orientation. As a result of studies of the structures and properties of interfaces it is now possible to (1) control the structure and chemistry of certain large angle grain boundaries and (2) produce single phase and multiphase materials having a crystal size in the order of a few nanometers. The methods to achieve these two general interface characteristics and their effects on some physical, mechanical and chemical properties of commercial metallic alloys, ceramics, semiconductors and metallic glasses, are presented.

KEYWORDS

Solute, process control; microchemistry; mechanical, annealing, chemical properties; nanocrystalline materials.

CONTROL OF STRUCTURE AND CHEMISTRY OF GRAIN BOUNDARIES

Watanabe (1984) introduced the concept of "grain boundary design and control", with the purpose of improving various properties in polycrystalline materials. This concept is based on the fact that grain boundaries having certain coincidence site lattice (CSL) orientations with relatively low Σ values have been found to exert beneficial effects on various properties in materials. The coincidence site lattice corresponds to a special rotation of one lattice relative to the other to form a 3-dimensional atomic pattern in which a certain fraction of lattice points coincide (Kronberg and Wilson, 1949). The Σ-value is defined as the reciprocal of this fraction. The low Σ grain boundaries (sometimes termed "special") generally possess well- ordered structures, and usually lower energies in a pure material, than other large angle grain boundaries, termed "random". Also , the low Σ boundaries are less susceptible to impurity segregation effects. Consequently, the control of the character and density of such grain boundaries may result in the generation of new and improved properties which are quite different from that of the usual polycrystal produced without any control of the grain boundaries.

Boundaries deviating from CSL relationships by angles greater than that defined by Brandon's criterion (1966), ie. $\Delta\theta=15\Sigma^{-1/2}$ degs, and those with $\Sigma>49$, are classified as "random" here. In the absence of grain boundary control, Warrington and Boon (1975) have demonstrated that only 7.5% of the grain boundaries in a random distribution would be expected to be of the CSL-type with $\Sigma \leq 25$ (excluding low angle grain boundaries ($\Sigma 1$) and twins ($\Sigma 3$).

One of the first experiments which indicated that the type of large angle grain boundary may be controlled was conducted by Aust and Rutter(1959). New grains, introduced by recrystallization, were allowed to grow competitively at the expense of the substructure of a melt-grown single crystal. The orientations of the largest recrystallized grains were determined at 300°C for zone-refined lead with varying tin (solute) additions. Generally, large angle random boundaries were obtained with <5 ppm by weight of tin and with 50-1000 ppm tin. However, with additions of 5-40 ppm tin, a preferred orientation was observed, corresponding to "special" boundaries near coincidence rotations of 28°, 38° and 47° <111> and 28° <100> i.e. = $\Sigma 13$, 7, 19 and 17 respectively. These results are due to the different solute (Sn) interactions with the different grain boundaries, producing differences in grain boundary migration velocities.

The effect of solute concentration on the frequency of CSL-related grain boundaries is also demonstrated in Fig. 1 (Palumbo and Aust, unpublished). The effect of sulphur concentrations (0.3-10 ppm) on the distribution of grain boundary "types" in high purity nickel (99.999%) following recrystallization and grain growth at 1450 K. (avg. grain size = 750 μm) is shown. Individual grain boundaries have been characterized using the selected area channeling method and classified in the range $\Sigma = 3$ to 49. The most pronounced effect of sulphur content is observed with the frequency of CSL boundaries in the range $\Sigma = 5$ to 25, which increases from a fraction of 0.178 with nickel containing 0.3 ppm S, to a value of 0.290 with Ni-3 ppm S. With a further increase in sulphur content to10 ppm, this fraction decreases to a value of 0.220.

Recently, Watanabe (1985, 1986) has undertaken a systematic investigation into the grain boundary character distribution in polycrystalline materials. This was carried out using the electron channeling pattern (ECP) technique for orientation determination. The frequency of the three types of boundaries, i.e., low angle boundary, coincidence boundary and random boundary, depends on the original orientation and the amount of prestrain in aluminum polycrystals produced from single crystals. For example, the frequency of coincidence boundaries (up to $\Sigma 29$) ranged from 20% to 30%, increasing with the compressive prestrain (60% to 80%). The frequency of random boundaries never exceeded 65% in this work on aluminum. However, it should be kept in mind that the frequency of coincidence boundaries in the recrystallization structure of zone-refined high-purity Al is strongly affected by solute atoms. This was shown by Belluz and Aust (1975), as seen in Fig. 2. Here, the frequency of coincidence boundaries (up to $\Sigma 17$) increased from about 20% to 50% by a 10 ppm addition of Sn, while no effect was observed for Ti addition and a smaller effect for Cu.

As shown in Fig. 3, there is some indication that the frequency of coincidence boundaries increases with decreasing grain size, based on studies by Watanabe (1975) and Grabski (1975) for recrystallized polycrystals of Fe alloys and Al. Rutter and Aust (1962) found that the frequency of coincidence boundaries in a pure metal is dependent on the annealing temperature. For example, Fig. 4 depicts the results of experiments on artificial nucleation and growth into substructured crystals of zone-refined lead, after annealing at 300°C and 175°C. The 300°C anneal resulted in random high angle grain boundaries while the 175°C anneal produced coincidence boundaries about <111>, <110> and <100> axes. These results are largely due to the observed greater mobility of coincidence boundaries over the random boundaries at the lower temperature and their similar mobilities at the higher temperature in the pure metal.

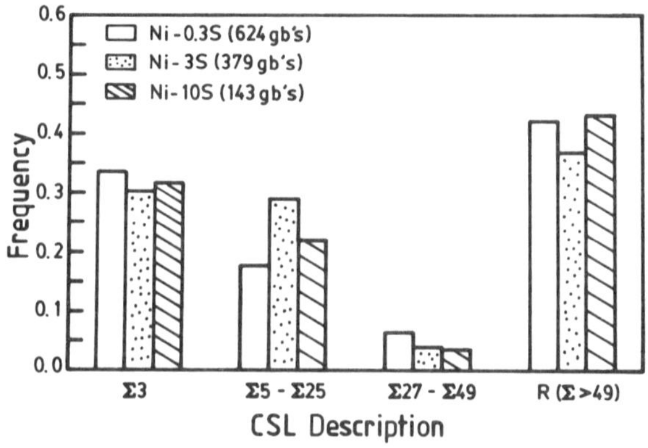

Fig. 1 Grain boundary character distributions in 99.999% Ni with varying S concentration.

In addition to the mobility of grain boundaries, the energy most also be considered. This arises in the formation of annealing twins which can occur when grain boundaries migrate. It was observed in pure metals such as Pb(Aust and Rutter, 1960), Al (Aust, 1961) and Cu (Ferran, Cizeron and Aust, 1963, 1967) that the frequency of coincidence boundaries can be increased by twinning. In materials with a low twin boundary energy, such as Cu and Pb, repeated twinning can occur which produces lower Σ grain boundaries for each twinning event. The driving force for the twinning is the lower energies of the Σ grain boundaries.

The role of twinning in grain boundary character distributions (GBCD) is demonstrated in Fig. 5, which shows the effect of sulphur content on the fraction of the CSL-related grain boundaries (Σ5-Σ49) in 99.999% Ni which are associated with twin boundaries (ie. Σ3) at triple junctions (Palumbo and Aust, unpublished). This fraction is a measure of the contribution of twinning in generating CSL-related grain boundaries, and is observed to increase with increasing specimen purity (ie. decreasing S-content), achieving a value of 92% with Ni-0.3 ppmS. The solute effects can be rationalized in terms of selective impurity segregation, whereby increasing impurity content results in diminished energy differences between CSL and non-CSL related interfaces.

Randle, Ralph and Dingley (1987, 1988) measured the orientation relationships of about 200 boundaries in a Ni-based superalloy (NIMONIC PE16) using an electron backscattering technique in conjunction with a SEM. As shown in Table 1, the frequency of coincidence boundaries increases with aging treatment. Almost half the boundaries are of the CSL type ($\Sigma \leq 49$) in the overaged samples. This overaging treatment reduces the amount of solute in solid solution and is expected to lead to reduced solute-boundary effects. This case may be analogous to the earlier Pb-Sn, Al-Sn and Ni-S results previously discussed, where too much solute leads to random boundaries while a suitable solute concentration favours coincidence boundaries.

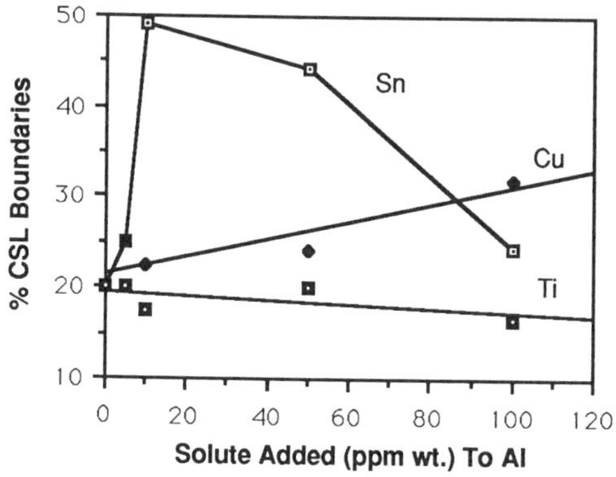

Fig. 2 The effect of Sn, Ti, and Cu additions on the frequency of coincidence boundaries in zone-refined Al.

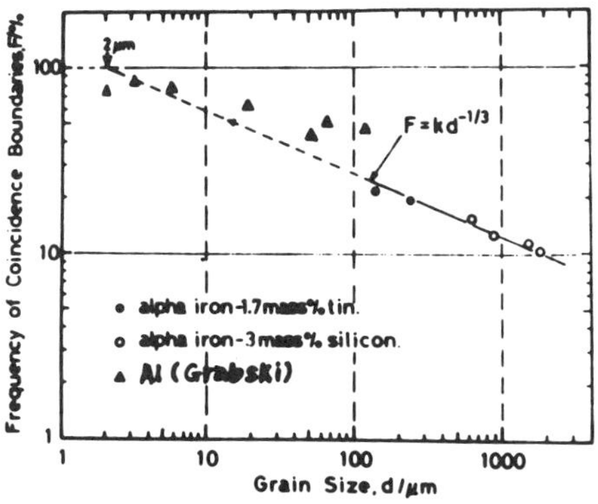

Fig. 3 The effect of grain size on the frequency of coincidence boundaries in Fe alloys and Al.

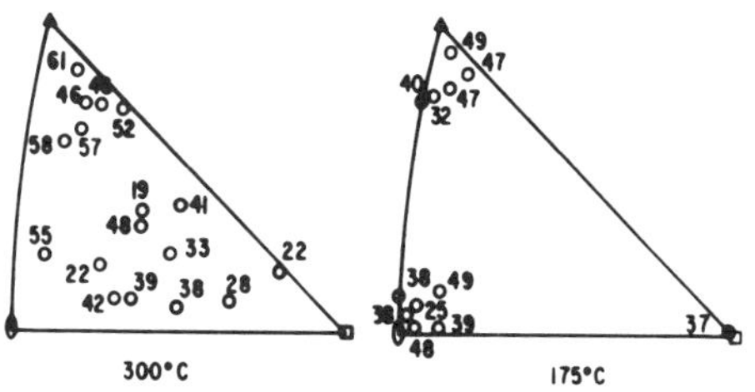

Fig. 4 Effect of annealing temperature on the frequency of CSL boundaries following artificial nucleation and growth in zone-refined Pb.

Hondros (1986) has discussed an important aspect of grain boundary design deriving from grain boundary microchemistry. He considered various approaches to the problem of control of bulk material properties by changing the grain boundary chemistry. One approach involves the dilution of the harmful species, by spreading out the small quantity over a very large grain boundary surface. This may be possible with the advent of highly microcrystalline alloys produced by rapid solidification techniques. The production of nanocrystalline materials, to be discussed later, is a similar approach to this problem. A second approach is that of "selective segregation", whereby a harmless surface active element having a strong repulsive interaction with a harmful segregant may expel the latter from the grain boundary. Hondros (1986) discussed several recent observations dealing with embrittlement which support this idea. In two cases it is shown that boron is enriched at the grain boundary and this combats grain boundary decohesion, a probable mechanism being the expulsion of the embrittling trace element.

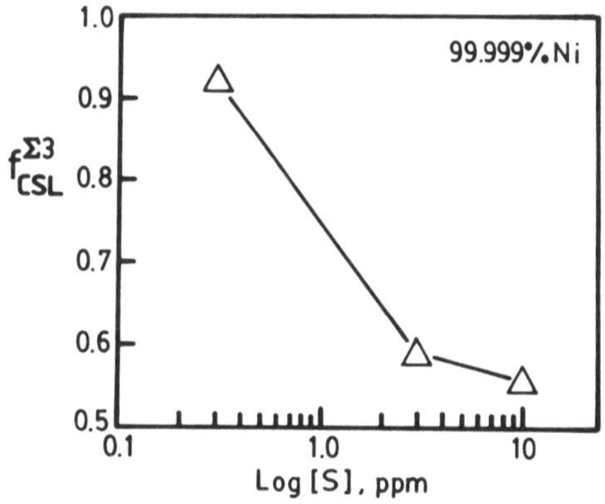

Fig. 5 The effect of S concentration on the fraction of CSL boundaries (Σ5-Σ49) in 99.999% Ni associated with twin boundaries (Σ3) at triple junctions.

TABLE I Effect of Heat Treatment on the CSL Distribution for Nimonic PE 16

Heat Treatment	No. of Boundaries Analyzed	CSL (%)	Non-CSL(%)
Unaged	43	18	82
Aged	74	27	73
Overaged	77	47	53

Another microchemical aspect of grain boundary design is to increase the solid solubility of the damaging species, thereby reducing the grain boundary concentration of the harmful solute. In this regard, there is some indication that high non-equilibrium solid solubilities may be achieved in microcrystalline systems by rapid quenching techniques, and by means of producing nanocrystalline alloys. The chemical stabilization of the harmful impurity elements through the formation of stable microprecipitates has shown some promise (Hondros, 1986). A somewhat different concept is that of "tying up" a harmful grain boundary segregant by means of separate, second phase particles at the grain boundaries (Aust and Palumbo, 1986). This is shown to be useful in the case of intergranular corrosion in a highly oxidizing environment (Aust, Armijo and Westbrook, 1966) and in environmentally- induced intergranular cracking (Thorpe, Ramaswami and Aust, 1984) of austenitic stainless steels.

SPECIAL PROPERTIES OF MATERIALS WITH A HIGH DENSITY OF CSL GRAIN BOUNDARIES

The presence of low Σ grain boundaries in materials has been found to exert beneficial effects in intergranular fracture and embrittlement associated with high-temperature creep and liquid metal embrittlement. For example, recent observations have indicated that low Σ coincidence boundaries (and low angle boundaries) are very resistant to high temperature grain boundary sliding and fracture, while high angle random boundaries slide and fracture easily (Watanabe, 1983). The stronger embrittlement of random grain boundaries is due to an increasing number of segregation sites with an increasing number of secondary dislocations in the interface (Gleiter, 1970).

Lim and Raj (1984) have shown that lower Σ coincidence boundaries in nickel polycrystals with smaller deviation angles from exact coincidence misorientations are more resistant than other boundaries to intergranular cavitation in low cycle fatigue at 573 K. Don and Majundar (1986) observed similar effects concerning intergranular cavitation of Type 304 stainless steel during creep tests at 866K under an applied stress of 172 MPa. In this study, drastic changes in cavitation across twin interceptions with grain boundaries were observed, and rationalized in terms of twinning effects on grain boundary character (ie. CSL).

Watanabe, Tanaka and Karashima (1984) have determined the character of interfaces in ß-brass polycrystals during studies of liquid Ga-induced intergranular fracture. It was found that the intergranular fracture preferentially occurred at high angle random boundaries. However, coincidence boundaries and low angle boundaries were difficult to break, as was also observed in bicrystal studies. In addition, in-situ SEM observations revealed that the change in fracture mode depends on the type of grain boundary which the propagating crack meets in ß-brass polycrystals. This is illustrated schematically in Fig. 6, which shows two fracture paths starting from the bottom of the figure. In case A, a crack nucleated at a random boundary propagates to a connecting coincidence boundary, but is prevented from further propagation along the boundary because of the high propagation resistance of the coincidence boundary, and then the crack enters the grain interior, changing the fracture mode. However, in case B the crack nucleated at a random boundary, keeps propagating, selecting random boundaries at every triple point. Consequently, in case B no change in the fracture mode is involved in the fracture process. Coincidence boundaries can be strong obstacles to the propagation of cracks so that intergranular fracture will stop at these special boundaries. These results seem to be independent of the orientation of the boundary plane.

It may be possible to improve some properties in ceramic materials by controlling the density of Σ coincidence boundaries. Not only have special boundaries in Al_2O_3 been found to exist, but they have also been found to be quite numerous (Carter, 1984). The analysis of several experimentally observed interfaces in Al_2O_3 have revealed that they can be regarded as special twin boundaries.

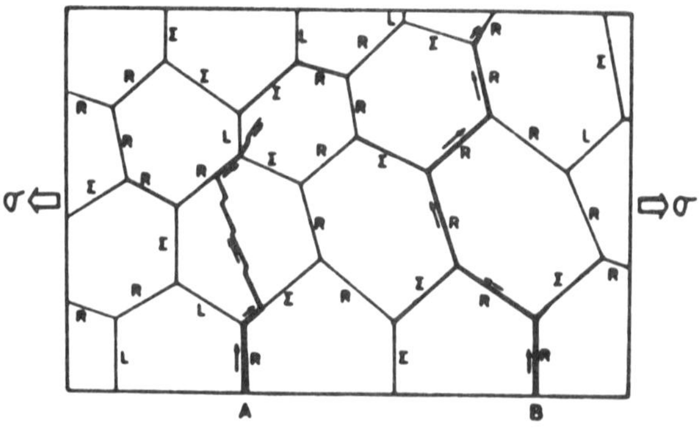

Fig. 6 Crack propagation paths in β-brass polycrystals embrittled by liquid Ga.

Wilbrandt and Haasen (1980), and Berger and co-workers (1983) have studied the initial stages of recrystallization in deformed single crystals of Cu and Al using HVEM and X-ray texture measurements. Recrystallization nuclei were observed to have orientations which were already present in the deformed microstructure. Nearly all observed orientations of the recrystallized grains can be described by coincidence relations to the deformed material. Also, twinning was found to lead to a lattice coincidence of a lower Σ value, or to a better approximation by the same coincidence orientation as before twinning. Typical Σ-values observed were near Σ =7, 11 and 13. Harase and co-workers (1986) have shown that the presence of grains with CSL boundaries after primary recrystallization was required to supply potential nuclei for secondary recrystallization in Fe-3% Si and thus obtain the Goss orientation texture. The latter texture is used commercially to obtain the desired magnetic properties in Fe-Si transformer materials. Work on copper (Makita, Hanada and Izumi, 1987) has also shown the importance of CSL boundaries in secondary recrystallization. It was found that the potential nuclei of secondary recrystallization already exist in the primary grains of copper with an extremely sharp cube orientation texture. These potential nuclei possess CSL orientations in the primary recrystallized matrix. The observed CSL boundaries correspond to Σ values of 7, 13, 21, 25 and 37. These preferred orientations of secondary grains in Cu are similar to those previously reported by Kronberg and Wilson (1949) and by Sharp and Dunn (1952).

It has been shown (Palumbo and Aust, 1988b) that susceptibility to the initiation of localized corrosion at grain boundaries in high purity (99.999%) Ni is strongly dependent on the nature of the boundaries (ie. CSL). Figure 7 summarizes the structure-dependence (ie. Σ, $\Delta\theta$) of the intergranular corrosion morphology observed at an applied electrochemical potential of 1150 mV in 2N H_2SO_4. Grain boundaries close to low Σ CSL relationships typically display a high resistance to the initiation of localized corrosion. In this study, special properties were not observed with grain boundaries having $\Sigma > 25$. The intergranular corrosion phenomena observed in this study cannot be attributed entirely to structural effects since it has been shown that structure-sensitive impurity segregation can influence corrosion properties even in materials of high purity (Aust and Iwao, 1974). Figure 8 (Palumbo and Aust, unpublished) shows the effect of sulphur content on the electrochemical potential limits for intergranular corrosion of 99.999% nickel in 2N H_2SO_4. Increasing sulphur content (0.3-50 ppm) results in reduced overpotential requirements for the initiation of grain boundary corrosion. The region bounded by the upper and lower curves in this figure (ie. comprehensive attack and immunity to localized corrosion) represents a potential range whereby grain boundary selectivity in the initiation of localized corrosion is observed. The magnitude of this range is thus a measure of the maximum difference in corrosion susceptibility of the grain boundaries. (ie. most susceptible to least susceptible.) This difference is observed to be greatest (ie. 175 mV) at intermediate sulphur levels of 3 and 10 ppm. With increasing sulphur content to 50 ppm, this potential range decreases to 125 mV. A decrease to 125 mV is also observed with the high purity material (ie. [S] = 0.3 ppm). These effects can be attributed to selective sulphur segregation at the intermediate impurity levels.

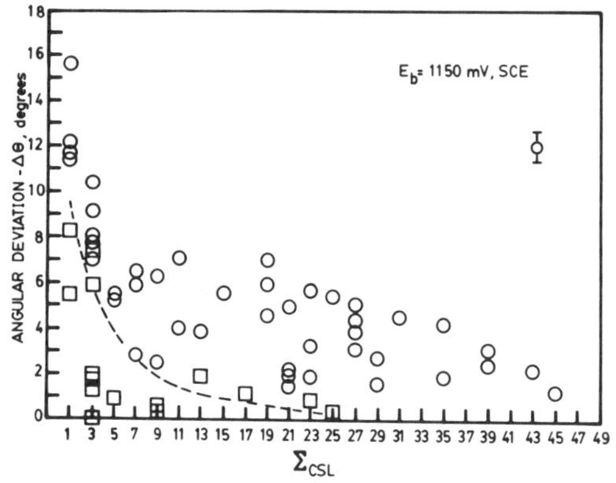

Fig. 7 Structure-dependence (Σ,$\Delta\theta$) of intergranular corrosion susceptibility in 99.999%Ni([S]=3ppm) in 2N H_2SO_4. ○, Corrosion; □, Immunity.

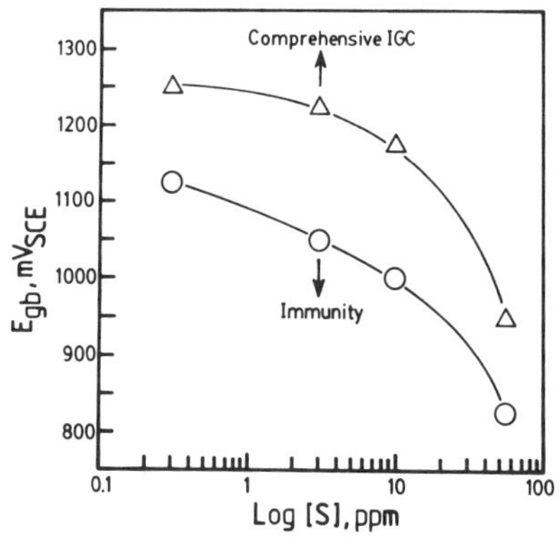

Fig. 8 Effect of S concentration on the potential limits for intergranular corrosion of 99.999% Ni in 2N H_2SO_4.

Susceptibility to intergranular corrosion can also be modified by the interaction of grain boundaries at triple junctions. It has been recently demonstrated (Palumbo and Aust, 1988a) that susceptibility to corrosion at triple junctions is dependent on the specific crystallographic arrangement of crystals at the junction, in accordance with Bollmann's criteria (1984) for a nodal balance of intersecting grain boundary dislocation arrays. This triple line character of interfaces, as given by Bollmann (1984), is another form of interface design producing different structures at triple junctions.

NANOCRYSTALLINE MATERIALS

Nanocrystalline materials are single phase or multiphase polycrystals, with a crystal size in the order of a few nanometers, typically 1 to 10 nm. As a result, about 50 vol. % of the material consists of grain boundaries or interphase boundaries, corresponding to about 10^{19} boundaries per cm^3. In principle, any method for preparing a polycrystalline material with a crystal size of a few nanometers and a random texture may be utilized to generate nanocrystalline materials. The experimental arrangement used by Birringer, Herr and Gleiter (1986) is a modified gas condensation method. In order to produce multiphase nanocrystalline materials, the inert gas condensation method may be used with slight modification (Birringer and Gleiter, to be published).

If the atomic structure of nanocrystalline materials differs from the structure of glasses and crystals, the structure-dependant properties of nanocrystalline materials are expected to be different from the properties of the chemically identical substances in the glassy or crystalline state.

The self diffusively in nanocrystalline Cu, with a crystal size of 8 nm and in the temperature range of 293 to 393 K revealed an enhancement of the self diffusivity by a factor of about 10^{19} in comparison to lattice diffusion (Horvath, Birringer and Gleiter, 1987). This remarkable enhancement may be understood in terms of the high grain boundary density which provides a connective network of short circuit diffusion paths. Also, an enhancement of the self diffusivity in nanocrystalline Cu in comparison to grain boundary diffusion of a factor of about 100 was noted. This enhancement appears to result from (I) the low impurity concentration in the nanocrystalline boundaries in comparison to boundaries in conventional polycrystals and (II) rapid diffusion along the connective network of grain boundary triple junctions. Two examples of enhanced solubility of solute in a solvent have been reported (Birringer and co-workers, in press). Hydrogen solubility in nanocrystalline Pd is about one to two orders of magnitude larger than in a Pd crystal at 292K. A second observed enhanced solid solubility is the increase of the solubility of Bi in Cu from $<10^{-4}$ at % in the crystal lattice to about 4% in nanocrystalline Cu.

The synthesis of a nanocrystalline ceramic ($Ti0_2$, rutile) has been carried out by Siegel and Hahn (in press) and by Karch, Birringer and Gleiter (in press) by producing small Ti crystals and then oxidizing them to $Ti0_2$ prior to compaction. It was observed that conventionally brittle ceramics become ductile permitting large (e.g. 100%) plastic deformation at low temperature (e.g. 293K), if a ceramic material was produced in the nanocrystalline form. The ductility seems to originate from the diffusional flow of atoms along the intercrystalline interfaces.

Koo and co-workers (1988) have found that a nanocrystalline-amorphous mixed phase of a Si-H alloy (25% H_2) can be produced at room temperature from amorphous Si-H thin films (approximately 100 nm thick) by using neutron irradiation with a dose of $2x10^{16}$ neutrons/cm^2. Figure 9 shows a neutron irradiated sample of the amorphous Si-H film with a high order of crystallinity. The crystals (black in Fig. 9) have a size of 10-15 nm. Subsequent annealing at 200-250°C of the irradiated film can produce a high density, uniform nanocrystalline material with a crystal size of 10-20 nm. This nanocrystalline form of hydrogenated silicon has recently drawn interest as a new material for applications in optoelectronics (Hamakawa, 1985; Imura and co-workers, 1984). The higher electrical conductivity and lower optical absorption coefficient of the nanocrystalline Si-H alloy as compared to amorphous Si-H material are attractive properties for solar cell applications.

Thorpe, Ramaswami and Aust (1988a) have recently produced nanocrystalline phases in Metglass 2826A ($Ni_{36.7}Fe_{31.5}Cr_{13.8}P_{12.1}B_{5.9}$) by annealing the metallic glass at 400 K for about $8x10^4$ min. As seen in Fig. 10, the FCC phases produced by the latter solid solution anneal showed a size distribution from 2 to 15 nm and were surrounded by regions of untransformed amorphous matrix (white featureless regions). According to

Thorpe, Ramaswami and Aust (1988b), this mixed nanophase-amorphous material produced an improvement in the general corrosion resistance and passivation, as compared to the original metallic glass. For example Fig. 11 shows a potentiodynamic polarization plot of the solid solution annealed metallic glass compared to the as-quenched, amorphous Metglass 2826A. It is evident that the passive current density of the nanophase-amorphous sample is nearly one order of magnitude less than the amorphous sample over the entire potential range. In addition, the degree of interfacial attack of the nanophase-amorphous material, as determined by electrochemical potentiokinetic reactivation (EPR) testing, was much less than that of primary crystallized Metglass 2826A (annealed 60 min at 873 K) and similar to that observed in a solution-treated 304 stainless steel (Thorpe, Ramaswami and Aust, 1987). This improvement in both the general and localized corrosion resistance of the nanophase-amorphous Metglass 2826A is attributed to the observed Cr enrichment in the passive film resulting from the enhanced interphase boundary diffusion of Cr. The more protective Cr film would lower the frequency of film breakup and increase the repassivation rate over active surface sites.

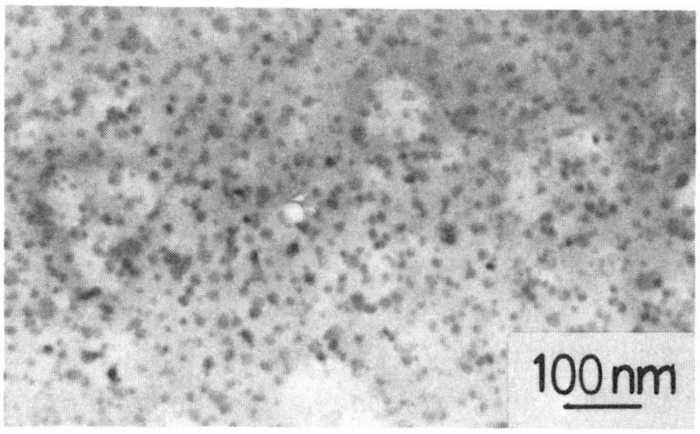

Fig. 9 STEM micrograph of a neutron irradiated sample of amorphous Si-25%H.

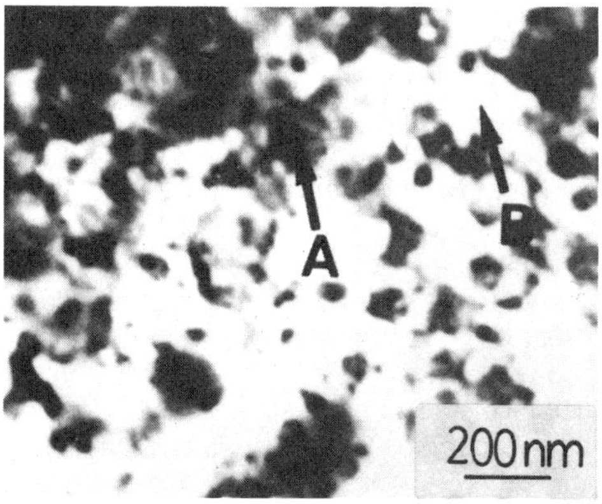

Fig. 10 STEM micrograph of Metglass 2826A after an SS anneal (400K, 8×10^4min). A, nanocrystals; B, amorphous.

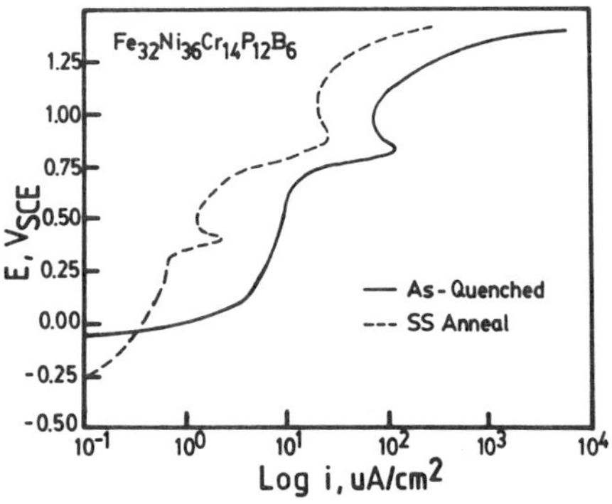

Fig. 11 Potentiodynamic polarization curves for as-quenched and SS annealed Metglass 2826A in 0.1M NaCl+0.1M Na_2SO_4 at 303K.

ACKNOWLEDGEMENTS

The authors would like to thank the Natural Science and Engineering Research Council of Canada for financial support. One of the authors (G.P.) would also like to acknowledge financial support through the Ontario Graduate Scholarship program.

REFERENCES

Aust, K.T., and J.W. Rutter (1959). Trans. Met. Soc. AIME, 215, 119-127.
Aust, K.T., and J.W. Rutter (1960). Trans. Met. Soc. AIME, 218, 1023-1028.
Aust, K.T. (1961). Trans. Met. Soc. AIME, 221, 758-761.
Aust, K.T., and J.W. Rutter (1962). Trans. Met. Soc. AIME, 224, 111-115.
Aust, K.T., J.S. Armijo, and J.H. Westbrook (1966). Trans. ASM, 59, 544-556.
Aust, K.T., and O. Iwao (1974). Localized Corrosion NACE-3, 62-75.
Aust, K.T., and G. Palumbo (1986). Trans. Japan Inst. Met. (Suppl.), 27, 995-1004.
Belluz, R.V., and K.T. Aust (1975). Met. Trans., 6A, 219-220.
Berger, A., P.J. Wilbrandt, and P. Haasen (1983). Acta Met., 31, 1433-1443.
Birringer, R., U. Herr, and H. Gleiter (1986). Trans. Japan Inst. Met. (Suppl.), 27, 43-52.
Birringer, R., and H. Gleiter. Adv. Mat. Sci. and Eng., Pergamon Press, to be published.
Birringer, R., H. Hahn, H. Hofler, J. Karch, and H. Gleiter. Diffusion and Defect Data, in press.
Bollmann, W. (1984). Phil. Mag. A, 49, 73-79.
Brandon, D.G. (1966). Acta. Met., 14, 1479-1484.
Carter, C.B. (1984). Mat. Res. Soc. Symp., in W. Krakow, D.A. Smith, L.W. Hobbs (Eds.), Electron Microscopy of Materials, Elsevier Sci. Publ. (North Holland), 31, 267-278.
Don, J., and S. Majundar (1986). Acta Met., 34, 961-967.
Ferran, G., G. Cizeron, and K.T. Aust (1963). Acad. Sci. Paris, 257, 3595-3597.
Ferran, G., G. Cizeron, and K.T. Aust (1967). Mem. Sci. Rev. Met., 54, 1067-1076.
Gleiter, H. (1970). Acta Met., 18, 117-121.
Grabski, M.W. (1985). J. de Physique, 46, C4-567 -C4-579.
Hamakawa, Y. (1985). SPIE, 543, 74-82.
Harase, J., R. Shimuzu, K. Kuroki, T. Nakayama, T. Wada, and T. Watanabe (1986). Trans. Japan Inst. Met. (Suppl.), 27, 563-570.
Hondros, E.D. (1986). Trans. Japan Inst. Met. (Suppl.), 27, 393-400.
Horvath, J., R. Birringer, and H. Gleiter (1987). Solid State Comm., 62, 319-322.
Imura, T., H. Kaya, H. Terauchi, H. Kiyomo, A. Hiraki, and M. Ichihara (1984). J. Appl. Phys., 23, 179-183.
Karch, J., R. Birringer, and H. Gleiter. Nature, in press.
Koo, Y.C., R. Perrin, K.T. Aust, and S. Zukotynski (1988). Met. Trans., 19A, 1345-1349.
Kronberg, M.L., and F.H. Wilson (1949). Trans. AIME, 185, 501-514.
Lim, L.C., and R. Raj (1984). Acta Met., 32, 1183-1190.
Makita, H., S. Hanada, and O. Izumi (1987). Trans. Japan Inst. Met., 28, 447-456.
Palumbo, G., and K.T. Aust (1988a). Scripta Met., 22, 847-852.
Palumbo, G., and K.T. Aust (1988b). J. de Physique (coll.), 49, in press.
Palumbo, G., and K.T. Aust, unpublished.
Randle, V., and B. Ralph (1987). J. Mat. Sci., 22, 2535-2541.
Randle, V., B. Ralph, and D. Dingley (1988). Acta Met., 36, 267-273.
Sharpe, M., and C.G. Dunn (1952). Trans. AIME, 4, 42-43.

Siegel, R.W., and H. Hahn, Nanophase Materials in M. Yussouff (Ed.), Current Trends in Physics of Materials, World Scient. Publ. (Singapore), in press.
Thorpe, S.J., B. Ramaswami, and K.T. Aust (1984). Acta Met., 32, 1297-1304.
Thorpe, S.J., B. Ramaswami, and K.T. Aust (1987). Corrosion, 43, 534-538.
Thorpe, S.J., B. Ramaswami, and K.T. Aust (1988a). Acta Met., 36, 795-803.
Thorpe, S.J., B. Ramaswami, and K.T. Aust (1988b). J. Electrochem. Soc., 135, 2162-2170.
Warrington, D.H., and M. Boon (1975). Acta Met., 23, 599-607.
Watanabe, T. (1983). Met. Trans., 14A, 531-545.
Watanabe, T. (1984). Res. Mechanica, 11, 47-84.
Watanabe, T. (1985). J. de Physique, 46, C4-555 - C4-564.
Watanabe, T. (1986). Trans. Japan Inst. Met. (Suppl.), 27, 73-82.
Watanabe, T, M. Tanaka, and S. Karashima (1984). in M.H. Kamdar (Ed.), Embrittlement by Liquid and Solid Metals, Proc. AIME Symp., AIME (Warrendale), 183-196.
Willbrandt, P.J., and P. Haasen (1980). Z. Metallk., 71, 385-395.

Passive Film Breakdown at Grain Boundary Junctions in High Purity Nickel

G. Palumbo and K. T. Aust

**Department of Metallurgy and Materials Science,
University of Toronto, Toronto, Canada M5S 1A4**

ABSTRACT

Electrochemical studies in 2N H_2SO_4 were conducted in order to assess the effect of crystallographic parameters on passive film stability at grain boundary junctions (triple lines) in high purity nickel. Orientation relationships between crystals were determined using the selected area channeling method (SACP), and represented in rotation matrix notation. Susceptibility to localized corrosion was found to be strongly dependent on the specific crystallographic arrangement of crystals at the triple line, in accordance with Bollmann's crystallographic criteria (Bollmann, 1984) for a nodal balance of grain boundary dislocations. Preferential corrosion (ie. relative to the adjoining grain boundaries) was only evident under conditions whereby these criteria were not satisfied. Results from this study are consistent with a mechanism whereby dislocations act as preferential sites for passive film breakdown, and provide experimental support for Bollmann's disclination treatment of triple lines (Bollmann, 1984; Bollmann, 1988a).

KEYWORDS

triple lines, localized corrosion, grain boundary dislocations, I-lines, U-lines, nearest neighbour relation, coincidence site lattice

INTRODUCTION

It has been shown by Bollmann (1970) that grain boundaries can be represented by grain boundary dislocation (GBD) networks, and that these networks can be calculated by means of the O-Lattice method. In the calculation of such networks, there arise many possible representations for transitions between neighbouring grains (Bollmann, 1982); however, the physically significant transition is that which leads to the lowest interfacial dislocation content (ie. most widely spaced GBD array). This transition, termed the "nearest neighbour relation" (NNR) yields the lowest interfacial energy (ie. for a given orientation relationship) (Bollmann, 1984).

In polycrystalline materials, typically three grain boundaries (3 GBD networks) meet in a line (ie. the triple line). Bollmann (1984; 1988a) has shown that these lines can be interpreted as disclinations, and can be classified as two types: I lines, and U lines. When the sequence of NNR's about the triple line leads to the identity matrix, adjoining GBD arrays form nodes where the Burger's vectors balance to zero (I lines). When the sequence of NNR's does not lead to the identity matrix, a nodal balance is not possible (Bollmann, 1984; Bollmann; 1988a).

The existence of GBD nodes at an I-type intersection of $\Sigma3$-$\Sigma9$ -$\Sigma27b$ grain boundaries in Cu - 3 at.%Si has been experimentally confirmed by TEM (Clarebrough and Forwood, 1987). We have recently reported (Palumbo and Aust, 1988a) on the existence of U-type intersections in high purity nickel, and on their susceptibility to localized corrosion. In the present work, further results are presented concerning the character dependence of localized corrosion at triple junctions.

EXPERIMENTAL

Electrochemical tests were conducted on deformation-free disc specimens of high purity nickel (99.999%, [C]<1ppm, [S]<1ppm) in deaerated 2N H_2SO_4 at 303K. The electrochemical test procedure has been described in detail previously (Palumbo and Aust, 1988a) and involves prepassivation of the nickel electrode at +750 mV_{SCE} (saturated calomel reference electrode), followed by a potential step to +1200 mV_{SCE} in order to locally destabilize the preformed film. A total anodic charge transfer(Q) of 5C/cm^2 was allowed in the destabilization step. This method was previously shown to result in preferential film breakdown at grain boundaries (Palumbo and Aust, 1988b).

Orientation relationships between adjoining crystals were determined by the selected area channeling method (Accelerating voltage: 30keV, Rocking angle: 10°), and assessed as anticlockwise rotations in matrix form. CSL classification (Σ= 1 to 49) of individual grain boundaries was by the rotation matrix method (Mykura, 1979), utilizing Brandon's criterion (Brandon, 1966) for the maximum allowable angular deviation ($\Delta\theta=15^o\Sigma^{-1/2}$).

Triple line character was determined using the procedure described in detail by Bollmann (1988a). Orientation relationships for the adjoining grain boundaries at the junction were redefined in terms of the co-ordinates of one of the adjoining lattices (see Figure 1). These rotation matrices (in orthonormal coordinates) were then converted to structural coordinates (R*) using the structure matrix for fcc systems (Bollmann,1982),

$$R^* = S^{-1} R\, S \quad \text{where } S_{fcc} = \begin{pmatrix} 0.5 & -0.5 & 0 \\ 0.5 & 0.5 & 0.5 \\ 0 & 0 & 0.5 \end{pmatrix} \qquad [1.0]$$

The unimodular transformations (U where det(U)=1) necessary to convert the general transformations (R) to NNR's (see Figure 1) were then determined using the method described in detail by Bollmann (1982). The general tranformation in structural coordinates (R*) was first used to formulate matrices representing internal (M^i) and external (M^e) crystal coordinates,

$$R^* = M^i_{(0<x<1)} + M^e_{(integer)} \qquad [2.0]$$

The column vectors of M^i (x_i) were then used to determine the square of the distance (d_i^2, where i = number of the column vector) between these columns and the corners (8) of the respective unit cell using the relation,

$$d_i^2 = [G(x_i-x_c)]^*(x_i-x_c) \qquad [3.0]$$

where $G=S^TS$, and x_c represents the coordinates of the unit cell corners. The values of d_i^2 were rounded to 3 significant decimal places, and the smallest value(s) for each column vector x_i determined. The cell corner coordinates yielding these minimum d_i^2 values were then added to the respective column vectors of M^e formulating the matrix U^{-1}. The resulting U matrix (ie. $(U^{-1})^{-1}$) was checked to ensure unimodularity (ie. det(U)=1). These U-matrices were then used to assess triple line character by evaluating the characteristic tensor (T_{123})for the triple junction using grain 1 as a reference lattice (see Figure 1). In cases where multiple equivalent U-matrices were determined for a given grain boundary, T-matrices were calculated for each combination. Triple junctions were classified as U-lines when none of the possible arrangements resulted in the identity matrix.

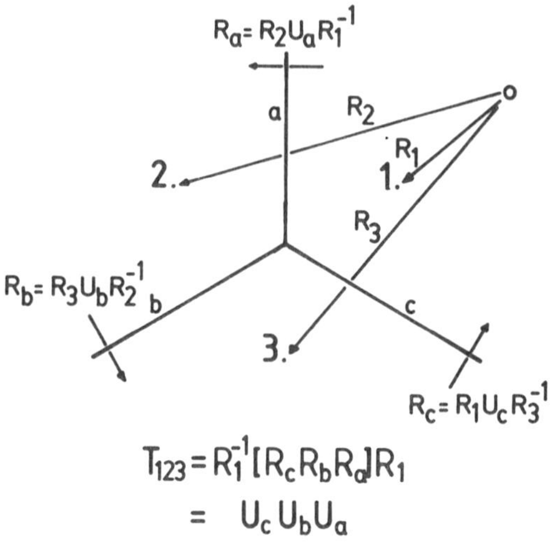

Fig.1 Schematic representation of the crystal orientation relationships, and the determination of the characteristic tensor (T_{123}) for the triple junction using grain 1 as a reference lattice.

RESULTS AND DISCUSSION

Figure 2a illustrates the corrosion morphology most typically observed at the triple junctions, whereby the extent of corrosion at the junction is consistent with that of the adjoining boundaries. The junction shown in Fig. 2a is formed by the intersection of a $\Sigma 3$, $\Sigma 27a$, and a non-CSL boundary ("random"). Orientation relationships in the system are shown in Fig. 1b and can be represented by rotations about a common $[01\bar{1}]$ axis with the $\Sigma 3$ (boundary b) and $\Sigma 27a$ (boundary c) represented by anticlockwise rotations of 71° and 32° respectively, and the non-CSL boundary (boundary a) by a clockwise rotation of 102°. Figure 2b also shows the calculated U-matrices resulting in nearest neighbour transitions between the adjoining crystals. Five equivalent NNR's were calculated for the $\Sigma 3$ boundary, with only one possible NNR determined for each of the $\Sigma 27a$ and non-CSL interfaces. (Note: The U-matrix for the $\Sigma 27a$ boundary is the identity matrix, indicating that this orientation relationship (ie. 32° $[01\bar{1}]$) represents a nearest neighbour configuration.) Five possible tensors ($T_{123}=U_cU_bU_a$) exist for this triple junction, one of which being the identity matrix (shown in Fig. 2b). The possibility of a nodal balance of GBD's at this junction leads to its classification as an I-line. It should be noted that in the study by Clarebrough and Forwood (1987) where a nodal balance of GBD's at a $\Sigma 3$-$\Sigma 9$-$\Sigma 27b$ intersection was confirmed by TEM, analysis of their crystallographic relationships yields 36 possible T-matrices, 2 of which being the identity matrix. As discussed by Bollmann (1988b), a nodal balance of GBD's at the junction , if crystallographically possible, would be the preferred state due to energetic considerations.

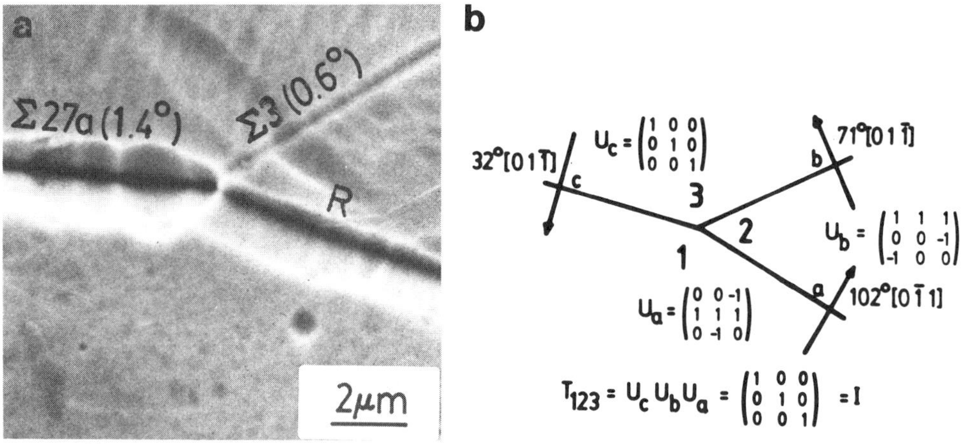

Fig. 2(a) Scanning electron micrograph of the corrosion morphology evident at a $\Sigma 3$-$\Sigma 27a$-'non-CSL' junction (Note: Values in brackets following Σ-value represent angular deviation from the exact CSL). (b) Corresponding orientation relationships and U-matrices leading to an I-arrangement.

Figure 3a shows an example of extensive corrosion at the triple junction formed by the intersection of a $\Sigma 11$ and two non-CSL related grain boundaries. The extent of corrosion at the junction is much greater than that evident along any of the adjoining grain boundaries. The crystallographic relationships are shown in Fig. 3b. Two possible NNR relationships are obtained with the $\Sigma 11$ boundary (boundary c), and one for each of the non-CSL boundaries (a and b). Neither of the two possible T-matrices for this system is the identity matrix, thus precluding the possibility of a nodal balance of GBD's at the junction (U-line).

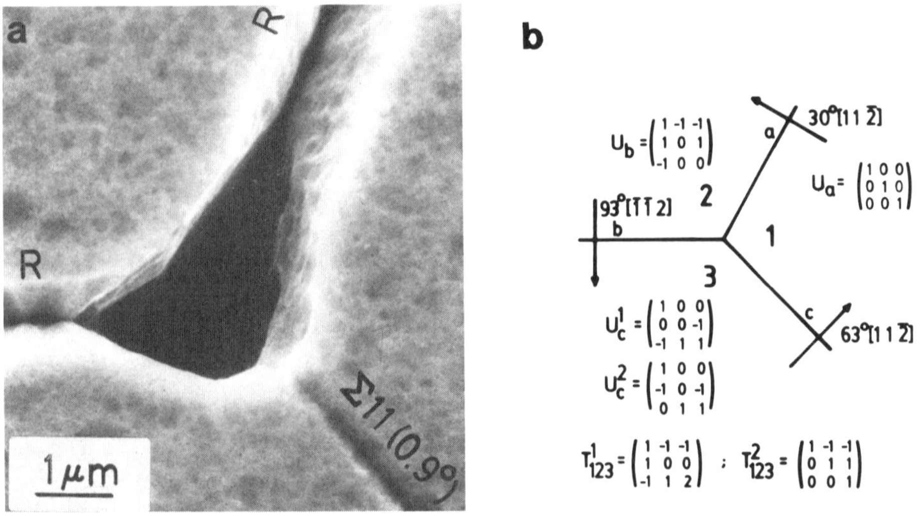

Fig. 3(a) Scanning electron micrograph of the corrosion morphology evident at the intersection of a $\Sigma 11$ boundary with two non-CSL related interfaces. (b) Corresponding orientation relationships and determined U-matrices demonstrating a U-line arrangement.

Table I summarizes the observations of this study. Of 68 triple junctions examined, 52 were determined to be the I-type. This preference for I arrangements in the triple line distribution is consistent with the energetic considerations discussed by Bollmann (1988b). Preferential triple line corrosion was **not** observed at any of the I arrangements examined. Of the 16 U-lines investigated, 13 displayed preferential corrosion relative to the adjoining grain boundaries.

Figures 4 and 5 further illustrate the preferential corrosion observed at U-type intersections. The corrosion morphology is typified by substantial penetration at the junction, with limited lateral spreading (ie. $\leq 2\mu m$). It has been previously shown (Palumbo and Aust, 1988b) that the depth propagation of localized corrosion (under the electrochemical conditions used in this study) is largely governed by differences in the rates of passive film breakdown and repair at crystalline defects, with lateral spreading largely governed by general surface dissolution rates.

TABLE I Summary of I and U Line Distributions and Corrosion Behaviour

	No. Observed (%)	No. Displaying Preferential Corrosion
I-Lines	52 (76.5%)	0
U-Lines	16 (23.5%)	13

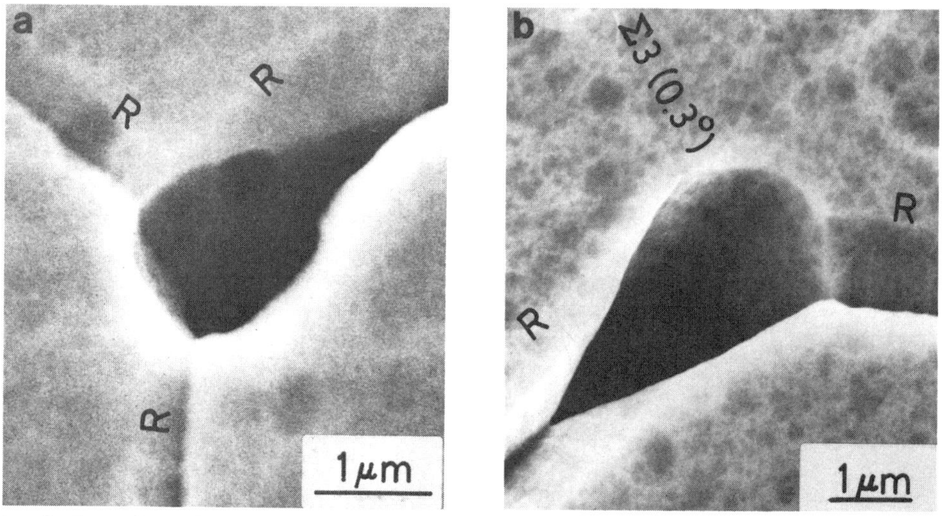

Fig. 4 Scanning electron micrographs of the corrosion morphology typically observed at U-type triple junctions. (a) The intersection of three non-CSL related interfaces. (b) The intersection of a $\Sigma3$ boundary with two non-CSL related interfaces.

Susceptibility to enhanced corrosion at U-type junctions can be attributed to the high dislocation content associated with these intersections (Bollmann, 1984). The degree of susceptibility of the U-lines would be expected to be dependent on the extent of "mismatch" at the intersection. Bollmann (1988a) has shown that the dislocation surplus at the intersection is described exactly by the characteristic tensor (T_{123}) of the triple line. Further work is necessary to clarify the T-dependence of U-line properties.

The role of dislocations in altering the stability of surface films on nickel has been previously demonstrated (Latanision and Opperhauser, 1970; Palumbo and Aust, 1988c); however, structural factors may not be the sole cause of preferential U-line corrosion. Aust and Iwao (1974) have shown that impurity segregation can be a significant factor in the intergranular corrosion behaviour of very high purity materials. Enhanced impurity segregation to U-type intersections must be considered since it has been shown (Gleiter, 1970; Michael and co-workers, 1988) that GBD's can act as preferential segregation sites. The surplus of dislocations expected at U-lines may result in greater solute enrichment at the junction than along the adjoining interfaces, resulting in a greater susceptibility to surface film destabilization at these sites.

SUMMARY

Preferential corrosion at triple junctions in high purity nickel has been found to occur exclusively at those junctions where the crystallographic arrangement of adjoining crystals precludes the possibility of a nodal balance of grain boundary dislocations (Bollmann's U-lines). The observed phenomena is attributed to the structural (dislocation content) and chemical (impurity sink) properties of U-lines, giving rise to preferential sites for passive film breakdown. The results of this study provide further support for Bollmann's (1984, 1988) disclination treatment of triple lines.

ACKNOWLEDGEMENTS

The authors would like to thank W. Bollmann for many useful comments and suggestions concerning this study. Financial support from the Natural Science and Engineering Research Council of Canada is gratefully acknowledged. One of the authors (G.P.) would also like to acknowledge support through the Ontario Graduate Scholarship Program.

REFERENCES

Aust, K.T. and O. Iwao (1974). Localized Corrosion, NACE-3, 62-65.
Bollmann, W. (1970). Crystal Defects and Crystalline Interfaces, Springer-Verlag, Berlin.
Bollmann, W. (1982). Crystal Lattices. Interfaces. Matrices, W. Bollmann (privately published), Geneva.
Bollmann, W. (1984). Phil. Mag. A, 49, 73-79.
Bollmann, W.(1988a). Phil. Mag. A, 57, 637-649.
Bollmann, W. (1988b). private communication.
Brandon, D.G. (1966). Acta Met., 14, 1479.
Clarebrough, L.M. and C.T. Forwood (1987). Phil. Mag. A, 55, 217-225.
Gleiter, H. (1970). Acta Met., 18, 117-121.
Latanision, R.M. and H. Opperhauser (1971). Corrosion, 27, 509-515.
Michael, J.R., C.H. Lin, and S.L. Sass (1988). Scripta Met., 22, 1121-1124.
Mykura, H. (1979). Grain Boundary Structure and Kinetics, ASM, Ohio, 445-456.
Palumbo, G. and K.T. Aust (1988a). Scripta Met., 22, 847-852.
Palumbo, G. and K.T. Aust (1988b). J. de Physique (coll.), 49, (in press).
Palumbo, G. and K.T. Aust (1988c). Microstructural Science, 17, (in press).

ROLE OF INTERNAL INTERFACES IN THE MECHANICAL BEHAVIOUR OF ENGINEERING MATERIALS

R. A. Varin

Department of Mechanical Engineering,
University of Waterloo,
Waterloo, Ontario, Canada N2L 3G1

ABSTRACT

The recent developments in the understanding of the role of grain boundaries and interfaces in the mechanical behaviour of engineering materials are briefly reviewed. It is shown that dislocation processes occurring at grain boundaries are important in controlling the mechanical behaviour of polycrystals. It is suggested that at elevated temperatures the annihilation of dislocations at grain boundaries may control the mechanical behaviour of ultrafine-grained engineering alloys. This process may also lead to substantial increase in ductility. Some aspects of boron segregation at grain boundaries in engineering alloys are discussed. The role of interfaces in metal matrix composites with special emphasis on the interfacial zones is also presented.

KEYWORDS

Grain boundaries; mechanical behaviour of polycrystals; dislocation processes at grain boundaries; boron segregation; interfacial zones in metal matrix composites.

INTRODUCTION

In the 1970's and early 1980's a significant development occurred in the understanding of the structure of grain boundaries and their role in the mechanical behaviour of polycrystals. In the last 10 years there has been a real explosion of research in the field of composite materials and again a special emphasis has been put on the structure and role of interfaces.

So far, however, there has been no attempt to provide some overview of the similarities and differences in the elementary mechanisms occurring at grain boundaries in polycrystals and interfaces in composites which are responsible for their mechanical properties. The aim of the present paper is to present a brief overview of these problems. The content of the present paper is restricted to metallic polycrystals and metal matrix composites (MMCs).

GRAIN BOUNDARIES IN POLYCRYSTALS

Ambient and Cryogenic Temperatures

A most typical representation of the effect of grain boundaries on the yield strength and flow stress of engineering polycrystals is the well-known Hall-Petch relationship: $\sigma_\varepsilon = \sigma_o^\varepsilon + k_\varepsilon d^{-\frac{1}{2}}$ at constant strain ε. The k_ε parameter represents the "average" strength of grain boundaries for propagating slip during plastic flow (yielding). A typical example of H-P relationship in AISI 316 steel is shown in Fig. 1. It is seen that the k parameter considerably increases with decreasing temperature into the cryogenic region. It means that grain boundaries become more resistant to slip penetration at cryogenic temperatures. Unfortunately, the behaviour demonstrated in Fig. 1 is not well-documented for many important conventional engineering materials and in particular, for advanced alloys. Very recently,

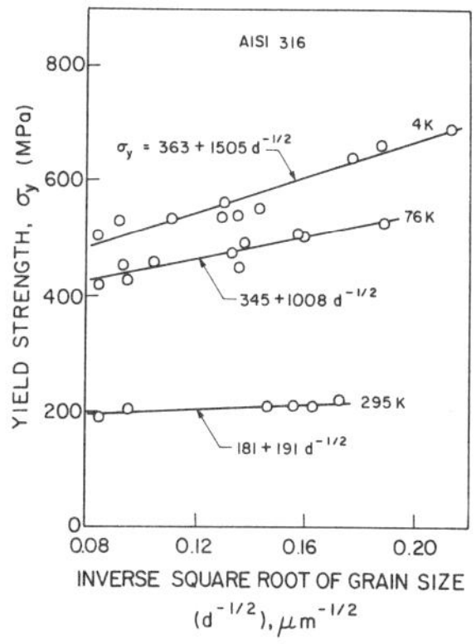

Fig. 1. Effect of grain size on the yield strength at three test temperatures for AISI 316 steel (Tobler, Beekman and Reed, 1983).

Chang and co-workers (1987) have reported some enhancement in strength of polycrystalline, boron-doped Ni_3Al intermetallic at cryogenic temperatures without, however, any detailed Hall-Petch type analysis. Apparently, much more research is needed in this area.

Processes occurring at grain boundaries can also affect ductility of structural alloys. On the one hand, it is well-known that grain boundary segregation of elements such as tin, antimony, phosphorus, sulphur, etc. renders many engineering materials, especially steels, totally embrittled (see reviews by Briant and Banerji (1978), and Olefjord (1978)). On the other hand, a segregation of boron to the grain boundaries considerably enhances ductility of some alloys such as polycrystalline Ni_3Al intermetallic (Liu and Sikka, 1986). The examples of both detrimental and beneficial nature of grain boundary segregation are well-known and documented in literature. What is, however, less-known is the fact that large number of precipitates at grain boundaries can promote inhomogeneous deformation (Lüders-type). This is illustrated in Fig. 2. It is seen that in ultrafine-grained AISI 316 and 304 steels the Lüders strain, ε_L, increases with the number of precipitates per unit area of grain boundary surface. However, the behaviour of low-carbon AISI 304 L steel seems to be independent of the grain boundary precipitates. In turn, grain boundary segregation of some elements, including boron, acts in opposite direction enhancing homogeneous deformation as shown in Fig. 3. The Lüders strain, ε_L, decreases with boron content up to ~15 ppm and then levels off (Fig. 3b). A

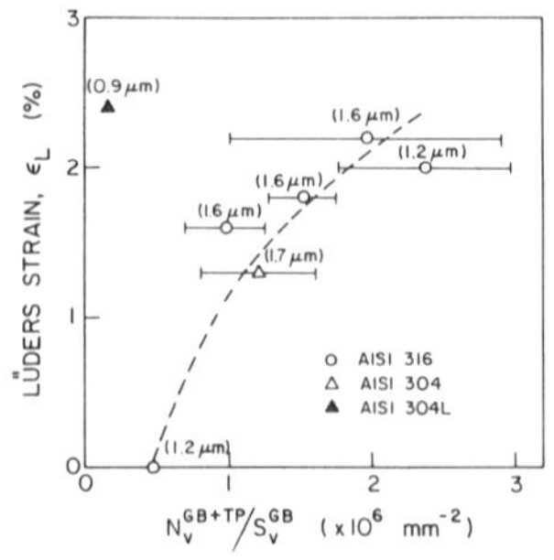

Fig. 2. Magnitude of the Lüders strain as a function of the number of particles per unit area of grain boundary surface (N_V^{GB+TP} - volume density of particles at grain boundaries (GB) and triple points (TP), S_V^{GB} - the grain boundary surface area per unit volume). Numbers in parentheses show grain size.

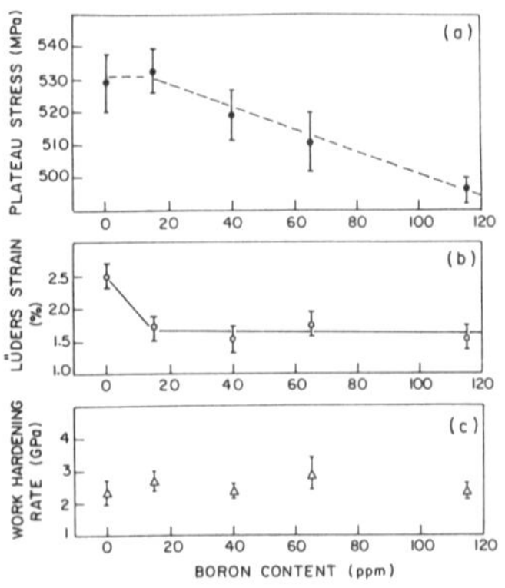

Fig. 3. Variations of (a) plateau stress, σ_P, (b) the Lüders strain, ε_L, and (c) the work hardening rate $\theta = d\sigma/d\varepsilon$ with boron content in ultrafine-grained AISI 304 L steel (Kurzydlowski, Varin and Tangri, 1988).

similar reduction of the Lüders strain by grain boundary segregation of phosphorus and tin in low-carbon steels was also observed by Hu (1983). It is also seen in Fig. 3a that the plateau stress (the stress to propagate Lüders bands) decreases linearly with the boron content above 15 ppm. The reduction in the plateau stress due to addition of 750 ppm boron was also observed by Schulson and co-workers (1986) in fine-grained Ni_3Al intermetallic. These effects are believed to occur as a result of enhanced generation of dislocations from grain boundaries by segregated elements (Hu, 1983) and/or the ease of slip accommodation at grain boundaries. Again, all the above effects are virtually unexplored in advanced alloys despite that some of them, such as Ni_3Al, are known to deform inhomogeneously (Schulson and co-workers, 1986).

An important question now arises: what is the elementary mechanism responsible for the effects of grain boundaries on the mechanical properties? An overwhelming experimental evidence, collected mainly by transmission electron microscopy, indicates that this mechanism is an interaction of dislocations with grain boundaries. Interaction in a very broad sense, including generation as well as transfer of slip dislocations across grain boundaries. These two micro-events usually occur simultaneously during straining as shown in Fig. 4 and discussed by Kurzydlowski, Varin and Zielinski (1984).

Elevated Temperatures

The best method of illustrating the effect of temperature on the role of grain boundaries in strength is to plot the variation in the k_ε parameter as a function of temperature. A typical example for AISI 316 steel is shown in Fig. 5. A significant drop of k_ε at high temperatures indicates a weakening of the locking effects at grain boundaries. Again this type of behaviour has not been systematically investigated in many conventional as well as advanced materials. The influence of grain boundaries on ductility can also be affected by temperature. However,

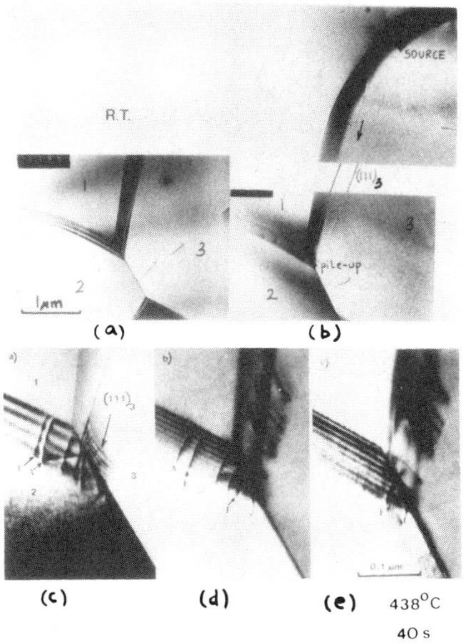

Fig. 4. Sequences of dislocation generation and slip transfer in an austenitic stainless steel during "in situ" straining of a thin foil in TEM. (a) Before generation. (b) Generation from the source on $(111)_3$ slip plane. (c,d) Transfer of slip by generation of dislocations in grain 2 at the tip of pile-up. (e) Spreading of dislocations which entered grain boundary 1-2 after annealing at 438°C for 40 s (Kurzydlowski, Varin and Zielinski, 1984).

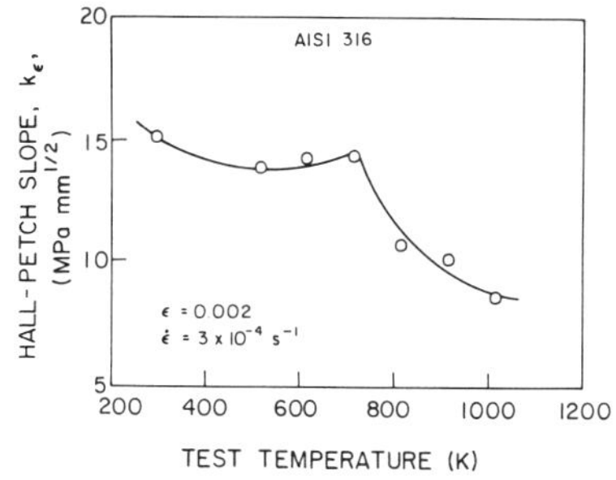

Fig. 5. Variation of k_ε with temperature in AISI 316 steel (Mannan, Samuel and Rodriguez, 1982).

usually with increasing temperature the strain rate effects come into play. The combined effect of grain boundaries, temperature and strain rate is illustrated in Fig. 6. It is seen that a considerable refinement of grain size enhances ductility at low strain rates (Fig. 6a). Completely opposite effect is observed for coarse-grained material (Fig. 6b). It is, therefore, important to find out the "threshold" value of grain size below which ductility is enhanced. Unfortunately, data of this type is lacking. An experimental evidence seems to indicate that elementary mechanism responsible for high temperature effects at grain boundaries is the annihilation (spreading) of trapped dislocations (extrinsic grain boundary dislocations - EGBDs) (Varin and Romanowska-Haftek, 1986; Varin, Kurzydlowski and Tangri, 1986). The mechanism of spreading is completely unexplored in advanced engineering materials but it may be very important in high temperature plastic behaviour of advanced intermetallics which are inherently brittle in polycrystalline form (Liu and Sikka, 1986). Any enhancement of their ductility by a certain combination of grain size, temperature and strain rate could lead to improvement in processing characteristics.

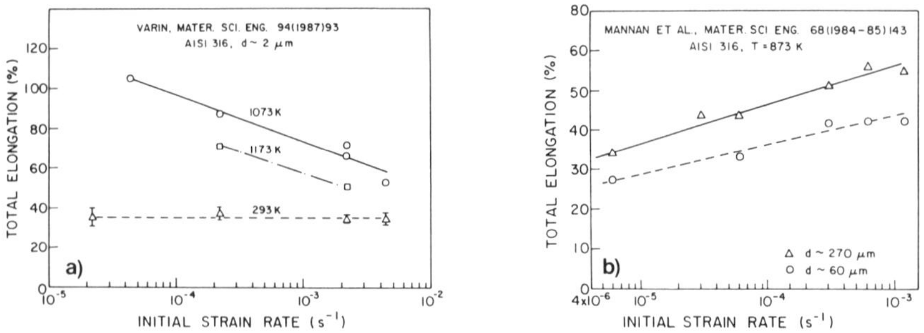

Fig. 6. Effect of grain boundaries, temperature and strain rate on ductility of AISI 316 steel. (a) Ultrafine-grained material (Varin, 1987). (b) Coarse-grained material (Mannan, Samuel and Rodriguez, 1984-85).

INTERFACES IN METAL MATRIX COMPOSITES

Ambient Temperatures

Interfaces in many composite systems contain a brittle layer composed of mostly intermetallic compounds, formed either during fabrication and/or service. A typical example of such a layer, formed in magnesium alloy/carbon fiber composite, is shown in Fig. 7. In the composite systems with the brittle layer strongly adhering to the fibre the strength of the composite decreases with increasing thickness of the brittle layer (Ochiai and Murakami, 1979; Ochiai and Osamura, 1987). This behaviour is

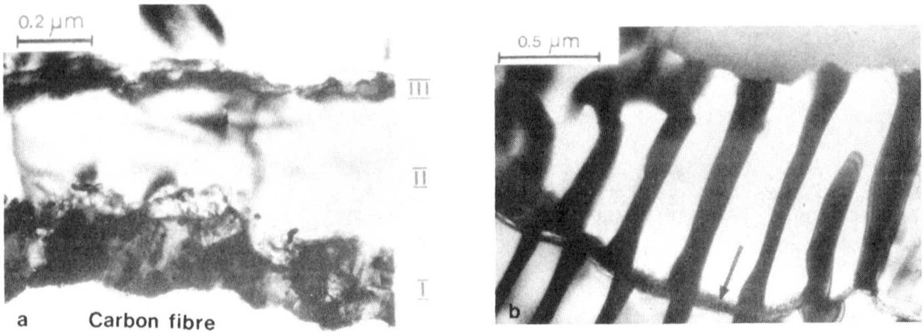

Fig. 7. Interfacial zones in Mg-Al/Ni-coated carbon fiber composite. (a) Triple-layer zone: I - carbides, II - pure Ni, III - microcrystalline or amorphous oxides. (b) Ni coating consumed in the eutectic reaction, the original Ni/Mg boundary indicated by arrow (Hall, 1987).

explained by the formation of microcrack equal to the layer thickness and its expansion into the reinforcement (fiber) by the notch effect (Ochiai and Murakami, 1979; Ochiai and Osamura, 1987). In the composite system where debonding between the reinforcement and the brittle layer takes place easily, it is observed that the apparent interfacial shear strength, τ_i, measured by pull-out test, is inversely proportional to the thickness of the brittle layer, Fig. 8. In this case the microcrack in the layer propagates along the interface layer/reinforcement. Interfaces without a brittle layer can control strength and ductility by a dislocation mechanism

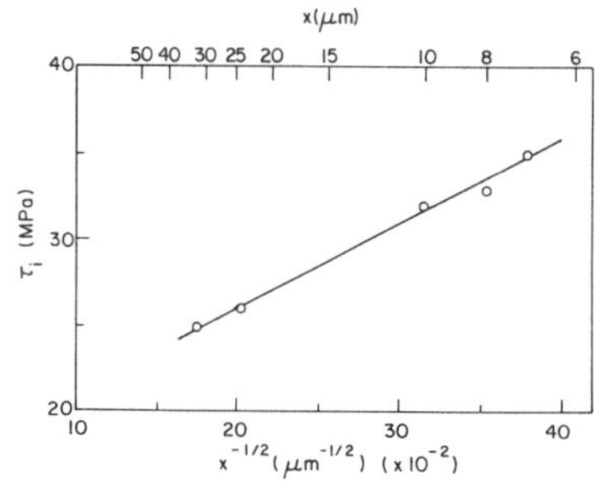

Fig. 8. Correlation between the apparent interfacial shear strength (τ_i) and compound layer thickness (x) in aluminum/stainless steel wire composite (Tiwari, Tiwari and Gopinathan, 1987).

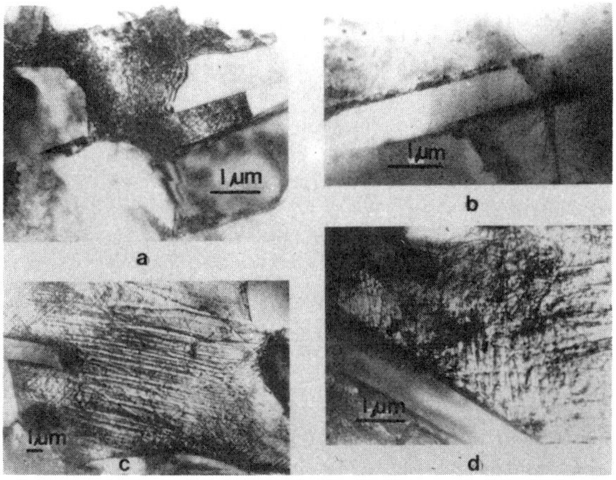

Fig. 9. (a) Transmission electron micrograph of high dislocation density in "as received" 6061 Al/20% SiC composite. (b) Heating to 650 K; most of the dislocation disappeared. (c,d) After cooling to 340 K; dislocations reappeared in the vicinity of the Al/SiC interface (Vogelsang, Arsenault and Fisher, 1986).

similar to the one occurring at grain boundaries. Vogelsang, Arsenault and Fisher (1986) have shown that dislocations can be generated at the Al/SiC whisker or platelet interface by the coefficient of thermal expansion (CTE) mismatch between matrix and reinforcement, Fig. 9. In soft matrices, e.g. annealed alloys of aluminum, the increased matrix dislocation density at the interfaces can significantly increase the strength of composite (Arsenault, 1984; Derby and Walker, 1988).

In discontinuously reinforced composites, e.g. by SiC whiskers, the sharp corner profile at whiskers ends is a site of severe stress concentration causing localized plastic flow at microscopic loads below the matrix yield stress. Increasing plastic strain results in microstructural damage in the form of void initiation and coalescence as shown in Fig. 10. These effects are sensitive to strain rates. As

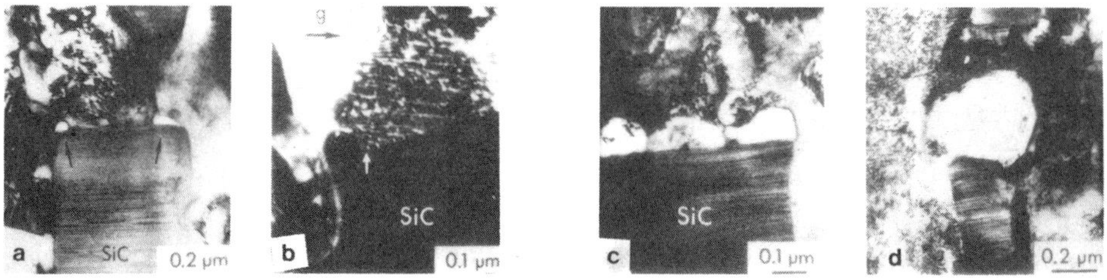

Fig. 10. (a) Electron micrograph (bright field) of void initiation at whisker end in 6061 Al/SiC composite. (b) A weak beam image (3 g) of the same region. Emission of dislocations from the whisker end is clearly visible.
(c) Intermediate stage of voids growth in which voids coalesce.
(d) Advanced stage of growth showing formation of a cavity and extensive matrix deformation (Nutt and Duva, 1986).

reported recently by Pickard and co-workers (1988) the failure strain of 2124 Al/SiC whisker composite increased with strain rate. At low strain rates failure occurred at the matrix/whisker interface. At high strain rates the major failure mechanism was by fracture through the whiskers along the shear bands formed in the matrix, rather than at the interface (Pickard and co-workers, 1988).

It is clear that the interfacial dislocation processes can also control the mechanical behaviour and properties of many MMCs. There is a need to study these processes to assess whether such MMC materials are intrinsically brittle or whether there are means of increasing and optimizing their ductility whilst retaining their high strength.

Elevated Temperatures

The fundamental data on the role of interfaces in the high temperature deformation régime of MMCs is rather scarce. Nevertheless, two effects have been suggested. On the one hand, Nardone and Strife (1987) argued that the attractive interaction of dislocations and the incoherent interface between the SiC whisker and the 2124 Al matrix, can be responsible for an increase in the resisting stress against creep, σ_R, and as a result, an increase in the stress exponent $n = 21$ during creep. On the other hand, a deformation under thermal cycling at low applied stress ($\sigma \simeq 20$ MPa) of Al/SiC whisker composite can lead to superplasticity and tensile elongations up to 300% (Wu and Sherby, 1984). These results are directly attributable to the generation of internal stresses due to the difference in the CTE of the matrix and the SiC whisker and enhanced generation of dislocations at the interfaces as shown in Fig. 9.

ACKNOWLEDGEMENT

This work has been supported by Grant GP6100 from the Natural Sciences and Engineering Research Council of Canada which is gratefully acknowledged.

REFERENCES

Arsenault, R. J. (1984). Mater. Sci. Eng., 64, 171-181.
Briant, C. L., and S. K. Banerji (1978). Int. Met. Rev. No. 4, 164-199.
Chang, K.-M., S. C. Huang, A. I. Taub, G.-M. Chang, and J. W. Morris, Jr. (1987). Metall. Trans. A, 18A, 1819-20.
Derby, B., and J. R. Walker (1988). Scr. metall., 22, 529-532.
Hall, I. W. (1987). Metallography, 20, 237-246.
Hu, H. (1983). Metall. Trans. A, 14A, 85-91.
Kurzydlowski, K. J., R. A. Varin, and W. Zielinski (1984). Acta metall., 32, 71-78.
Kurzydlowski, K. J., R. A. Varin, and K. Tangri (1988). Unpublished results.
Liu, C. T., and V. K. Sikka (May 1986). J. Metals, 38, 19-21.
Mannan, S. L., K. G. Samuel, and P. Rodriguez (1982). Proc. ICSMA-6, Ed. R.G. Gifkins, Pergamon Press, Vol. 2, 637-642.
Mannan, S. L., K. G. Samuel, and P. Rodriguez (1984-85). Mater. Sci. Eng., 68, 143-150.
Nardone, V. C., and J. R. Strife (1987). Metall. Trans. A, 18A, 109-114.
Ochiai, S., and Y. Murakami (1979). J. Mater. Sci., 14, 831-840.
Ochiai, S., and K. Osamura (1987). Metall. Trans. A, 18A, 673-679.
Olefjord, I. (1978). Inst. Met. Rev. No. 4, 149-163.
Pickard, S. M., B. Derby, J. Harding, and M. Taya (1988). Scr. metall., 22, 601-606.
Schulson, E. M., T. P. Weihs, I. Baker, H. J. Frost, and J. A. Horton (1986). Acta metall., 34, 1395-99.

Tiwari, S. N., A. N. Tiwari and V. Gopinathan (1987). J. Mater. Sci., 22, 2680-84.
Tobler, R. L., D. H. Beekman, and R. P. Reed (1983). Austenitic Steels at Low Temperatures, Eds. R. P. Reed and T. Horiuchi, Plenum Press, 135-157.
Varin, R. A., and E. Romanowska-Haftek (1986). Metall. Trans. A, 17A, 1967-75.
Varin, R. A., K. J. Kurzydlowski, and K. Tangri (1986). Mater. Sci. Eng., 80, L5-L9.
Varin, R. A. (1987). Mater. Sci. Eng., 94, 93-107.
Vogelsang, M., R. J. Arsenault, and R. M. Fisher (1986). Metall. Trans. A, 17A, 379-389.
Wu, M. Y., and O. D. Sherby (1984). Scr. metall., 18, 773-776.

INTERFACIAL ZONES IN ALUMINUM MATRIX COMPOSITE CONTAINING Ni_3Al-TYPE INTERMETALLIC RIBBONS

M. P. Metelnick*, R. A. Varin* and Z. Wronski**

*Department of Mechanical Engineering, University of Waterloo, Waterloo, Ontario, Canada N2L 3G1

**Metals Technology Laboratories, CANMET, 568 Booth Street, Ottawa, Ontario, Canada K1A 0G2

ABSTRACT

An aluminum matrix composite containing rapidly solidified $Ni_{75}Al_{23}B_1Zr_1$ (at .%) ribbons has been fabricated by casting at 700, 715, 730 and 875°C. The interfacial zones formed in the composite have been investigated by scanning electron microscopy combined with standardized EDX spot analysis. The sequence of layers in the specimens fabricated at 700, 715, 730 is as follows: AlNi → Al_3Ni_2 → the outer layer between Al_3Ni and eutectic. Composite specimen fabricated at 875°C contains two types of interfacial zones, namely, a single-layer AlNi and a triple-layer zone. The thicknesses of individual layers have been measured as a function of casting temperature.

KEYWORDS

Metal matrix composite; aluminum matrix; $Ni_{75}Al_{23}B_1Zr_1$ (at .%) rapidly solidified intermetallic ribbons; casting; interfacial zones.

INTRODUCTION

Amorphous metallic ribbons for reinforcement of composites are attractive due to their good mechanical strength at ambient temperature, in particular, much better transverse properties than the fiber-shaped reinforcements (Ruutopold, Varin, Wronski, 1988). The development of metal matrix composites reinforced with metallic ribbons is however hindered by the deterioration of the ribbon's strength with increasing temperature owing to the crystallization of its initially amorphous (glassy) structure (Gilman, 1984). A possible way to overcome this problem is to employ metallic ribbons whose mechanical properties would increase with increasing temperature. A typical example of such behaviour is the intermetallic compound Ni_3Al with boron additions to improve its ductility (Aoki, Izumi, 1979; Liu, White, Horton, 1985; Liu, Sikka, 1986). The temperature peak strength of Ni_3Al can be considerably enhanced up to 1000-1200 MPa by certain alloying additions such as Hf and Zr (Ochiai and co-workers, 1986; Bellows, Tien, 1987).

The aim of this work is to study the effect of various casting temperatures on the interfacial zones in aluminum matrix (1xxx series) composite with incorporated rapidly solidified ribbons of the Ni_3Al-type intermetallic.

EXPERIMENTAL

The metal matrix composite was fabricated by casting granulate aluminum, 99.3% purity, with rapidly solidified $Ni_{75}Al_{23}B_1Zr_1$ ribbons (composition given in at .%). Boron was added to ductilize this initially brittle intermetallic. The obtained ribbons were tested for 180 deg. bend ductility as proposed by Taub, Huang and Chang (1984). The ribbons could be bent to the full 180 deg., with no signs of fracturing.

The ribbons were rapidly solidified on a berrylium copper wheel using the free jet single-roller melt spinning method at a surface speed of 28 ms^{-1}. The fabricated ribbons were approximately 2 mm wide, up to 6-8 cm long and about 100 μm thick. The crystallographic structure of the ribbons, analysed by the Debye-Scherrer X-ray diffraction method using nickel filtered copper radiation, was found to correspond to Ni_3Al intermetallic

For casting the ribbons were pre-mixed with the granulated aluminum in the proportion: 60 g of aluminum and 3 g of ribbons. Casting was carried out at four temperatures: 700°C, 715°C, 730°C, 875°C. The slag was removed twice during the fabrication process. The second slag removal was done just prior to incorporation of the ribbons. The ribbons were stirred for about 2-3 minutes after being put into the melt and then air cooled for 20-30 minutes. The molds were graphite coated silica (Salamander A-1).

The ingots were sectioned and then polished using various grades of sand paper and alumina powder, and finally a diamond lapping wheel with 2.5 μm diamond paste. Optical microscopy and scanning electron microscopy (JEOL JSM-840) combined with standardized EDX spot analysis (using Al and Ni standards) were employed to investigate the samples.

RESULTS AND DISCUSSION

In general, the overall structure of samples fabricated at various temperatures appeared similar. Fig. 1 is a scanning electron micrograph of the 700°C sample representing a typical example of the structure observed in "as-cast" samples. It is seen, that the aluminum matrix contains relatively fine second phase particles. The standardized EDX analysis has revealed that these particles contain from 6 to about 19 at .% Ni depending on the casting temperature. These nickel contents

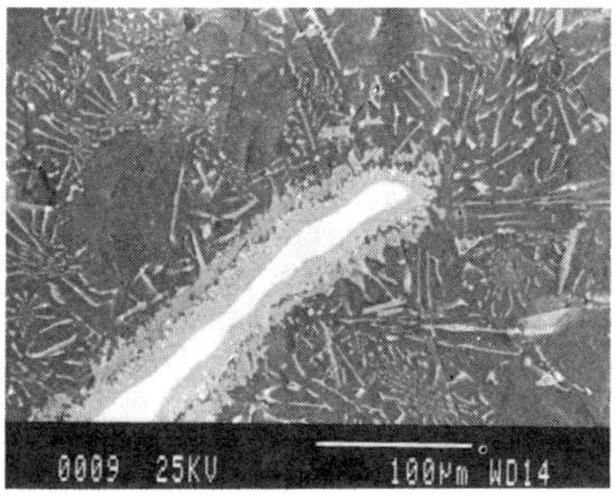

Fig. 1. SEM micrograph of the structure of specimen cast at 700°C.

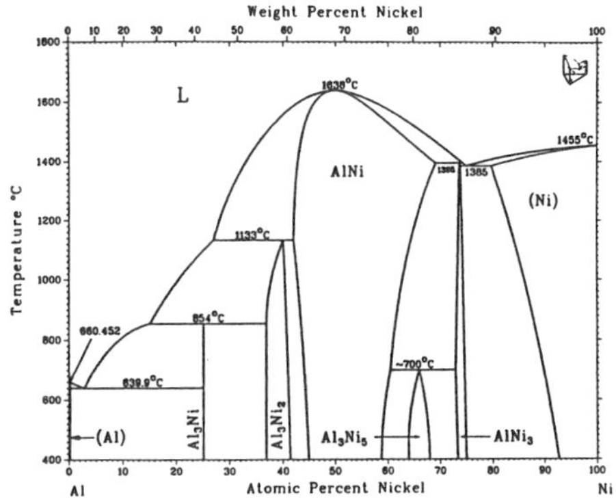

Fig. 2. Al-Ni phase diagram (Massalski, 1986).

correspond to the compositions between the eutectic and Al_3Ni in the Al-Ni phase diagram (Fig. 2). The origin of the formation of these particles is probably related to a partial dissolving of ribbons during stirring in the molten aluminum. This process enriches small areas of aluminum melt with nickel. According to Fig. 2, upon cooling the nickel enriched liquid undergoes the eutectic reaction at 640°C forming eutectic particles in the aluminum matrix. As also seen in Fig. 1, the interfacial zones have developed around the ribbons. The standardized EDX spot analysis was carried out to investigate the chemical composition of these zones (Al and Ni contents). As seen in Fig. 3 the interfacial zones are multi-layer zones and typical locations of the electron beam for each layer are designated in Fig. 3.

According to the results of the EDX analysis the general sequence of the diffusion layers in the interfacial zone can be represented schematically as shown in Fig. 4. In all the specimens, irrespective of the casting temperature, the first layer directly adjacent to the ribbon is always AlNi compound. It is interesting to

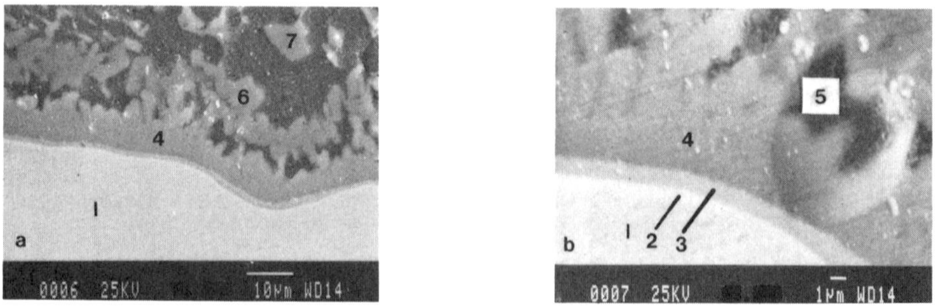

Fig. 3. Locations of the electron beam during the EDX analysis of the 700°C specimen. (a) Lower magnification. (b) High magnification showing the individual layers in the interfacial zone. 1: 75 at .% Ni ($AlNi_3$); 2: 54-59 at .% Ni (AlNi); 3: 32-33 at .% Ni (Al_3Ni_2/Al_3Ni); 4: 22 at .% Ni (Al_3Ni); 5: 7.8 at .% Ni (~Eutectic); 6: 13 at .% Ni (Al_3Ni/Eutectic); 7: 19 at .% Ni (Al_3Ni/Eutectic).

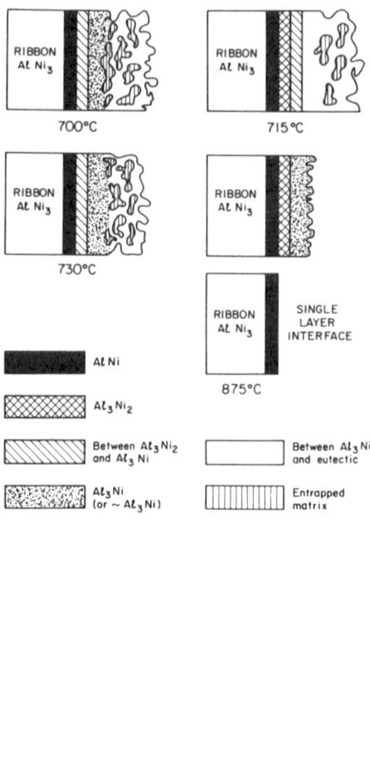

Fig. 4. Schematic sequence of layers observed in the interfacial zones of samples cast at various temperatures.

point out that there is no distinct interphase boundary between the AlNi layer and the ribbon. The second layer, again irrespective of the casting temperature, is always a very distinct layer of either Al_3Ni_2 compound (715 and 875°C samples) or a layer with a composition between Al_3Ni_2 and Al_3Ni (700 and 730°C samples). The 715°C specimen exhibited an extra diffuse layer with a composition between Al_3Ni_2 and Al_3Ni compounds, without distinct interfaces on both sides. The thick and irregular outer layer in the 700, 715 and 730°C specimens has a composition between Al_3Ni and the eutectic. It has no clearly delineated boundary between the area enriched in Al_3Ni and the rest of the layer (Fig. 3). It seems that a part of this layer has been formed by the eutectic reaction at 640°C (Fig. 2).

The sample fabricated at 875°C is different than the others. First, a large portion of interfaces around many ribbons contain only a single layer of AlNi compound (Fig. 5a). Second, some interfacial zones contain a triple-layer zone but the outer, finger-like layer in this zone, corresponds to the Al_3Ni compound and is much thinner than the outer zone in other specimens (Fig. 5b).

The thicknesses of the outer, Al_3Ni_2 and AlNi layers were measured as a function of casting temperature and the average values with standard deviations are plotted in

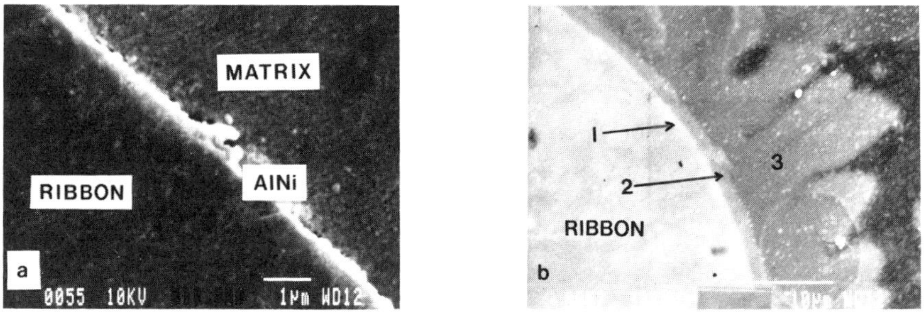

Fig. 5. SEM micrographs of the interfacial zones in the 875°C sample. (a) Single layer of AlNi. (b) Triple-layer zone. 1: 60 at .% Ni (AlNi); 2: 36 at .% Ni (Al_3Ni_2); 3: 22 at .% Ni (Al_3Ni).

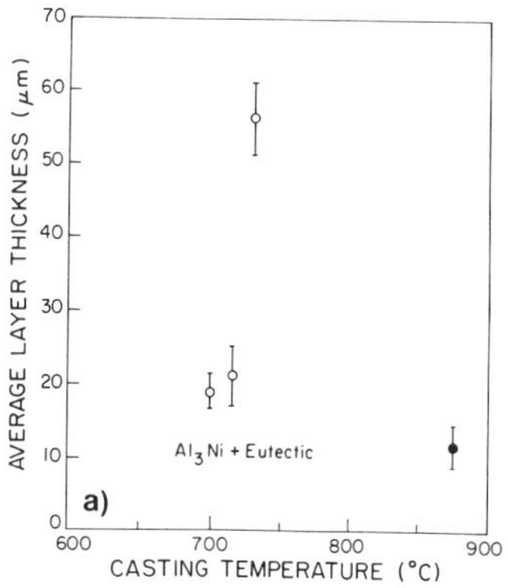

Fig. 6. Average thicknesses of various layers as a function of casting temperature. (a) Outer layer (Al_3Ni + eutectic). (b) Al_3Ni_2. (c) AlNi.

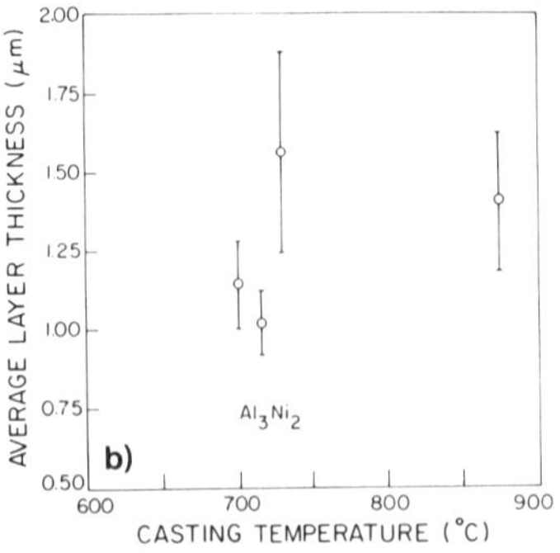

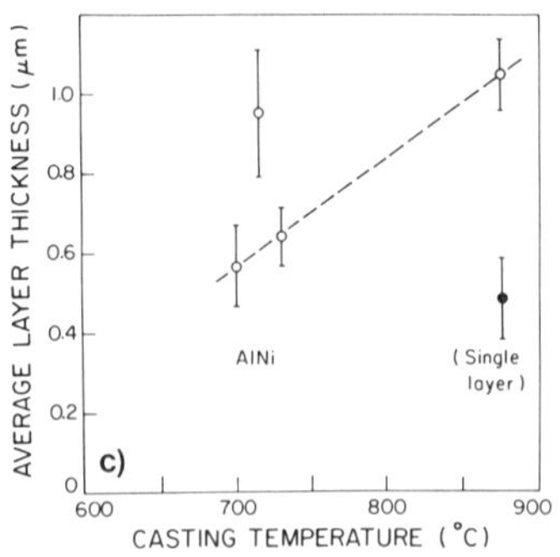

Figs. 6a, b and c, respectively. Ten readings were taken for each average value. As seen in Fig. 6a the overall thickness of the outer layer in specimens fabricated at 700-730°C (labelled "Al_3Ni plus eutectic") increases drastically with the increasing casting temperature. However, the outer layer in the triple-layer interfacial zone of the specimen fabricated at 875°C is the thinnest one. This probably results from the fact that this layer consists of only Al_3Ni compound. Constructing a tie-line at 875°C in the Al-Ni phase diagram (Fig. 2) one can see that the nickel content in the layer of liquid aluminum, being in equilibrium with the Al_3Ni_2 diffusion layer, would be about 17 at .%. Upon cooling between 875°C and 854°C this liquid aluminum layer is slightly depleted with nickel to about 15 at .% and a part of it undergoes a peritectic reaction at 854°C. As a result of this reaction a layer of Al_3Ni may be formed on the previously formed Al_3Ni_2 diffusion layer. According to the lever rule at 854°C the amount of Al_3Ni formed peritectically will be small and correspondingly its layer will be rather thin. If the cooling rate of remaining liquid during subsequent cooling below 854°C was high enough it probably could prevent formation of a continuous eutectic layer on the previously formed thin Al_3Ni layer. The suggested mechanism seems to explain qualitatively the experimental finding that the outer layer in the 875°C sample is the thinnest one. A problem, however, still remains open for discussion why the triple-layer interfacial zone is being formed at some parts of the ribbon/matrix interface but the other interfaces contain only a single-layer zone.

As seen in Fig. 6b within the experimental error the average thickness of Al_3Ni_2 layer seems to be relatively independent of the casting temperature. This result is interesting because it has been observed that the thickness of Al_3Ni_2 layer strongly depends on temperature during one-step aluminizing of nickel (Hickl and Heckel, 1975). Apparently, the Al_3Ni_2 layer formed during casting does not follow the behaviour observed during pack aluminizing.

The average thickness of AlNi layer formed as a part of multi-layer interfacial zones also increases with the increasing casting temperature (Fig. 6c). However, the average thickness of AlNi layer in a single-layer interfacial zone of the 875°C sample is only 0.49 ± 0.1 μm. As mentioned before, there is no distinct boundary between the AlNi layer and the ribbon in the multi-layer interfacial zone as opposed to a single-layer of AlNi in the 875°C sample (Fig. 5). This morphological difference seems to indicate that the formation of AlNi layer in a single-layer zone occurred by a different mechanism than the formation of this layer in a multi-layer zone. The hypothesis can be taken into consideration that in the early stage of the interfacial zone formation at 875°C both AlNi and Al_3Ni_2 were formed according to the phase diagram in Fig. 2. Subsequently, the Al_3Ni_2 layer was dissolved due to diffusion of nickel into growing AlNi layer. Such a mechanism is well-documented during the homogenization of the surface layer obtained by the aluminizing treatment of nickel and its alloys (Hickl and Heckel, 1975). There are, however, two difficulties with this hypothesis. First, in the homogenization treatment of the aluminized layer there is the lack of an aluminum surface input which, in turn, reduces the concentration gradient in the Al_3Ni_2 layer growth (Hickl and Heckel, 1975). In the present work, there is a continuous contact of liquid (and solid) aluminum matrix with the ribbon. Second, a gradual consumption of the Al_3Ni_2 layer by the AlNi layer in the aluminized layer leads to a substantial thickening of the AlNi layer although the total coating thickness increases only slightly (Hickl and Heckel, 1975). However, the AlNi layer in the single-layer interfacial zone is much thinner than its counterpart in the triple-layer interfacial zone of the 875°C sample (Fig. 6c). Therefore, the exact mechanism of the formation of single-layer AlNi interfacial zone in the 875°C specimen is not well understood. Further experiments are now in progress to shed some light on this problem.

CONCLUSIONS

(a) The interfacial zones in $Ni_{75}Al_{23}B_1Zr_1$ ribbon/aluminum matrix composite fabricated by casting at 700, 715 and 730°C consist of several layers with various contents of nickel and aluminum.

(b) The first layer in the above samples, directly adjacent to the ribbon, is a thin layer of AlNi compound without a distinctly delineated interface between the ribbon and the layer. The second layer is a very distinct and also relatively thin layer of Al_3Ni_2 compound (or a layer with a composition between Al_3Ni_2 and Al_3Ni). The outer layer is an irregular and rather thick layer with a composition between Al_3Ni and eutectic.

(c) Composite specimen fabricated at 875°C contains two different types of interfacial zones, namely, a single-layer of AlNi and a triple-layer zone.

(d) The first two layers in the triple-layer zone of the 875°C specimen are exactly the same as those in the 700-730°C specimens, i.e. AlNi and Al_3Ni_2. The irregular outer layer corresponds very close to the Al_3Ni composition.

(e) The thickness of AlNi layer increases continuously with the increasing casting temperature.

(f) Within the experimental error the thickness of Al_3Ni_2 layer seems to be relatively independent of the casting temperature.

(g) The thickness of the outer layer in the specimens fabricated at 700-730°C (Al_3Ni plus eutectic) increases with casting temperature. However, the outer layer in the 875°C specimen (Al_3Ni) is much thinner than the others.

ACKNOWLEDGEMENTS

This work has been supported by Grant GP6100 from the Natural Sciences and Engineering Research Council of Canada which is gratefully acknowledged.

REFERENCES

Aoki, K., and O. Izumi (1979). J. Jap. Inst. Met., 43, 1190.
Bellows, R. S., and J. K. Tien (1987). Scr. metall., 21, 1659.
Gilman, J. J. (1984). Metallic Glass Materials, Industrial Materials Science and Engineering, Marcel Dekker, New York, p. 1.
Hickl, A. J., and R. W. Heckel (1975). Metall. Trans. A, 6A, 431.
Liu, C. T., C. L. White, and J. A. Horton (1985). Acta metall., 33, 213.
Liu, C. T., and V. K. Sikka (May 1986). J. Metals, 38, 19.
Massalski, T. B. (Ed.) (1986). Binary Alloys Phase Diagrams, Vol. 1, American Society for Metals, Metals Park, p. 140.

Ochiai, S., Y. Mishima, M. Yodogawa, and T. Suzuki (1986). Trans. JIM, 27, 32.
Ruutopold, A., R. A. Varin, and Z. Wronski (1988). Mater. Sci. Eng., 98, 547.
Taub, A. I., S. C. Huang, and K. M. Chang (1984). Metall. Trans. A, 15A, 399.

METAL NON-METAL INTERFACES

J. Shirokoff, J.S. Sprenger and U. Erb

Department of Metallurgical Engineering
Queen's University, Kingston, Ontario K7L 3N6 Canada

ABSTRACT

Recent results of studies to detect interfaces of low energy in metal/non-metal systems using the sphere plate technique are reviewed. The findings are discussed in terms of the present models for such interfaces.

KEYWORDS

Orientation relationships, preferred orientations, epitaxy, interfacial energy.

INTRODUCTION

Interfacial studies are important in understanding structure-property relationships of materials. For crystalline/crystalline systems, the structure and properties are strongly controlled by orientation relationships between the two crystals and their corresponding defect structure at the interface, (Chadwick and Smith, 1976; Gleiter and Chalmers, 1972). In the past, several geometric models to describe special interfaces (i.e. interfaces showing properties that are different from those observed for general interfaces) have been proposed and many experimental observations could be explained by these models. However, as more and more experimental data become available, the general applicability of such geometric models is now being questioned. For example, in the case of interfacial energy, Sutton and Balluffi (1987) suggest that when certain principles, eg., low reciprocal volume density of coincidence site lattice (CSL) sites (Σ) (Kronberg and Wilson, 1959), high planar CSL site density (Γ) (Brandon and co-workers, 1964), high Γ at constant interplanar spacing/or large interplanar spacing (Wolf, 1985), and high density of lock-in rows of atoms (Fecht and Gleiter, 1985) are applied to experimental results there is general disagreement that interfacial energy is directly related to these parameters. It is now generally felt that variations of the interfacial energy as a function of crystal misorientation can only be fully understood if other parameters such as the electronic structure, details of the bonding at the interface, etc. are taken into account.

Interfaces in amorphous/crystalline systems are less well understood than interfaces in crystalline/crystalline systems. Most of the models for amorphous/crystalline interfaces are derived from the study of liquid/solid interfaces and concern themselves mainly with the atomistic structure at the interface and a few monolayers on either side. Jackson's (1958) theory deals with a nearest neighbour bond model which can predict faceting for atomically smooth interfaces or non-faceting for atomically rough interfaces. More recently, high resolution electron microscopy studies have confirmed Jackson's theory for solid/liquid interfaces in InSb (Saka and co-workers, 1985). Cahn's theory (1960) deals mainly with the thermodynamics of solid/liquid interfaces while Monte Carlo simulations (Bonnisent, Gauthier and Finney, 1979) are concerned with the atomic structure at different distances from the interface.

With respect to the orientation-dependence of solid/liquid interfacial energies, little information is available so far. For the liquid/solid fcc metal system it is known that interfaces having the (111) or (100) planes parallel to the interface have somewhat lower energies than interfaces having other crystal faces parallel to the interface (Basterfield, Miller and Weatherley, 1970). However, more work is needed in this field, especially for the case of "solid" amorphous/crystalline systems.

The sphere-plate technique is one of several experimental techniques used to study the structure and properties of interfaces. Although criticized on several occassions (Goodhew and Allen, 1977; Sutton and Balluffi, 1987), this technique has been an extremely powerful tool over the past 15 years to detect special interfaces of low interfacial energy in a variety of systems. These include grain boundaries in pure metals and alloys (Herrmann and co-workers, 1975) and interphase interfaces in metal/non-metal systems (Erb, Abel and Gleiter, 1982; Fecht and Gleiter, 1985; Shirokoff and Erb, 1986; Shirokoff and Erb, 1987; Shirokoff, Hui and Erb, 1988; Shirokoff and Erb, 1988). The purpose of the present paper is to review some of our recent findings on crystalline metal/crystalline non-metal and crystalline metal/amorphous non-metal interfaces using the sphere-plate technique. These results will be discussed in view of the models mentioned above.

EXPERIMENTAL DETAILS ON THE SPHERE-PLATE TECHNIQUE

The sphere plate technique involves the study of the rotation of tiny single crystal spheres sintered onto a substrate into orientation relationships of low interfacial energy. Since its first introduction (Herrmann and co-workers, 1975) several modifications have been made to the sphere plate technique (Erb, Abel and Gleiter, 1982; Shirokoff and Erb, 1986). The results reported in the present study are based on the latest modificaton. In these experiments sphere plate samples are obtained by annealing vapour deposited thin films on a substrate. During annealing these films first break up into islands before forming single crystal spheres (Fig. 1). Upon further annealing, these spheres rotate into orientation relationships of low interfacial energy which are detected by x-ray diffraction techniques.

In this paper we present results of sphere-plate experiments for the following systems:

- Ag on NaCl single crystals
- Ag on Si single crystals
- fcc metals on amorphous SiO_2
- the complex interfacial system $Ag/SiO_2/Si$

INTERFACES IN THE Ag/NaCl SYSTEM

The first results of sphere-plate experiments for the Ag/NaCl [100] system were published in 1982 (Erb, Abel and Gleiter, 1982). Two orientation relationships of low interfacial energy were detected one of which is the cube/cube orientation, i.e. an orientation relationship for which the three [100] directions of the Ag lattice are parallel to the [100] directions of the NaCl lattice (epitaxial orientation). The second orientation of low energy was

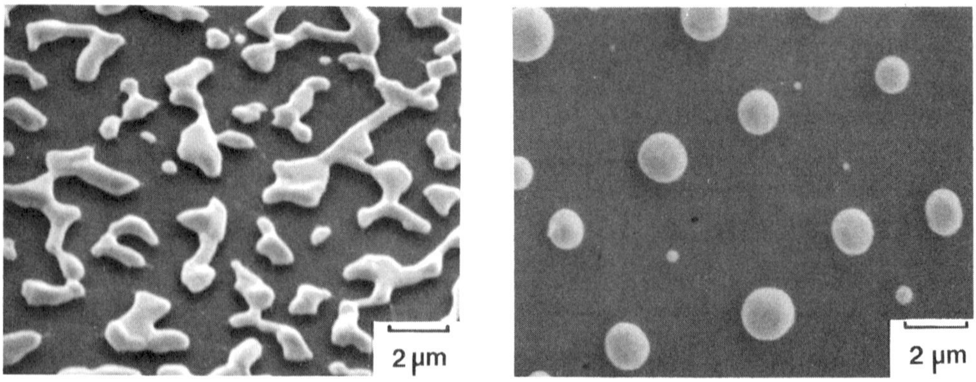

Fig. 1: Sphere formation of Ag on amorphous silica.
Left: intermediate stage, right: final stage.

described by a 45° rotation of the Ag lattice with respect to the NaCl lattice about the common [001] direction (i.e. Ag(100)//NaCl(100) and Ag[001]//NaCl[011]. When analysing the atomic arrangement at the interface for both orientation relationships, the mismatch for the former case is in the order of 30% while it is only about 3% for the latter orientation. Thus, it was concluded, the atomic matching at the interface seems to play a minor role in establishing low energy interfacial orientation relationships for this system.

The two preferred orientations of Ag on NaCl found by Erb, Abel and Gleiter (1982) represented only two of the ten possible preferred orientation observed previously by Matthews (1966) in thin film growth experiments. In order to study this discrepancy more in detail, sphere-plate experiments were repeated (Shirokoff and Erb, 1986) using the modified sphere plate technique described in the previous chapter which essentially yields interfaces which are cleaner than those produced in Erb, Abel and Gleiter's (1988) study. Figure 2 shows the (111) pole figures of silver (top half) and sodium chloride (bottom half) for a fully annealed sphere plate sample. The pole figure is analyzed in terms of parallel planes and directions for both phases (see table 1) and was also checked for twin orientations for each preferred orientation of silver since many of the spheres are expected to be twinned owing to the low stacking fault energy of silver. This analysis yielded a total of 12 cusps (orientations A, B, and C in table 1) in the interfacial energy vs misorientation curve for Ag/NaCl interfaces at 700°C. Ten of these orientation relationships (orientatons A and B) were also found by Matthews (1966) in evaporated films before coalescence of individual particles. Orientations A correspond to the same two cusps detected by Erb, Abel and Gleiter (1982).

In addition to Matthew's work (1966) two new preferred orientations were found which correspond to orientations C in table 1. This has been explained by way of the superior statistics inherent in the sphere-plate technique when compared to TEM studies used by Matthews (Shirokoff and Erb, 1986).

In summary, the major findings for the Ag//NaCl(100) interfaces are as follows:

i) When using the sphere-plate technique to detect cusps in the interfacial energy versus misorientation curve, careful contamination-free sample preparation is essential in order not to obscure cusps which are sensitive to impurity atoms.

ii) At 700°C, there is a total of 12 cusps in the interfacial energy versus misorientation curve.

iii) All reported preferred orientations in film growth studies can be explained in terms of low interfacial energy.

iv) Atomic matching (i.e. the geometry of the interface) is of minor importance for low energy interfaces.

TABLE 1: Preferred Orientation Relationships Between Ag and NaCl (100)

Orientation Relationship	Parallel planes	Parallel directions	Intensity in x-ray pole figure	Orientation relationship in the work by J.W. Matthews
A	Ag(100)//NaCl(100)	Ag[010]//NaCl[010]	strong	present
	Ag(100)//NaCl(100)	Ag[010]//NaCl[011]	very strong	present
B	Ag(111)//NaCl(100)	Ag[$\bar{1}$10]//NaCl[01$\bar{1}$]	strong	present
	Ag(111)//NaCl(100)	Ag[$\bar{1}$10]//NaCl[010]	strong	present
C	Ag(110)//NaCl(100)	Ag[$\bar{1}$10]//NaCl[011]	strong	not present

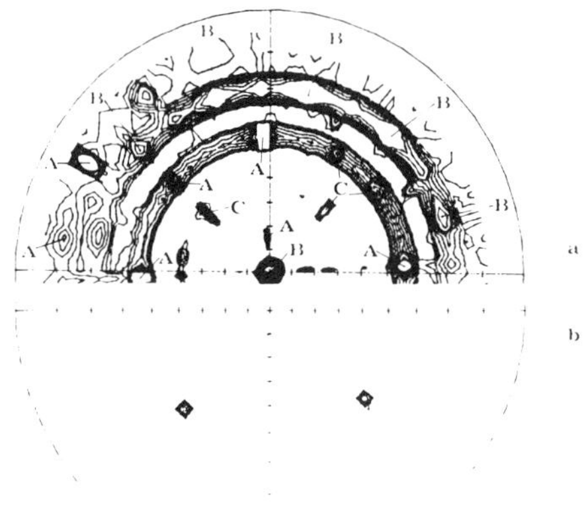

Fig. 2: (111) pole figures for a) Ag, b) NaCl

INTERFACES IN THE Ag/Si SYSTEM

As for the case of the Ag/NaCl, preferred orientations for Ag/Si(100) and Ag/Si(111) systems were measured at 700°C (Shirokoff, Hui and Erb, 1988; Shirokoff and Erb, 1988). The orientation relationships of low interfacial energies are listed in tables 2 and 3. The six preferred orientations may be classified into two categories:

a) A,B and F, for which close and second close packed planes of the Ag spheres are parallel to the substrate surface without any directional alignment, and

b) C,D and E for which the spheres have also well defined directional alignment.

TABLE 2: Preferred Orientation Relationships Between Ag and Si(100)

Orientation relationship	Parallel planes	Parallel directions	Intensity in x-ray pole figure	Orientation relationship in the work by LeGoues et al.
A	Ag(111)//Si(100)	no preferred directional orientation	strong	not present
B	Ag(100)//Si(100)	no preferred directional orientation	weak	not present
C	Ag(100)//Si(100)	Ag[011]//Si[011]	strong	present

TABLE 3: Preferred Orientation Relationships Between Ag and Si(111)

Orientation relationship	Parallel planes	Parallel directions	Intensity in x-ray pole figure	Orientation relationship in the work by LeGoues et al.
D	Ag(111)//Si(111)	Ag[110]//Si[110]	strong	present
E	Ag(111)//Si(111)	Ag[110]//Si[114]	weak	present
F	Ag(111)//Si(111)	no preferred directional orientation	weak	not present

While orientatons A, B and F are probably due to the native oxide layer on the Si crystals (Shirokoff and Erb, 1988) (see below) orientations C and D correspond to the epitaxial orientations previously reported (LeGoues, Liehr, Renier and Krakow, 1988) for thin film growth studies. As far as orientation E is concerned, it is presently unknown whether this is a separate preferred orientation or the result of twinned particles in orientation D.

FCC METAL - AMORPHOUS SILICON DIOXIDE

The study of orientation relationships for fcc silver on amorphous silicon dioxide (silica) using the sphere-plate technique has revealed two cusps in the interfacial energy vs misorientation curve (Shirokoff and Erb, 1987). These are identified as the two intensity maxima labelled A and B on the pole figure in Fig. 3. They correspond to orientation relationships for which close packed (111) and second close packed (100) planes of the silver spheres lie parallel to the interface without any preferred rotational alignment with respect to the SiO_2 substrate. These are the same preferred orientations as previously reported for silver films as grown from the vapour phase on glass substrates (Gittis and Dobrev, 1985). Preferred orientations for various other face centred cubic metals on amorphous SiO_2 are found in table 4 (Sprenger, Shirokoff and Erb, 1988).

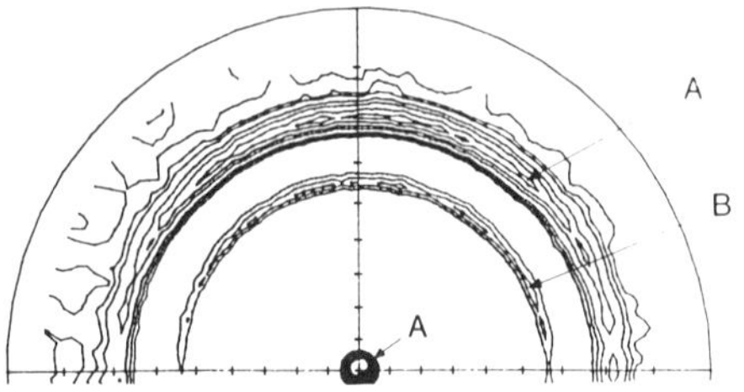

Fig. 3: (111) pole figure of Ag on amorphous silica.

TABLE 4: Preferred Orientations of Various Face Centered Cubic Metals on Amorphous SiO_2

System	(111)	(200)
Ag/Silica	strong	weak
Au/Silica	strong	----
Cu/Silica	strong	----
Ni/Silica	weak	strong
Pb/Silica	strong	weak

In summary, all fcc metal/amorphous SiO_2 systems studied so far seem to follow the general trend that close packed or second close packed planes are parallel to the substrate surface as predicted by Basterfield, Miller and Weatherly (1970). However, when comparing different fcc metals, it appears that factors other than the atomic packing density of the crystalline phase must be considered to predict which of the two orientations actually occurs in any given system.

MORE COMPLEX INTERFACES: SILVER-SILICON DIOXIDE-SILICON

From the studies mentioned earlier for Ag on Si interfaces, three preferred orientations, namely A, B and F correspond to orientations for which close packed planes and second close packed planes of the Ag phase lie parallel to the substrate without any preferred directional alignment. Since it is known from the study of Ag on amorphous silica that Ag prefers the (111) and to a lesser extent the (100) orientation without any preferred rotational alignment, it is believed that these orientations A, B and F are due to the thin interfacial layer of (native) oxide which separates the Ag phase from the Si single crystal substrate. This of course presumes that orientation relationships C, D and G, which are not observed on glass substrates, must be due to the influence of the silicon substrate through the amorphous interfacial layer. In fact the epitaxial orientation, C, was observed for even thicker (180 Å) amorphous interfacial layers on the Ag/Si(100) system. Figure 4 illustrates the x-ray

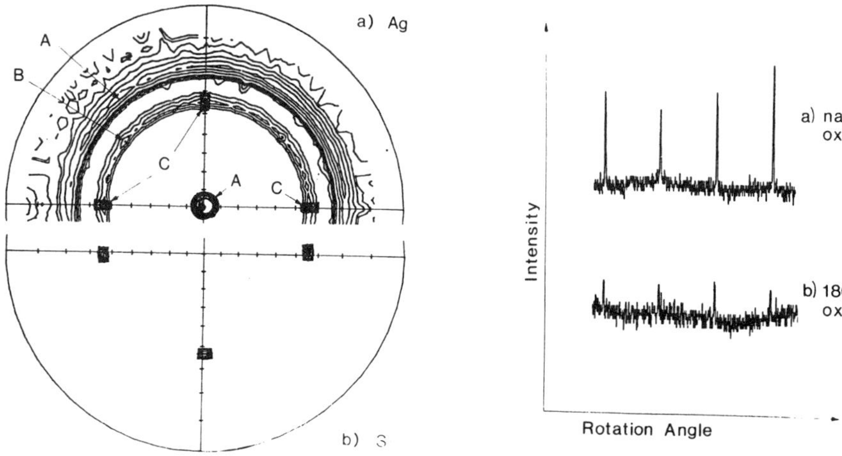

Fig. 4: Left: (111) pole figure for a) Ag and b) Si. Right: X-ray intensity scans for Ag at constant sample inclination of 54°.

intensity (i.e. number of spheres which rotate into orientation C) of the epitaxy orientation for native oxide and built-up oxide layers on silicon (Shirokoff, Hui and Erb, 1988). Although fewer spheres rotate into the epitaxial orientation on built-up oxides, this orientation still occurs indicating that information transfer from the Si substrate to the Ag spheres exists even through fairly thick oxide layers. This has been discussed in terms of long range forces. However, at the present time it is unknown whether this long range information transfer is due to electrical forces or stresses in the interlayer, and this will be the subject of future studies.

CONCLUSIONS

Cusps were found in the interfacial energy versus misorientation curve for both crystalline/crystalline and crystalline/non-crystalline interfaces.
When analyzing the orientation relationships for crystalline/crystalline interfaces it appears that the crystallographic geometry at the interface is of minor importance. However, the atomic packing density seems to play an important role for preferred orientations in amorphous/crystalline interfaces.
When amorphous interfacial layers separated two crystalline phases, the resulting orientation relationships which develop after annealing give further evidence that long range forces play an important role in preferred orientations through amorphous interlayers.

ACKNOWLEDGEMENT

The financial support of the Natural Sciences and Engineering Research Council of Canada is greatfully acknowledged. One of the authors (J.S.) is the recipient of an OGS scholarship and a Queen's University School of Graduate Studies and Research CTG.

REFERENCES

Basterfield, J., W.A. Miller and G.C. Weatherley (1970). Anisotropy of interfacial free energy in solid-fluid and solid-solid Systems. Can. Metall. Quart., 8, 131-143.

Bonissent, A., E. Gauthier and J.L. Finney (1979). Monte Carlo simulation of the crystal-melt interface of a Lennard-Jones substance. Phil. Mag. B, 39, (1), 49-59.

Brandon, D.G., B. Ralph, S. Ranganathan and M.S. Wald (1964). A field ion microscope study of atomic configuration at grain boundaries. Acta Metall., 12, 813-821.

Cahn, J.W. (1960). Theory of crystal growth and interface motion in crystalline materials. Acta Metall., 8, 554-562.

Chadwick, G.A. and D.A. Smith (1982). Grain Boundary Structure and Properties. Academic Press, New York.

Erb, U., W. Abel and H. Gleiter (1982). The significance of atomic matching for the structure of interphase boundaries. Scripta Metall., 16, 1317-1319.

Fecht, H.J., and H. Gleiter (1985). A lock-in model for the atomic structure of interphase boundaries between metals and ionic crystals. Acta Metall., 33, (4), 557-562.

Gittis, A., and D. Dobrev (1985). Development of textures during the growth of silver films condensed in vacuum on glass. Thin Solid Films, 130, 335-340.

Gleiter, H., and B. Chalmers (1972). Structure of Grain Boundaries. In B. Chalmers, J.W. Christian and T.B. Massalski (Eds.), High Angle Grain Boundaries. Prog. Mat. Sci., 16, Pergamon Press, Oxford, Chap. 1, pp. 1-12.

Goodhew, P.J. and R.M. Allen (1977). Comments on: Low energy grain boundaries in metals. Scripta Metall., 11, 37-39.

Herrmann, G., H. Sautter, G. Baro and H. Gleiter (1975). Measurements of grain boundary energy by sintering experiments. In J.L. Walter, J.H. Westbrook, and D.A. Woodford (Eds.), Grain Boundaries in Engineering Materials. Proc. 4th Bolton Landing Conf. June 1974, Claitor's Publ. Div., Baton Rouge, LA., pp. 43-54.
Jackson, K.A. (1958), Mechanism of growth. In Liquid Metals and Solidification. ASM, Cleveland, OH., pp. 174-186.
Kronberg, M.L. and F.H. Wilson (1949). Secondary recrystallization in copper. Trans. Am. Inst. Min. Engrs., 185, 501-514.
LeGoues, F.K., M. Liehr, M. Renier and W. Krakow (1988). Microstructure of epitaxial Ag/Si(111) and Ag(Si(100) interfaces. Phil. Mag. B., 57, (2), 179-189.
Matthews, J.W. (1966). Growth of Face-centered-cubic metals on sodium chloride substrates. J. Vac. Sci. Techn., 3, (3), 133-145.
Saka, H., A. Sakai, T. Kamino and T. Imura (1985). In situ HREM observation of solid-liquid interface. Phil. Mag. A., 52, (3), L29-L32.
Shirokoff, J., and U. Erb (1986). Detection of the complete set of preferred orientations of silver on sodium chloride. Scripta Metall., 20, 1607-1611.
Shirokoff, J., and U. Erb (1987). Measurement of preferred orientations in the silver-silica interface. Thin Solid Films., 151, (1), 65-70.
Shirokoff, J., and U. Erb (1988). The role of interfacial energy in the development of preferred orientation of silver on silicon. Submitted.
Shirokoff, J., K.C. Hui, and U. Erb (1988). The study of long range forces in epitaxy systems: the Ag/Si interface. In S. Sass and R. Raj (Eds.), Interface Science and Engineering '87. J. Physique-Colloq., Paris, in press.
Sprenger, J.S., J. Shirokoff and U. Erb (1988). To be published.
Sutton, A.P., and R.W. Balluffi (1987). Overview No. 61: On geometric criteria for low interfacial energy. Acta Metall., 35, (9), 2177-2201.
Wolf, D. (1985). On the relationship between symmetric tilt, twist, "special", and "favoured" grain boundaries. J. Physique, Paris, 46, (C4), 197-211.

FUNDAMENTAL CONCEPTS OF REINFORCED POLYMERS

M.R. PIGGOTT

ADVANCED COMPOSITES PHYSICS & CHEMISTRY GROUP
DEPARTMENT OF CHEMICAL ENGINEERING
UNIVERSITY OF TORONTO
TORONTO, ONTARIO M5S 1A4

Abstract

The properties and uses of fibre composites are briefly reviewed. The Strength and Young's modulus of these materials obey relatively simple relationships when the fibres are aligned and continuous, and modified equations can be used for strength when the fibres are short and not all in the same direction. The properties of laminates made from aligned fibre laminae are particularly advantageous when their light weight and corrosion resistance are taken into account, and this has made them the material of choice for aerospace, and sports applications. These high performance composites are only a small part of the fibre composites business. Lower performance composites, while not necessarily cheap, are cost effective due to ease of molding, reduction in numbers of parts needed etc., and are widely used in automobiles and domestic products, etc.

Keywords:

Fibre composites; strength; stiffness; toughness

1. INTRODUCTION

1.1 Brief History

Fibre reinforced polymers are a relatively recent development, although man has used nature's cellulose fibre reinforced lignin (i.e. wood, and also cane) since time immemorial. In the early part of this century wood fibre and cotton fibre reinforced thermosets (e.g. phenol formaldehyde) were invented, but it was not until the early 1940's that reinforced polymers were developed which could withstand significant loads.

This development required the production of suitable reinforcement. Griffith in 1921 [1] showed that glass was weak because it had flaws in it, which could easily propagate because of the small work of fracture of the glass. It was difficult to toughen the glass, but relatively easy to ensure the almost total absence of a significant flaws, by making glass fibres having small diameters (e.g. 10 μm). They needed to be carefully protected, but this could be effected by coating them with a relatively soft organic material. Thus it was possible to produce fibres with a Young's modulus about equal to aluminum, and about the same density, but having strengths of 2 GPa or more.

During the second world war the need for an insulating but strong material arose, for use as a radar dome for military aircraft. It was found that combining glass fibres with a thermosetting polyester resin produced a material with the desired properties, was easily moldable into complex shapes, was light, and was corosion resistant. Thus the modern reinforced plastics industry was born.

Glass reinforced polyesters quickly found use in boats and sports equipment, were gradually accepted by the chemical industry for building chemical plant because of their excellent resistance to acids and other highly corrosive materials, and have been making major inroads into the transport industry, for trucks, trains, and small aircraft, but still have not affected the automobile industry very greatly. They are at present mainly used for sports car bodies, and the smaller parts of mass-production cars, but are expected soon to make steel obsolescent for leaf springs, hoods, doors, and wheels, etc.

1.2 The Fibres

It was quickly realized that fibre composites would have an even greater impact on the structural materials industries if lighter fibres with higher Young's moduli could be manufactured. This led to the development of boron fibres (see table 1 for properties) but these have proved to be too expensive, and have been almost totally eclipsed by carbon. Two typical carbons are included in table 1, but as fig. 1 shows, there are presently at least 49 types of carbon fibre on the market. The limit of properties at the moment is roughly given by the line drawn in fig.1, which fits the equation

$$\sigma_{fu} E_f = 2000\ (\mathrm{GPa})^2 \qquad (1)$$

Here σ_{fu} is the fibre strength and E_f is its Young's modulus.

A wide range of other fibres have been developed, including polmers such as Kevlar and highly orientated polyethylene; also silicon carbide and alumina. Table 1 has details of these. Note the low densities of the fibres, especially the carbons and polymers.

Another possible reinforcement in fibre form is the single crystal whisker. These are typically 1 μm diameter and 1 mm long. So far, however, it has not proved possible to make reinforced polymers with them economically, despite their potential strengths (up to 21 GPa) and moduli (up to 2250 GPa)

1.3 The Polymers

Developments in polymers have followed closely on the progress made with fibres. A wide range of thermosetting polyesters is now

TABLE 1
Properties of Selected Fibres and Wiskers

Fibre	Diamter (μm)	Young's modulus (GPa)	Strength (GPa)	Density (Mgm^{-3})	Maximum Temperature (^oC)
Kevlar[1]	12	130	3.6	1.45	200
Polyethylene[1]	38	119	2.6	0.87	100
E Glass	15[2]	72	3.4	2.54	550
Boron	100	420	3.5	2.60	700[3]
HM Carbon[4]	6	400	2.6	1.82	2500[3]
HS Carbon[4]	8	250	4.4	1.80	2000[3]
Alumina/Zinconia	20	390	2.1	4.2	1400
Silicon carbide	10	470	4.8	2.8	1300
Silicone nitride	10	300	2.5	2.5	1200
Al_2O_3[5]	1	2250[6]	15	4.0	1200
Sic[5]	1	840[6]	21	3.2	1600

Notes

1. 24% (Kevlar) and 47% (polyethylene); higher modulus versions available.
2. diameters ranging from about 8 to 22 μm available.
3. 500°C in oxidising atmosphere.
4. wide range of carbons available; see fig. 1.
5. whiskers
6. maximum value, with most favourable crystal orientation.

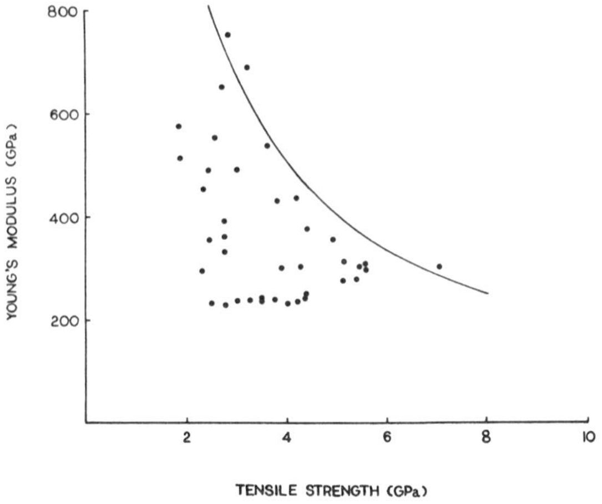

Fig. 1. Strength vs Young's Modulus of carbon fibres. Line is strength x modulus = 2000 $(GPa)^2$.

available, ranging from soft and rubbery to hard and brittle. Flame retardments and other fillers are often included.

For higher performance, epoxies are now used. Again, a wide range of types are available, with special emphasis being put on developing tough ones, by the use of very finely divided elastomers. For higher temperatures, polyimides are used. These extend possible use temperatures up to about 300°C.

Soon after development of the first reinforced thermosets it was realized that thermoplastics, such as nylon, could be mixed with short glass fibres and extruded and molded to produce a wide range of products having complex shapes, good surface finish, etc. The short fibres, made shorter during the extrusion, did not increase the strength very greatly (and sometimes not at all) but did increase the Young's modulus of the material, and decrease its thermal expansion coefficient, making it more dimensionally stable over a range of temperatures, and also made it less easily deformed at elevated temperatures.

Recently, special high temperature thermoplastics have been developed which make it possible to build fibre composite structures which can be used at temperatures above 300°C. These include polyetheretherketone (PEEK) and polyetherketone ketone (PEKK), polyether sulphone, (PES) etc. Although it is more difficult to make fibre reinforced structures with them than for example with epoxies, because of the high temperatures and pressures needed for fabrication, they have the advantage of being very tough, as well as having mechanical properties which are as good as the epoxies

1.4 Forms of Material Used

Fibres and resins can be combined while a structure is being made, such as in the various hand lay up processes used for making boats, etc. In these the fibres, often in woven form, are laid on or in a mold, and the liquid resin is brushed on, and worked in with a roller. Large and small storage and pressure vessels and pipes are made by filament winding, with the fibres passing through a resin bath before being wound out the mold. If large parts are being produced, it is common to use a room temperature curing resin. Continuous lengths of special shapes, e.g. I beams, can be produced by pultrusion, i.e. by combining fibrs and liquid resin, and then pulling them through a heated die to harden the resin. These processes use continous filament, and sometimes woven cloth, and more rarely, random chopped glass mat.

Alternatively, fibres can be combined with the resin prior to use. For example, with thermoplastics like nylon, a chopped fibre-resin mixture is extruded, and then cut into pellets for molding ,later. With the cheaper thermosets, such as the polyesters, short fibres are combined with the partly cured polymer, in the form of sheet (sheet molding compounds, or SMC's) on in bulk form (BMC's). The molding compound must be kept cool to prevent the resin hardening before molding to final shape. These compounds, involving relatively short fibres, in random orientation, generally produce low performance products, since the amount of fibres in the mixture is generally not very great (5-20%) and the shortness of the fibre results in inefficient use of its strength and modulus (these are typically 30-300 times those of the polymer)

Prepregging is a method of combining fibres with resin to produce a material with potential for high performance. Prepregs use either woven fibres, or continuous fibre strands. If a thermoset is being used, the fibre web is coated with the liquid resin mixture, and then partly cured so that it is still drapeable. Usually the resin has been brought to the gel stage and so will not flow away. With thermoplastics, the resin is finely divided, and held together by slight fusing or some temporary binder. Again the material is soft and flexible so that it can be draped over a mold.

1.5 Uses of Reinforced Polymers

With the advent of higher performancee fibres, these composites have been employed in aircraft and space craft and sports equipment which make very high demands on a material in terms of specific mechanical properties.

One of the first major uses of reinforced polymers was in boat building; small boats are almost all made with glass reinforced polyesters at present. Ease of molding and resistance to corrosion were very important reasons for this. Use in chemical plant for large vessels as well as small quickly followed.

Uses in the automobile have not developed so quickly, except for sports cars (some are almost wholly made of fibreglass) because of the relatively long time needed for the resin in large moldings to harden. However, with the development of reinforced reaction injection moldings (RRIM) it is now normal to make smaller parts of cars (e.g. the hood) with reinforced plastics. The parts generally weigh less than the corresponding metal parts, and the total number of parts needed is much reduced because of the complex shapes that can be accurately and reproducibly molded. Other sections of the ground transportation industry make much use of glass reinforced e.g. for the cabs of trucks, and for railway vehicles etc.

Glass reinforced thermosets and thermoplastics are much used in household items such as refrigerators, bath tubs and showers, and cases for drills, typewriters, micromputers, etc. They generally reduce the cost of a part without loss of properties, feel warm to the touch, etc.

The advent of the higher performance fibres led to the use of composites in aerospace, fig. 2. Modern military aircraft contain large amounts of carbon and Kevlar fibre. Civilian aircraft presently use them for fairings, but it is likely that they will be used for load bearing structures in the near future. The main strucures of helicopters nowadays are made almost wholly with reinforced plastics, again usually with Kevlar and carbon fibres. The peculiar advantages of light weight and high modulus make it advantageous to use carbon fibre reinforced polymers in high speed machinery such as the weaving loom.

Medical uses are also important. Splints can be designed so that their properties mimic bone, by choice of the amount of fibre to match the Young's modulus of the bone. Their light weight and warm feel also make their use for prostheses widespread. They are also important as supports for x-ray examination; such use of carbon ensures that the x-ray absorbtion of the support is very small. In these applications the high cost of the high performance fibre is not a serious disadvantage, because it is overshadowed by other costs.

Applications where high cost is not a major problem are found in sports equipment, and reinforced polymers are widely used in tennis racquets, golf clubs, skis, etc.

Fibre reinforced polymers are thus entering most areas of materials use. It is likely that in the not too distant future they will be used in almost all applications where heat, erosion and abrasion resistance is not required.

Fig. 2. The Voyager,which circumnavigated the world without refuelling was made almost entirely with carbon-epoxy, and had a structural weight of only 427 kg (Courtesy Hercules Aerospace Products Group)

2. COMPOSITES PROPERTIES

2.1. Elastic Constants

In a well-made composite these depend on fibre lengths and orientation. The simplest case is the continous aligned fibre composite which obeys a Rules of Mixtures expression in the fibre direction [2]:

$$E_1 = V_f E_f + V_m E_m \qquad (2)$$

where the E's are Young's moduli and the V's are volume fractions with subscripts f and m respresenting fibre and matrix, and 1 represents the fibre direction and signifies that this is the composite modulus in that direction. (For a well made composite with no air voids $V_f + V_m = 1$.) In the 2 and 3 directions normal to the fibres and normal to each other, we have an inverse Rules of Mixtures:

$$E_2 = E_3 = (V_f/E_f + V_m/E_m)^{-1} \quad (3)$$

and the same is true for the in-plane shear modulus, G_{12}, i.e.

$$G_{12} = (V_f/G_f + V_m/G_m)^{-1} \quad (4)$$

Densities of composites, ρ_c, are also given by rules of mixtures so long as the composite is well consolidated, so that no significant voids are present:

$$\rho_c = V_f\rho_f + V_m\rho_m \quad (5)$$

The attractiveness of fibre composites is readily apparent when appropriate values are put into these equations. With aligned fibre composites $V_f = 0.6$ is readily achieved, so using table 1, we can see that when the HM carbon is used, a Young's modulus of 240 GPa may be attained in the fibre direction (the modulus of the resin, usually 1-4 GPa can be neglected). This is greater than steel, and gives a modulus/density ratio of 154 compared with steel having 27 $kNmg^{-1}$. The comparison is even more favourable when specific resistance to flexure i.e. $E^{1/3}/\rho$ is the basis used.

At right angles to the fibre direction the properties are dominated by the polymer. We can neglect E_f in equation 2 for a fibre like carbon, and so

$$E_2 = E_3 \cong V_m/E_m \quad (6)$$

Thus using an epoxy resin matrix with $E_m = 2.5$ GPa we estimate $E_2 \cong$ 6.2 GPa.
Thus E_2 and $E_3 << E_1$. So if a composite sheet is to be used, as for example in an aircraft wing, it is important to have the fibres going in more than one direction.

Since we want fibres in more than one direction, and we want a high fibre volume for the best properties, we make our aircraft wings etc. with laminates. These have many layers or thin sheets of aligned fibres reinforced polymer (laminae) with alignment directions being different in the different laminae. We can, in fact, design our material, putting fibres only in desired directions, to give a material having maximum efficiency for the structure and stresses envisaged.

The properties of the laminate are an average of the properties of the laminae, taking fibre directions, and numbers of fibres in a given direction into account. We can illustrate this with a cross ply laminate, in which the laminae are aligned in two directions at right angles. If we have equal numbers of fibres in each direction, then in the two principal directions, the Young's moduli are the mean of those given by equations 1 and 2. In our example this gives about 80 GPa. There is some fall-off in modulus in intermediate directions, as indicated in fig. 3. For more information about how to make laminate calculations, see Jones [3]

2.2 Tensile and Shear Strengths

The aligned fibre composite is highly anisotropic with respect to strength as well as modulus. A similar expression to equation 2 is used for strength, σ_{1u}, in the fibre direction:

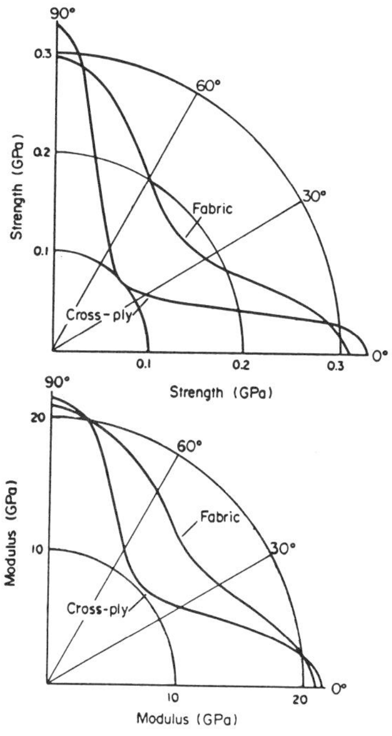

Fig.3. Polar plots of strength and modulus of glass-epoxy laminates. When carbon fibres are used instead of glass the strength in the 0° and 90° directions is about 600 MPa and the modulus is about 80 GPa. Variation with angle is very much the same as for fibres.

$$\sigma_{1u} = V_f\sigma_{fu} + V_m E_m \sigma_{fu}/E_f \qquad (7)$$

where the σ's are stresses, and the subscript u indicates the ultimate value of the stress (i.e. the strength). We do not use the matrix strength in equation 7 because the fibre breaking strain is normally smaller than that of the polymer. So the composite breaks when the strain reaches the breaking strain of the fibres, i.e. σ_{fu}/E_f, at which point the matrix stress is $E_m\sigma_{fu}/E_f$.

In the other directions the strength is determined by the matrix, so we use:

$$\sigma_{2u} \cong \sigma_{3u} \leq \sigma_{mu} \tag{8}$$

The reason for the less than sign is that if the adhesion between the fibres and matrix is not good enough, we can get failure at the interface before the applied stress reaches the matrix strength. The in-plane shear strength, τ_{12u} is determined by the shear strength of the polymer:

$$\tau_{12u} \cong \tau_{mu} \tag{9}$$

but in this case we can usually exceed the matrix strength by a small amount. This is because the fibres often bridge the shear failure crack, and have to be broken, and because the shear failure crack is not strictly planar.

Once again, in the fibre direction, the contribution of the polymer can be neglected, so we can achieve strengths, with the carbon fibres composite discussed previously, of 1.5 GPa in the fibre direction. σ_{2u} and σ_{3u} are often only 50 MPa or so, but the cross ply laminate will have strengths in the two principal directions of about 0.6 GPa. See fig. 3. for the variation of strength with orientation.

Some structures are designed to give particular strengths. For example pressure vessels are filament wound, and the amount of fibres needed and their orientations are determined by the strength requirements. In the case of a cylindrical vessel the hoop stress is twice the axial stress, so a winding angle of 55° is used. The structure must be symmetrical, so the fibres are wound back and forth at +55° and -55°.

2.3 Compression Properties

The compressive strength of a fibre composite does not have any single governing equation, and early attempts to explain it in terms of buckling of straight fibres [4] were not successful since the values predicted were too high. Fibre, matrix and interface properties all affect the strength, and can also affect the Young's modulus in compression, which is often lower than the tensile modulus [5].

Compressive properties are difficult to measure, since premature buckling is difficult to prevent, and grip failure is also a problem. The most difficult composite to test has aligned fibres, and even the test methods recommended by the ASTM tend to give low results in this case. Instead, the sandwich beam, test see fig. 4, should ideally be used, although it is costly and difficult to carry out.

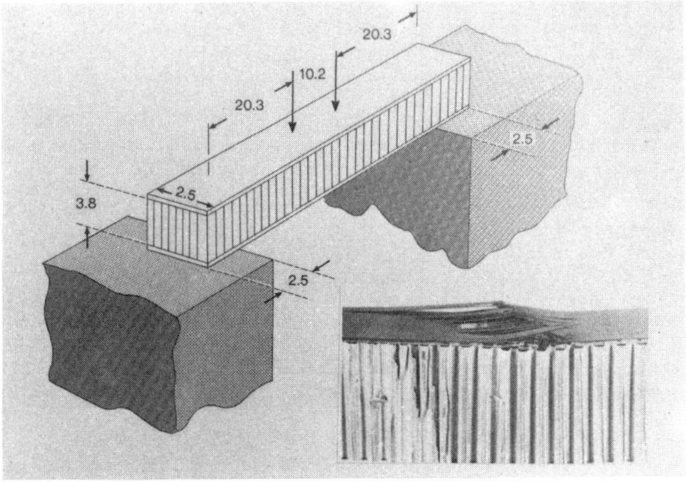

Fig. 4. Sandwich beam test used to measure the compression strength of laminae and laminates.

That fibre properties directly affect compressive strength is demonstrated by steel and Kevlar.The correlation between fibre compressive strength and composite compressive strength can be demonstrated with steels with different hardnesses. In the case of Kevlar, the fibre appears to be weak in compression, so that the compressive strength of an aligned continuous Kevlar fibre reinforced epoxy is about one twelth of the corresponding tensile strength. The Young's modulus in compression of Kevlar fibres also appears to be low, i.e. about a half of the tensile modulus, so that the composite has a correspondingly low modulus.

Fibre straightness is very important. Thus woven materials give composites with lower compressive strength and modulus than expected (i.e. less than a half that for the corresponding aligned fibre composites). See table 2.

Small fibre divagations probably account for the effect of the fibre-matrix interface. If the adhesion there is weak, the compressive strength is reduced, almost certainly as a result of fibre buckling being facilitated. In addition ,the polymer yield and tensile strength affect the compressive properties. Again this is likely to be through easier buckling in the case of a weak matrix.

Compression properties of composites have been reviewed recently and the reader is referred to the review [6] for further details and literature references.

TABLE 2

Compressive Strength and Toughness of Some Unidirectional Fibre Reinforced Epoxy Resins

Fibre	Compressive Strength (GPa)	Shear Strength (MPa)	Work of Fracture Across fibres kJm^{-2}	Work of Fracture Delamination kJm^{-2}
Boron	2.59	-	18	-
Carbon	1.31	110	40	1.56
Glass	0.83	76	130	3.24
Kevlar	0.21	52	140	-

2.4 Short Fibre Composites

When the fibres are short, they are usually not aligned, so we have to take into account both fibre length and fibre alignment. Thus we use, for modulus

$$E_c = \chi_1 \chi_2 V_f E_f + V_m E_m \qquad (10)$$

Here χ_1 is the alignment factor which is 1.0 for aligned fibres, about 0.375 for fibres which are random in a plane, and 0.2 for fibres which are random in all directions. When some alignment is present, χ_1 can be estimated by taking into account the number of fibres in each direction. (In this case, of course, E_c will have different values for different stressing directions.)
χ_2 is the length factor, which is also less than one since a significant part of the fibre is involved in stress transfer, and hence the fibre is not being used to maximum efficiency. It is given approximately by

$$\chi_2 \cong \left[1 - \sqrt{E_f/E_m}\ /s\right] \qquad (11)$$

Here s is the fibre aspect ratio (length/diameter). More exact evaluation may be obtained from the literature [7]. However, it should be noted that theoretical estimations of χ_2 are not very well supported by experimental data. There has been very little experimental work done on this, and in the most systematic study up to the present χ_2 was found to be less than the value given by equation 11, especially for shorter fibres [8]. If fibre lengths are variable χ_2 has to be evaluated by weighted summation, taking the length distribution into account

A similar expression is used for strength:

$$\sigma_{cu} = \chi_3 \chi_4 V_f \sigma_{fu} + V_m \sigma_{mu} \qquad (12)$$

where χ_3, the direction factor, takes the same values as χ_1 for the random cases, but the summation uses different governing equations when some orientation is present[8]. The fibre length factor is

$$\chi_4 = (1 - s_c/2s) \tag{13}$$

where s_c is the critical fibre aspect ratio. s_c can be measured by making a fibre composite which contains only one fibre, and, so long as the polymer breaking strain is at least three times that of the fibre, it is found that the fibre can be broken into small pieces by stressing the polymer. When the strain in the polymer reaches about twice the breaking strain of the fibre, fibre breaking ceases, and the broken fibre lengths range from $s_c d/2$ to $s_c d$ where d is the fibre diameter. Equation 12 has been approximately verified for aligned carbon fibre composites using this technique for measuring s_c [8].

If the fibre length is variable χ_4 must be determined from an appropriate weighted sum using equation 13.

Since χ_1, χ_2, χ_3, and χ_4 are all less than one, and since it is not possible to achieve very high fibre volume fractions if the fibres are not continuous and aligned, the mechanical properties of short fibre composites are less good, and often much worse than those of the aligned continuous fibre composites.

The compression strength of short fibre composites does not appear to have been investigated systematically.

2.5 The Interface and Fracture Properties

The interface between fibres and matrix plays a key role in the transfer of stress in a fibre composite, and hence controls some properties. These include shear and off-axis strength, and compression strength, as mentioned above. Also, the strength and modulus of short fibre composites are affected by the interface. The governing factors are the interface shear stress τ_i and work of fracture, G_i

The shear stress at the interface determines the critical aspect ratio which, in turn, affects the strength of short fibre composites (see equation 13). Thus, by equating the internal forces in the fibre, when it is about to break, and the external surface forces, we can show [2] that

$$s_c = \sigma_{fu}/2\tau_i \tag{14}$$

so that τ_i can be estimated from s_c measured using the method described in the previous section.

Recent experiments in which single fibres were embedded in resins, and then pulled out suggest that the interface fails by a brittle fracture process, at least for thermosets such as epoxy and polyester. The work of fracture G_i is very low, in the range 10-200 Jm^{-2} [9]. This indicates that τ_i is a post-debonding frictional shear stress. Hence

$$\tau_i = \mu P \tag{15}$$

where P is the pressure exerted on the polymer by the fibre, and μ is the coefficient of friction at the interface. We can use this information to explain some of the processes which contribute to the fracture toughness of composites.

Again we consider the simplest composite, with fibres continuous and aligned. The work of fracture G shows great anisotropy:

$$G_1 \cong V_f G_f^* + V_m G_m^* + G_{1i} \tag{16}$$

$$G_2 \cong G_3 \sim G_m \tag{17}$$

where the subscript i refers to processes taking place at the interface, and the asterisks indicate that the properties G_f and G_m are modified by the presence of the other component in the composite. Thus

$$G_f^* > G_f \tag{18}$$

because the fibre is supported by the polymer, and this inhibits brittle fracture. This effect is unimportant with fibres such as glass, carbon, and the other ceramics, since they are very brittle so that, even for $G_f^* = 3G_f$, the contribution of this term is negligible. In the case of Kevlar and other polymer fibres there is a significant contribution from G_f., though, so this effect must then be taken into account.

In the case of the matrix

$$G_m^* << G_m \tag{19}$$

because the fibres limit the size of the worked zone at the fracture surface. Thus there is little to be gained by using a tough polymer to enhance G_1, which tends to be rather low, see table 2.

The interface work, G_{1i} can be influenced favourably however. When a composite is broken at right angles to the fibres, the fibres do not usually break in the crack plane. Instead, they can break at a distance equivalent to many fibre diameters away. If the average distance is L, corresponding to an aspect ratio (L/d) of s_1, then G_{1i} has two major components

$$G_{1i} = V_f d \tau_i s_1^2/6 + V_f G_i s_2 \tag{20}$$

where s_2, multiplied by d, is the average debonded length of fibre, and the right-hand term is the contribution from the work of fibre debonding. The other term is the work of fibre pull out against frictional forces. Unless $s_2 >> s_1$ the contribution from G_i is usually negligible This is because s_1 is normally much greater than one, and $\tau_i d/6$ is of the same order of magnitude as G_i. (Typical values for τ_i are 20 MPa [10] and d is in the range 6-22 μm, with the lowest values corresponding to carbon, which also has the lowest G_i)

In order to increase G_{1i} it is necessary to increase s_1. This can be effected by reducing the pressure at the interface [11], or by reducing the bonding between fibres and matrix [12]. Weakening the bonding, as indicated earlier, reduces other properties, so control of the pressure, by reducing the shrinkage of the polymer is the preferred method.

G_2 represents the resistance to delamination. This is controlled by the matrix as indicated by equation 17, and so is enhanced by using tough thermoplastics, instead of the rather brittle thermosets. With the thermosets $G_2 \cong 2.6\ G_m$ due to the fracture surface not being entirely in the matrix, while with the thermoplastics $G_2 \sim G_m$ [13] see fig. 5. Very low works of fracture

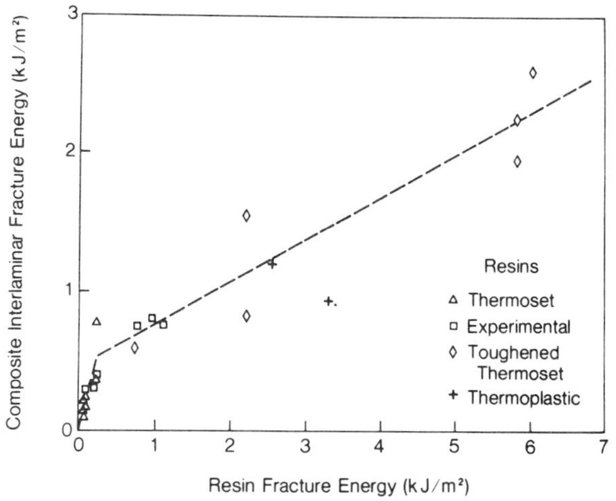

Fig. 5. Composite interlaminar fracture energy (double cantilever beam specimen plotted against neat resin Mode-I fracture energy for various materials. Dashed lines indicate general trends in data: slopes are 2.3 at low fracture energies and 0.31 at higher fracture energies.[13] (courtesy D.L. Hunston and ASTM).

are involved in delamination, as can be seen from table 2. This is a major shortcoming of laminates.

2.6 Water Absorption and Diffusion

Water can diffuse through the polymer and attack the interphasial region that exists between the fibres and the polymer. This interphasial region is present because the fibres are coated before use. This coating is partly to protect the fibres from damage during mechanical handling, and partly to enhance the adhesion between fibres and matrix. In the case of glass fibres it has an important role in protecting the fibre surface from attack by water. This is characterized by leaching of ions from the glass, with the consequent development of osmotic pressure, and cracking, and spalling.

For a composite with a water resistant interface, the weight gain on immersion in water, W_c, obeys Fick's Law with respect to rate, and a simple mixture rule with respect to the saturation weight gain, i.e.

$$W_{sc} = V_f W_{sf} + V_m W_{sm} \qquad (21)$$

where the subscript s indicates the saturated weight again. (This equation is not obeyed when the interface is susceptible to attack.) For glass, carbon, and other ceramic fibres W_{sf} is negligible, but it is significant with polymer fibres such as Kevlar. The polymer matrix has W_{sm} typically a few percent of the total polymer weight.

Diffusion in aligned fibre composites obeys a mixture rule for transport along the fibre direction:.

$$D_1 = V_f D_f + V_m D_m \qquad (22)$$

where the D's are diffusion coefficients, and for ceramic fibres, D_f is negligible. In the other directions [14] we have

$$D_2 = D_3 = \left[\left(\frac{2}{\sqrt{1-c^2}} \tan^{-1} \sqrt{\frac{1+c}{1-c}} - \frac{\pi}{2} \right) / \left(D_m - D_f\right) + \left(1 - \sqrt{4V_f/\pi}\right) / D_m \right]^{-1} \qquad (23)$$

where

$$c = (1 - D_f/D_m) \sqrt{4V_f/\pi} \qquad (24)$$

with the interphase having very little effect, although diffusion along the interphase is significantly faster than in the polymer. Deviations from equation 23 that have been observed appear to be due to voids [14].

CONCLUSION

Fibre composites are very useful materials, since they are stiff, strong and light, as well as being highly corrosion resistant. Although the interactions between the fibres and the polymer are not yet fully worked out the mechanical properties of aligned continuous fibre, composites, and laminates made therefrom, can be predicted using relatively simple equations. These composites, however, are not very tough.

Composites made with short fibres have properties which are strongly dependent on the interface or interphase between the fibres and matrix, and do not obey simple expressions for Young's modulus, although their strengths appear to do so. These short fibre composites do not have such good mechanical properties as those containing continuous fibres because it is not usually possible to achieve such high fibre loadings, and because stress transfer is not so efficient, due to the smaller surface area of the fibres.

Fibre reinforced polymers absorb water, and unless steps are taken to ensure that the fibre (especially glass) has a protective coating on it, loss of properties can ensue.

The mechanical properties which can be achieved with reinforced polymers compare very favourably with metals, and when their corrosion resistance, light weight and ease of molding into complex shapes are taken into account, they are the material of choice in a wide range of applications, including chemical plant, automobiles and household goods such as refrigerators. They also serve in situations which take advantage of their unique properties of light weight and high modulus, such as in modern high speed weaving looms, and aircraft and spacecraft.

ACKNOWLEDGEMENT

The author is grateful to Professor M. Munro for inviting him to present this paper at the 1988 CIM meeting and to J. Clifford who always ensures that written papers are well prepared and presented.

REFERENCES

1. Griffith, A.A., (1921), Phil. Trans. Roy Soc,Vol A221 163.
2. Hull, D., (1981), An. Introduction to Composites,, (University Press, Cambridge, U.K.), Chpt. 5.
3. Jones, R.M, (1975), Mechanics of Composite Materials, (McGraw Hill, New York), Chpt. 2.
4. Hancox, N.L., (1975), J. Mater. Sci., 10 , 234.
5. Pigggott, M.R., and Harris, B., (1980), J. Mater Sci. 15, 2523
6. M.R. Piggott, (1985),.Developments in Reinforced Plastics, G. Pritchard, Ed., Elsevier, London, U.K.) Vol. 4, Chpt. 4.
7. Piggott, M.R., (1980), Load Bearing Fibre Composites, (Pergamon, Oxford,), Chpts 4 & 5.
8. Sanadi, A., and Piggott, M.R.,(1985), J. Mater Sci. 20, 421.
9. Piggott, M.R., and Chua, P.S., (1986), Ind. Eng. Chem. - Product R/D, 26 , 672.
10. Piggott, M.R., Sanadi, A., Chua, P.S., and Andison, D. (1986), Proc.1st Int.Conf. on the Composite Interface ,(Ed. H. Ishida & J. L Koenig, Elsevier, New York), 109.
11. Lim, J.T., Piggott, M.R., and Bailey,W.J., (1984), SAMPE Quarterly, 15, No. 4 , 75.
12. Harris, B., Beaumont, P.W.B., and De Ferran, E.M., (1971), J. Mater Sci. 6, 238.
13. Hunston, D.L., (1984), Comp. Tech. Rev. 6, 176.
14. Woo, M., and Piggott, M.R., (1988), J. Comp. Tech. Res. 10, 20.

PULTRUSION PROCESSING - GROWTH IN KEY TECHNOLOGY AREAS YIELD PRODUCT SOPHISTICATION

Jeff Martin and Joseph E. Sumerak

Pultrusion Technology, Inc. 1670 Enterprise Parkway
Twinsburg, Ohio

ABSTRACT

Pultrusion is a linear profile process where reinforcing fibers are first pulled through a resin matrix, then into a shaping and forming guide system and finally into a die where the product is generally heated and cured to its final dimensions. The majority of profile products are produced with fiberglass reinforcement and thermoset polyester or vinyl ester resins. Recently, both higher modulus polyester, nylon and high density polyethylene fibers have also been incorporated in profiles to further enhance the versatility of this composite process.

The result of this versatility has been the ability of pultruded profiles to participate competitively in markets like; electrical, corrosion, construction, transportation, as well as aerospace and defense. The number of applications in each of these markets continues to expand at growth rates faster than those for traditional materials. This growth is principally due to:

1. The large number of new composite materials available
2. More creative and sophisticated processing techniques
3. The successful application of pultruded profiles in existing markets
4. The awareness of end users and designers to the process and its versatility

One of the ironies of the pultrusion process is that of the three key technology areas involved in successful processing, two of these occur prior to the pultrusion machine. These three key areas are; the formulation of the resin matrix, the forming of the impregnated fibers just ahead of the die, and temperature management - the control of heat going into the die and the control of the rate of removal of heat after the exothermic reaction.

This paper will discuss these three key technology areas in depth and differentiate the technology in the machine from the technology involved in these key areas.

INTRODUCTION

It is estimated that the pultrusion process is nearly 40 years old. In its first 30 to 35 years the pultrusion process and associated markets grew at reasonable rates, but it was far from being the most active of composite processes. In the last five years pultrusion has taken a leadership position in the composites industry and is now experiencing double digit growth in the United States and worldwide. Yet in many respects pultrusion is one of the least understood of all of the composite processes and still the best kept secret in our composite industry.

PROCESS DESCRIPTION

Pultrusion is a linear profile process consisting of six principal elements; creels for reinforcement dispensing, resin impregnation area, material forming, die heating/curing station, pullers and cut-off saw.

Reinforcing fibers are first pulled from creels where unidirectional and multidirectional reinforcements are organized prior to going into a resin bath. They are impregnated with a resin matrix and precisely formed just prior to entering a heated die. The steel die is machined to the final dimensions of the part desired. A continuous pulling system alternately clamps and pulls on the part, pulling the material from the creels through the resin station and the forming guide area into the die and ultimately through a cut-off saw which automatically cuts the part to the desired manufactured length. In the past we were only able to make very simple structural profiles. Recently, new technology has allowed us to expand into very complex shapes. Both small complex profiles and large thin wall, even hollow complex profiles have been easily accomplished.

PULTRUSION MARKET GROWTH

In 1970, in the U.S. market there were only 20 pultruders producing 10 million pounds of product. In 1987 there were over 75 pultruders producing more than 110 million pounds of product. There are basically four reasons for this dramatic growth:

1. There are a large number of new composite raw materials available to the processor.
2. There are more creative and sophisticated processing techniques available.
3. The end-user and designer in the market place has become more aware of the process and the versatility of the material to meet his end-use requirements.
4. Those existing pultrusions have become very successful and have been proven in extensive use.

Probably the most significant of these has been the explosion of new materials available to the pultruder. There are better polyester resin systems, vinyl ester resins, processable epoxies and now methyl methacrylate resins that have increased our ability to pull at faster speeds and meet higher mechanical properties and service temperature characteristics. Carbon fiber, aramid fiber, and new reinforcing fibers such as polyester, nylon and high density polyethylene fibers have also been incorporated in profiles to further enhance the versatility of composites in this process. These materials as well as better processing technology have enabled the pultruded parts to increase their performance level while having tighter tolerances and better overall economics when compared to traditional materials.

KEY TECHNOLOGY AREAS

Pultrusion as a composite process appears to be quite simple. Yet it is as complex as any of the composite processes. An individual or company that embarks on a project to establish a pultrusion processing facility must learn and understand the uniquenesses of the process of the process. Many underestimate the technology and misunderstand where the real technology is in the process.

As an equipment supplier we frequently get asked if our machines can produce one shape or another. There are those who believe that purchasing a pultrusion machine automatically enables you to produce any porfile. When you first buy a car you do not get driving instructions. It takes time to learn the handling aspects of the car. The same is true in pultrusion. Purchasing equipment gives you the capability to heat, pull and cut a lineal profile. You want to purchase a reliable machine that will efficiently perform these tasks 24 hours a day with little maintenance required. Beyond the machine's capabilities the key technology to successful pultrusion processing is in three areas; material formulation, material preforming or shaping and process temperature management. It is ironic that two out of these three technologies occur before you get to the machine. This point confirms that the process is not machine dependent but relies heavily on material and process engineers direct involvement.

Resin Selection and Formulation

The first and critically important area is in the formulation of the resin. The resin selection itself is important as it governs the mechanical characteristics, operating temperature range, electrical insulation, corrosion profile. In addition, it governs process speed by its reactivity and can significantly control product aesthetics and tolerance capability. The resin matrix can be altered by chemical additives and fillers which will enhance its ability to handle higher temperatures, better electrical insulation and corrosion resistance as well as lower flame and smoke propagation. It is essential to the success of any pultruded profile that the polymer matrix be properly engineered to meet the desired end use properties while taking into account the processing characteristics necessary to meet the economic aspects of the application.

Recently new resins have come available to the pultrusion processor that has enabled him to meet even higher property requirements and enter new market/application areas. Epoxy resins with low viscosity and a quick curing resin system based on a bisphenol A/epichlorohydrin has been introduced to the market. It processes in fiberglass roving or other axial reinforcement systems with relatively high line speeds, low back pressure and good surface quality. This resin with a high heat distortion temperature has stability in many corrosive environments and exhibits excellent fatique resistance.

Methacrylate-based resins are being produced in several European countries. These resins use a methyl methacrylate monomer versus styrene in the crosslinking reaction. These resins reportedly process at speeds 2 to 3 times faster than conventional polyesters. Because of their inherent low viscosity these resins can take higher filler loadings. Higher concentrations of retardants such as alumina tribydrate should result in improved flame spread ratings and lower smoke than styrene-based systems. Profiles made with methacrylate resins show higher flexural strength, modulus and short beam shear than polyesters. Surface finish is excellent even when compared to low profile resins.

Determining the proper amount of resin, catalyst, fillers and other additives is extremely important to any success further down the processing stream. Each pultruder should have a polymer engineer, chemist or very practical, experienced, hands-on production man to provide this key resin formulation technology.

Material Preforming and Shaping

The second key technology area is in material forming. Any continuous reinforcement can be selected to meet a mechanical property. Usually, reinforcements are only used to enhance this characteristic since reinforcements tend to reduce electrical insulation and corrosion resistance. In many applications, this requires compromises in order to meet all the properties desired. Pultrusion is the one composite process that allows selective reinforcement of portions of a profile using different materials. This can be done on a continuous basis with good consistency and repeatability. The only disadvantage is that what is done must be done identically reinforced.

Material forming and preshaping is accomplished usually after impregnation although some initial steps of forming can be accomplished during impregnation. Forming guides are usually attached to the pultrusion die in order to ensure positive alignment of the formed materials to the die cavity. In the case of tubular products, a mandrel support is necessary to extend the mandrel in a cantilever fashion through the pultrusion die and yet resist the forward drag on the mandrel. Materials must form sequentially around the mandrel in an alternating fashion to prevent weak areas due to ply overlap joints. Forming guide slots and holes must be accurately sized with proper clearances to prevent excess tension on the relatively weak wet fibers while removing sufficient resin to prevent too high of a viscous drag at the die entrance.

An alternative method of impregnation and forming is the use of direct injection of resin into the forming guide or die after the dry materials have been formed. While this technique minimizes the problems and mess associated with the wet bath systems, some limitations exist with regard to wet out, air entrapment, and maximum filler content. A combination of techniques may be the answer for a specific profile based upon its complexity.

Materials commonly used for forming guides include Teflon (TM), ultra high molecular weight polyethylene (UHMWPE), chrome plated steel, and various sheet steel alloys. The pultrusion processor that who employs a craftsman capable of converting sheet metal and plastic stock into forming guides with precise control is most likely to be successful in processing complex shapes.

An old system called constant velocity forming is finding new favor recently as it enables the pultruder to produce a shape with a sheet metal guide that forms the shape in one action, taking the mat from a flat plane to a finished form in one step. More complex shapes are now possible with a combination of this constant velocity system along with carding plates and resin injection. This is allowing pultruders to compete on shapes that they could not previously produce.

Temperature Management

The third key technology is control of product temperature. In thermoset resins, it is important to control the rate and level to which heat is added to the matrix. It is also important to control the rate at which heat is removed from the profile. If too little heat is put into the die, the composite material will not reach its desired exotherm temperature resulting in an incomplete cure and less than designed properties in the finished profile. Too high a temperature can induce thermal stress cracking as the profile cools resulting in a poor electrical insulator or corrosion resistant material.

A number of different methods are employed for positioning the pultrusion die and applying the heat necessary to initiate the reaction. The use of a stationary die frame with a yoke arrangement that allows the die to be fastened to the die frame is the simplest arrangement. With this yoke arrangement, heating jackets utilizing hot oil or electrical resistance strip heaters may be positioned around the die at desired locations. Thermocouples placed in the die control the level of heat applied and multiple, individually controlled zones may be configured in this fashion. This approach is well suited to single cavity set ups but becomes more complex when the number of dies used simultaneously increases because each die requires a heat source and thermocouple feedback provision. The use of standard heating jackets and heating plates can be designed to accommodate multiple dies to help alleviate this limitation. Most of the older processes use one of these techniques. Pultrusion Technology, Inc. has developed a new heat management system that utilizes heated platens which have fixed zones of heating control with thermocouple feedback from within the platen. The advantage of this method comes from the ability to heat all dies uniformly with reduced temperature cycling as changes in temperature are detected early at the source of heat rather than at the load. When provided with the means to automatically separate the platens, the advantage of quick set-up and replacement of dies can lead to increased productivity through reduced down time. This system utilizes bottom platen height adjustment to exactly align the die centerline to the pulling mechanism. This eliminates product distortion and bowing associated with die/puller misalignment.

The use of a source of cooling water or air is essential in the front of the die at start-up and during temporary shut-down periods to prevent premature gellation of the resin at the tapered or radiused die entrance. This is accomplished either as a removal jacket or a self contained zone within the heating platen. Alternatively, the first section of the die is unheated and cooling through convection is relied upon. Knowledge and control of the die heating profile is the most critical pultrusion process control parameter as it determines the rate of reaction, the position of reaction within the die, and the magnitude of the peak exotherm. The use of heat sinking zones at the end of the die or auxiliary cooling may be necessary to remove heat prior to the product exiting the die.

The ability to deliver heat into the material prior to entering the die via Radio Frequency (RF) preheating, induction heating or conventional conductive heating of resin or reinforcements has been recognized as desirable for increasing process rates and the reduction of temperature differentials which contribute to thermal cracking in large mass products. Such heating devices are available as either integral units or stand-alone devices which can be positioned before the die entrance.

The Gelstar® Thermal Analyzer has been developed for process optimization and quality control. This instrument allows the tracking of external die temperature profiles and internal product temperatures as a function of die position during the curing process in a convenient graphic format. The data collected at a specific process speed becomes a graph of the steady state process conditions to be used for quality control, process engineering, and quality assurance documentation. This is a most significant development. Further process control developments of this nature will provide improved process capability and production efficiency enabling pultruders to meet even higher properties and faster processing speeds resulting in the pultruded profile being more competitive with traditional materials.

CONCLUSION

To be a successful pultruder one needs to understand where the key technologies are in processing. Once known, all pultruders need to strive to develop new approaches and refinements to resin formulation, material forming and temperature management. The pultrusion machine, by itself cannot assure you of excellent quality. You must use quality raw materials and quality engineering at the front of the machine to be sure that quality profiles come off the end of the line.

Pultruded profiles are the fastest growing segment of the composites industry. This is due to the solid engineering approach that most processors have taken to process management. We can all be assured of continued market growth as long as we continue to focus our efforts in the key technology areas.

PULTRUDED PROFILES - Where can I use them?

Jim M. Stanley and W. M. Matto

Pultrusions Canada Limited
247 Armstrong Avenue
Georgetown, Ontario, Canada
L7G 4X6

ABSTRACT

Pultruded profiles are finding increasing applications in many new and varied environments.

This discussion will describe two pultrusion applications by following each from design and production through to end use. The products to be discussed in detail are snap-on coverboards for the power rails on a light rail transit system and a moveable pool floor system used in areas where a variable depth of water would be advantageous.

Discussion will cover structural requirements and raw material selection, which were common to both products, along with comments on specific areas concerning each individual profile

Relating to coverboards, other topics covered were electrical testing, shape design,processing considerations, aging tests and field assembly. For the pool floor system additional areas evaluated were corrosion properties, structural testing, and long term under water acceptability.

A number of additional applications will be mentioned to demonstrate a variety of areas where pultruded products have succeeded.

With close co-operation, the user's requirements and the pultruders capabilities can be united to yield excellent products of consistently high quality

INTRODUCTION

Pultrusions has advantages over other reinforced plastic processes when the product application requires a composite with a uniform cross-section and certain other specific requirements such as close tolerances, optimum flexural strengths, good surface appearance, or variable lengths. The applications to be reviewed required these characteristics in varying degrees, and pultrusion became the "material of choice".

Although these examples may not relate directly to your specific needs, many of the design considerations will. Use them to stimulate your imagination to explore the versatility of pultrusion for your own unique applications.

POWER RAIL PROTECTION COVERBOARD

REQUIRMENTS

Overall design of the power rail protection coverboard was the responsibility of the supplier. Studies had been conducted by the transit authority in conjunction with their research agency which indicated a general design direction that would be acceptable. The approach was to have the fiberglass reinforced plastic coverboards for the light rail transit system applied directly to the power rail in a "snap-on" fashion. There were several reasons for this decision.

Electrical Protection for Safety:

With the cover in place three sides of the power rail would be enclosed thus leaving only the actual power transfer surface exposed.

Thermal Insulation:

Yielded 40% higher de-icing efficiency for the heater cable inserted into the power rail.

Indepedent Assembly:

Good overall appearance of the total system resulted because there were no alignment problems between the rail and the cover,which was attached directly to the rail.

Snap-on Instalation:

This reduced costs for the total coverboard system which included both product cost as well as field installation costs.

Thermal Movement:

Independent movement of the cover permitted unimpeded longitudinal movement of the power rails due to ambient temperature changes.

Variable Lengths:

On several occasions the field spacing of the insulator chairs was not accurate and required longer or shorter than standard length boards. Quantities of extra long or short pieces could be supplied on a pro-rated cost basis to maintain overall project costs.

DESIGN

The overall profile configuration was arrived at by meeting the following requirements:

1. The coverboard must fit inside the envelope given by the customer
2. It must provide a smooth upper contour to shed dirt and water
3. It must permit easy field installation onto the power rail

Designing within the envelope and having a smooth appearance presented little difficulty as one always prefers smooth curves in a pultrusion profile.

Design also involved determining the various angles and radii in order that the cover could be rotated onto the power rail without interference from friction points over the total 2 metre length. Several hand lay-up models were made and the final shape touched at all required contact points to fit correctly on the rail.

In the final design an angled surface was located on the bottom of the coverboard so that during installation, when the top was hooked on the power rail, the bottom angle assisted movement of the "leading edge" over irregularities on the power rail surface.

TESTING

The coverboards were subjected to a series of performance tests.

1. Drop Impact Test:
   This test became the major governing factor in the structural design of the profile. The test required a 70kg weight to be dropped from a height of 300mm at the centre of a 2m board. No cracks or crazes were to appear anywhere along its length.
2. Load Cycle Test:
   A 70kg load was cycled onto the upper curved portion of the part and removed for 50 cycles.
3. **Accelerated Aging Test:**
   The product was subjected to -4°C for 4 days and then 70°C for 7 days, after which electrical and mechanical tests were performed.
4. Dielectric Strength Test:
   Random testing of 5% of production gave results of 350 to 400 volts per mil - (well above the specification of 200 v/mil minimum).
5. Flame Retardance:
   Random samples were subjected to the UL horizontal flame spread test.

These power rail coverboards have been installed and have operated trouble free for over two years. Additional boards have been supplied for track extensions and maintenance.

SWIMMING POOL FLOORING SYSTEM

REQUIREMENTS

Several years ago we were approached by a firm in the process of developing a moveable floor system for use in commercial swimming pools.

This type of system had several overall advantages:

1. It would enhance therapeutic facilities by providing easy access for wheel chairs, elderly and handicapped users.
2. In recreational facilities it could provide non-swimmers with a safe environment whereby the depth could be increased to keep pace with aquired skills.
3. Overall there would be greater utilization of existing pool facilities.

The company already had considerable experience with this product, but solely with aluminum profiles. The history of coated aluminum profiles immersed in chemically treated pool water was totally unsatisfactory. The aluminum extrusions simply corroded away in a very short time, therefore, other materials had to be evaluated.

Again pultrusion became the "material of choice".

DEVELOPMENT and DESIGN

Some preliminary structural design and testing was done to determine general feasibility. this was followed by field evaluation with candidate composite samples submerged in an operating pool. These field trials lasted more than a year.

The results were satisfactory and the program was implemented. Several parameters were established as the basis for the final shape and size of the floor member with the history of the aluminum planks providing the major source of data. The principal design parameters governing the profile configuration were as follows:

1. The maximum deflections were derived from load/deflection tests performed on the original aluminum planks.

$$I = 1.28 \text{ in}^4 \text{ (Aluminum)}$$
$$E = 10.5 \times 10^6 \text{ (Aluminum)}$$
$$EI = 13.44 \times 10^6 \text{ (Aluminum)}$$
$$EI = 2.5 \times 10^6 \times 7.20 = 18 \times 10^6 \text{ (Pultrusion)}$$

A 34% higher value than the original aluminum extrusion was ultimately used as the customer felt a design change to a stiffer plank would be beneficial.

2. The maximun weight of a plank could not exceed 150 pounds.
3. The pool floor area required a 5% open area for water flow during raising and lowering operations.
4. The floor design loads were:
   - 25 pounds per square foot
   - 200 pound concentrated load
5. The profile had to provide an easy method of attachment to supports.
6. The colour and appearance of the floor surface was to be visually flawless for its entire length.
7. Continuous lengths of up to 53 feet would be required.

A return channel section was developed as an ideal shape which gave the required structural properties and as well provided an easy method of installation. Thicker return flanges were designed to incorporate a higher percentage rovings in order to balance the load carrying capacity of the top and bottom flanges. This also increased overall stiffness. The profile also permitted the use of a simple cross-bar clamping mechanism to hold each individual plank in place. A closed cell foam contained inside the channel gave an increase in total floor buoyancy.

Production of the profile was accomplished using an isophthalic resin system, a combination of glass mat and roving, and a surface layer of thermoplastic polyester veil.

ADDITIONAL PULTRUSION APPLICATIONS

1. The use of standard structural sections to fabricate such items as:
   - platforms and various support structures for corrosion and electrical applications.
   - flat surface stiffening members in other FRP fabrications
2. Sewage sludge scraper flites for use in primary sewage separation installations where traditional wooden flites are being replaced.
3. Large high voltage structures in the electrical industry where a combination of high strength and electrical insulation properties have replaced bulky steel frames and brittle ceramic insulators.
4. Suspended ceiling systems for sanitary locations where aluminum, coated steel and vinyl have proved inadequate.

Still other applications include oil well sucker rods, corrosion resistant gratings, ladders, cable trays, building panels, bus and trailer components, door and window frames, and lighting poles.

CONCLUSION

Pultrusion growth continues at a healthy rate. In order to acheive ultimate design satisfaction new products must be viewed not as replacement copies of existing products, but rather as new designs that improve and enhance overall product performance.

Our design versatility, using the pultrusion process, has increased and we hope that by presenting our experiences we will encourage more designers to consider pultrusion as their "material of choice".

Experience in the Design and Development of Composite Structures

P. McConnell, Canadair Inc., Montreal, Canada

ABSTRACT

Canadair has been active in the field of fiber-reinforced composites as it has advanced over the past decade. We have developed methods for characterizing the structural properties of composites so that secondary and primary structures can be safely produced with them. Many secondary components have been produced, particularly for the CL-601 Challenger business jet. Primary structures have been produced for surveillance drones and in one case for the Challenger. The certification procedures for these types of structure are continually being refined.

KEYWORDS

Composite materials, composite structures, material qualification, aircraft certification.

INTRODUCTION

The successful introduction of new material technology requires that i) the technology provide real improvements over the status quo while being economically practical, ii) sufficient understanding of the material exists to allow its efficient use with confidence, and iii) it be possible to demonstrate the safety of the technology to regulatory authorities.

Generally, an advance in one of these areas motivates complementary advances in the others. They must progress together, because the level of technology available for use is constrained by the least developed of the three.

Composite materials provide a good example of the way that these three areas progress in a complementary manner. Throughout the 1970's, extensive efforts were made to build experience with the new technology of graphite/epoxy composites. Studies such as Adsit and Waszczak (1979) found that the materials were limited by a lack of toughness, which led to the development of new, tougher, materials. We are now in a second period of experience building with these new materials (see, for example, Williams, O'Brien, Chapman, 1984). Thus, we have gone from a situation where a new technology required improved understanding, to one where that understanding motivated an improved technology, and then back to a situation of trying to completely understand the technology.

These ongoing changes in the technology have been accompanied by the development of certification methods under the guidance of regulatory authorities. In the United States, the FAA has provided Advisory Circular AC 20-107A, setting out a means of showing that composite materials in aircraft comply with the Federal Airworthiness Regulations. In Canada, direct consultations between industry and the Ministry of Transport are taking place to define certification requirements.

DESIGN METHODS FOR SECONDARY STRUCTURE

Table 1 lists most of the secondary composite components that have been produced at Canadair.

Table 1 Secondary Composite Structures

Component	Material	A/C	Date
Radome, MAD Boom, Rudder & Fin Inserts	Fiberglass	CL-28	1954
Wing Tips, Horizontal and Vertical Stab. Tips	Fiberglass	CL-215	1972
Wing/Fuselage Fairings, Nose Cone, Tail Cone, Nose L/G Doors, ADG Doors, Misc. Access Doors, Misc. Fairings, Flap L/E, Cowling Vert. Stab. L/E Side Panels, Misc. Ducting, Cockpit Panels, Wheel Bins	Kevlar	CL-600/601	1977
Belly Fairing, Cabin Floor, Stabilizer Fairings, Fuel System Reservoirs, Fuel Tank Access Panels, Wheel Disks, Winglet Tips	Graphite	CL-600/601	1977-1980

The approach used for the design of secondary composite structures is outlined in Fig. 1.

The test data required for these designs consists of material stiffness and ultimate strength values, fastener strength data, and sandwich panel data.

Strength values are B-basis allowables.

The component configuration is based on the designer's assessment of the loading situation and the overall part geometry, while allowance is made for manufacturing capabilities. The initial layup is based on the designer's experience and simple calculations.

When specifying the layup, the designer is guided by some simple rules. Nearly all of these components are made out of balanced fabrics, so the choice is limited to θ/-θ plies. For tractability and by tradition, the selection is usually limited to 0/90 and ±45 plies arranged in a symmetric layup. The percentage of each type of ply is assigned after an initial assessment of the relative magnitudes of the normal and shear stresses.

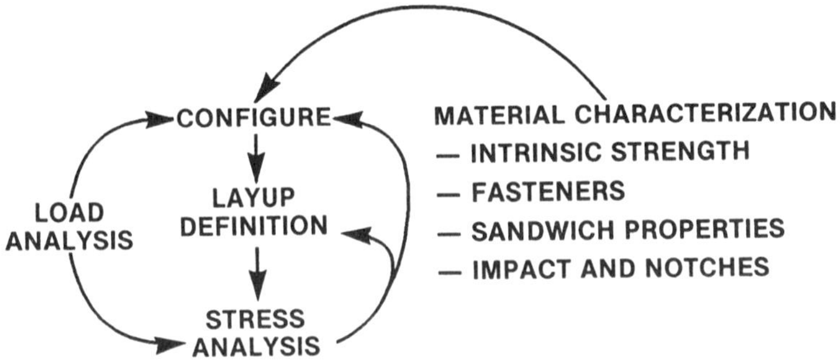

Fig. 1. Design flowchart for secondary structure.

When the initial design is ready, a stress analysis is performed. This is usually done with the aid of a finite element analysis package, unless the loading and geometry lend themselves to hand calculation. The results of the analysis are used in conjunction with the material data and a failure criterion to derive margins of safety for each ply. The margin of safety is usually calculated with respect to the ultimate load, which is taken as 1.5 times the limit load (the highest load that the component is expected to experience in its lifetime). For secondary structure the stiffness is often the governing consideration in design, resulting in high margins for material failure.

Margins of safety with respect to other relevant failure modes are also calculated. These may include buckling, fastener failure, impact damage propagation, and various sandwich construction failure modes. Some of these failure modes, such as impact damage propagation, are accounted for by applying a knockdown factor to the basic material allowable.

If any margin of safety is found to be negative, the design is modified, and the analysis is repeated. This continues until all margins of safety are positive.

For parts with small structural loads, like some cabin ducting, the choice of configuration and layup is based on the requirements of handleability and installation. Only limited strength analysis is performed in these cases.

The CL-601 main landing gear fairing is an example of composite secondary structure. This component was originally designed and produced in aluminum, but was later redesigned in graphite/epoxy for weight savings. The material system used in the fairing is Celion 3000 graphite fabric in a balanced 24 yar-per-inch 8-harness satin weave reinforcing BASF/Narmco 5208 Rigidite epoxy resin. Mechanical design data for this material system was generated by an extensive testing program which is described in the next section on primary structure.

The pattern of stiffeners on the component was chosen by a designer to match the anticipated load paths. A 0/90 layup was chosen because significant shear was not expected, and because the skin is only two plies thick in some areas. The layup could not have been symmetrical in those regions if both 0/90 and ±45-type plies had been included.

The success of this design is shown by the following figures: comparing the metal to the composite design, the part count was reduced by 126, the cost was reduced by 40%, and the weight was reduced by 34%.

DESIGN METHODS FOR PRIMARY STRUCTURE

Table 2 lists all of the primary composite components that have been produced at Canadair.

Table 2 Primary Composite Structures

Component	Material	A/C	Date
Experimental			
Wing preliminary design exercise, stiffened panels manufactured.	Graphite	--	1985
Pre-Production			
Outboard Flap Vane	Graphite	CL-601	1982
Production			
Rotor Blade	Glass, Kevlar	CL-227	1983
Drone Wings	Kevlar	CL-289	1975

The approach used for the design of primary composite structures is outlined in Fig. 2.

An extensive program of coupon tests is used to determine the intrinsic strength of any material system that is to be used in the design of primary structures. The strength quantities of interest are: longitudinal tension and compression, shear, flexure, short beam shear, and transverse tensile and compression strength. Tests are carried out at high and low temperature, and the effects of chemical exposure are assessed. A and B-basis allowables are derived, these are the population's lower 1 and 10 percentiles respectively, given with a 95% confidence level.

A-basis allowables are used in the design of primary structure.

Studies are carried out to obtain design data on the material's fastener-bearing, fatigue, and impact properties, and its notch sensitivity.

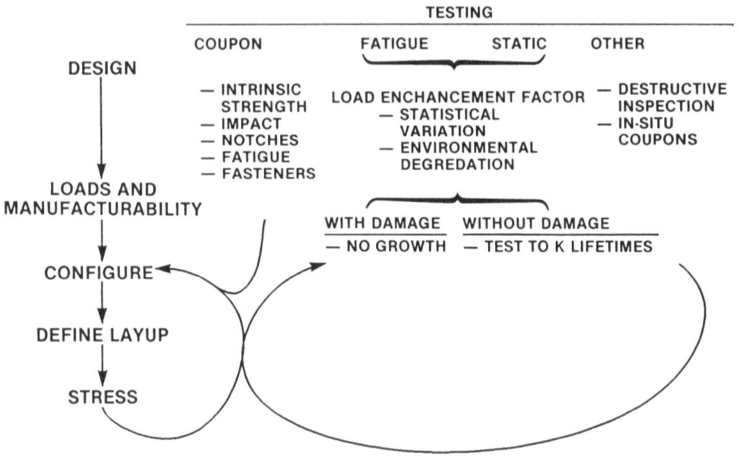

Fig. 2 Design flowchart for primary structure.

The design of the component is based upon the coupon-level data, plus loading data and current manufacturing capabilities. Selection of the layup requires consideration of buckling, edge stresses, and the strength envelope of the layup. As was described for secondary structures, margins of safety are calculated for all potentially critical failure modes. Material strength knockdown factors are used to account for some failure modes such as impact damage propagation.

Static, fatigue, and damge-tolerance tests are normally carried out on full-sized production components.

The static test normally proceeds to failure, and is considered successful if the failure load is somewhat greater than ultimate. The amount by which the test should exceed ultimate is determined by material variability (the test part is assumed to be fabricated from unusually strong material), and by environmental degradation of material strength (the material is assumed to realize its 'unconditioned' strength during the test). The test component is therefore assumed to be stronger than the "weakest" possible production component that has experienced environmental degradation.

The fatigue test is also run at enhanced loads, for multiple lifetimes. The full-sized component fatigue test is normally run at constant-amplitude, so studies of the fatigue spectrum are made to determine an equivalent constant-amplitude fatigue load.

The damage tolerance test is carried out to show that flaws induced by normal service will not grow under service loading. The flaws are usually 'barely visible' impact artifacts, although saw cuts have been used in the past. The loading is usually the same as that used in the fatigue test.

CL-227 ROTOR BLADE

This component is a 50 inch long fiberglass/Kevlar/graphite/epoxy structure with a foam core.

The analysis of the blade proceeded in two basic steps: a beam-type model was first used to model dynamic behavior, and later a plate-type model analyzed the stresses. The Tsai-Wu failure criterion was used to predict the margins of safety with respect to material failure.

Coupon tests were used to define the strength of proposed skin lay-ups, and design strains were restricted to account for impact damage. Fatigue studies of candidate skins led to the selection of an optimum layup.

The static test was run up to 153% of ultimate to account for material variability and environmental degradation. The fatigue test was run at a constant amplitude equal to 100% of limit load. Damage tolerance tests, which have yet to be completed, will be performed on both a static and a fatigue article.

Additional testing was performed to determine the compressive strength of test coupons taken from an actual production blade, verifying the production process.

A unique aspect of the design of this structure was the interrelationship between the manufacturing process (closed mold/expandable core) and the material properties. The manufacturing methods used to make coupons for material characterization tests did not necessarily simulate the curing environment of the actual component. This was the motivation for testing coupons taken from the actual production item.

CL-601 FLAP VANE

This component is a graphite fabric/epoxy structure 3.2 meters in length.

This was a relatively simple component to analyze. Some of the preliminary work was done treating the component as a beam with distributed loads. Later, refined modelling was carried out using finite elements with plate-type elements.

Static, fatigue, and damage tolerance tests were carried out on full-sized test components.

The structure was subjected to two static loading tests, one to design limit load, and another to failure. Failure occurred at 133% of design ultimate load.

Fatigue testing was carried out to simulate 5 lifetimes of takeoff and landing cycles. One lifetime was taken to be 190000 cycles, and testing was carried out in blocks of 9500 landing cylces followed by 9500 takeoff cycles. No changes in the structure were observed during the fatigue test.

The damage tolerance test consisted of a fatigue test carried out after a saw cut had been made in the region of maximum bending moment. It was found that no growth of damage occurred during fatigue when the flaw was smaller than the size at which it failed catastrophically.

The structure was also tested in flight. Between December 1984 and June 1985 the vane saw 31 flights, accumulating 60.3 hours of unrestricted flight. No changes in the condition of the vane occurred during the flight tests.

CL-289 WING

The CL-289 wing is a Kevlar/epoxy structure approximately 45 cm in length.

Kevlar/epoxy was chosen as the structural material due to its radar transparency, and because of its weight advantage over aluminum.

The wing was designed according to static strength and stiffness criteria. Fatigue and damage tolerance issues were neglected because of the limited life of the components. Each vehicle requires four wings, two of which are destroyed on each flight, meaning that each wing is only used for two flights.

The pre-production design was tested statically for stiffness and strength. The production design will be further tested at high and low temperature with absorbed moisture, and for erosion due to rain and dust, for the effects of salt fog, and for strength under shock loading.

WING DESIGN EXERCISE

This project was undertaken to identify the technologies that Canadair would have to develop to allow the production and certification of large primary composite structures.

A material selection procedure was undertaken to identify the optimum available material system based on certain desired material properties. These properties were, in order of importance: toughness, density, delivery reliability, shelf life, hot/wet properties, difficulty of fabrication, flow during cure, technical support, purchase price, toxicity, compatibility with other resins, cure cycle complexity, qualification time and cost, repairability, exothermic properties, and flammability. As a result of this analysis, Hercules IM6/Narmco 5245C was selected as the preferred material system for external areas where low-velocity impact was a concern. Union Carbide T300/Narmco 5208 was selected for use in internal areas where such impact events were not a concern.

The configuration of the wing was selected after trade-off studies of three alternatives: all honeycomb, multi-spar, and two-spar multi-rib. The final selection favored the two-spar design due to considerations of design, manufacturing, inspection, and repair. "Double Z" stiffeners were selected over hat stiffeners after consideration of manufacturability, inspectability, and airworthiness certification requirements.

An extensive series of coupon tests was conducted to confirm the assumed design strain of .5%. A total of 248 specimens were tested to determine the effects of stacking sequence, spectrum loading, notches, low velocity impact, and environmental exposure.

It was found that the design strain was limited to .42% by compressive strength after impact.

Sub-component tests were carried out on stiffened panels that represented typical compression surface skin areas. Analysis methods were developed to predict buckling levels and post-buckling behavior.

Compression tests of the skin sections showed that failure occurred by separation of the stiffeners from the skin. This failure mode was investigated using a fracture mechanics approach, and testing was continued after the area was reinforced. It was found that the reinforcement elevated the failure load, and changed the mode from stringer separation to material compressive failure.

One of the benefits of this exercise was the understanding it gave of the effort required to design and certify a large primary composite structure for a civilian aircraft. In particular, the material characterization effort was, not surprisingly, found to be much greater than would be required for an equivalent design effort using aluminum, since all common aluminum alloys are well characterized.

CONCLUSIONS

We have designed and brought into production a great number of secondary structural components fabricated out of advanced composite materials. The design of these structures required a complete understanding of the properties of the material systems used, an understanding which was accumulated by several years of basic research.

The amount of primary structure designed has been much less, mostly because of the great effort required in terms of material characterization and component testing that is required for certification. In the future, as the efficiency of the material systems improve, and the certification procedures become established, the design of flight vehicle structures in composites should become more common.

REFERENCES

Adsit, N.R., and J.P. Waszczak (1979). Effect of Near Visual Damage on the Properties of Graphite/Epoxy. Composite Materials: Testing and Design (Fifth Conference) ASTM STP 674, 101 - 117

Williams, J.G., T.K. O'Brien, and A.J. Chapman III (1984). Comparison of Toughened Composite Laminates Using NASA Standard Damage Tolerance Tests. Proceedings of ACEE Composite Structures Technology Conference, Seattle, Washington, August 13 - 15 1984, 51 - 73.

ADVANCED CONCEPTS FOR THE FILAMENT WINDING OF FIBRE REINFORCED POLYMERIC COMPOSITES

M. Munro

Department of Mechanical Engineering
University of Ottawa
Ottawa, Ontario, Canada, K1N 6N5

ABSTRACT

The basic process of the filament winding of continuous fibre reinforced polymeric composite components is first described. The critical manufacturing parameters are then identified and their effects are discussed. The major advanced concepts of thermoplastic tape winding, hybrid processes which include helical filament winding, and computer-aided design and manufacturing of filament wound fibre composite components are presented in detail.

KEYWORDS

Filament winding; fibre reinforced; composites; hybrid processes; computer-aided design and manufacturing.

INTRODUCTION

During the last twenty years fibre reinforced polymeric materials have been very successful in replacing metallic materials in existing structures. This has ranged from high volume, low cost applications (eg. fibreglass pipe and tanks) to high cost and high performance applications (eg. graphite fibre composite wing sections of the F-18). Their use has also permitted the construction of components that were not feasible with conventional metallic materials. The most visible example for Canada is the use of the highest modulus graphite fibre available at that time (Celanese GY-70) in the stiffness critical upper and lower boom arms of the Canadarm.

For high performance composites in which the fibre orientation, the volume fraction of fibres and the void content are critical manufacturing concerns there are a number of advanced processes available. These processes are generally automated or are being automated and include various forms of prepreg layup (manual, robotic transfer and gantry), pultrusion, braiding and filament winding. Filament winding is an attractive process for a number of reasons (relatively low cost, complex structures capability, etc.), however, its greatest advantage may be that it can be successfully combined with many other advanced fibre composite manufacturing processes. This aspect will be discussed in more detail later in the paper.

Since the introduction of filament winding in 1947 by M.W. Kellogg, a large variety of components have been fabricated. The simplest to fabricate are rings (with the fibres oriented circumferentially) suitable, for example, for high energy density fibre composite flywheels. Helically wound tube-type components have ranged from fibreglass pipe to the graphite fibre composite tubes used for the MX missile and the solid rocket booster casings which are under development for the Space Shuttle. As well, fibre composite automotive and helicopter drive shafts have been placed in service. Both spherical and elongated fibre composite pressure vessels have been produced in the thousands of components. These range from units that are again very large (e.g. the 2.3 m. dia. Kevlar 49 fibre composite pressure vessel used for the 1st Stage Motor of the Peacekeeper rocket) to small vessels that can be held in the hand yet are capable of storing large amounts of energy on a unit weight basis (Fig. 1). Other axisymmetric yet more complex structures include rectangular and square box sections, truncated cones and ellipsoids. Non-axisymmetric structures have included pipe tees (Fig. 2) and elbows, L, S and T-shaped air ducts, connecting rods, leaf springs, steering wheels and bumper frames. Other non-axisymmetric structures that have been built include a grain hopper car, storage silos, aircraft randomes, helicopter blades and boom sections, wind turbine blades and aircraft fuselages. Thus, the complexity of the component does not seem to be a limit of the filament winding process. The main purpose of this paper is to highlight aspects of the filament winding process that can possibly be used to manufacture these and even more complex components. A short introduction of the filament winding process is first given.

FILAMENT WINDING PROCESS

The process of the filament winding of fibre reinforced polymeric composites consists of placing resin impregnated bands of fibres under controlled tension through a payout eye on a rotating mould (mandrel) in a prescribed geometric path (Fig. 3). For rotationally symmetric components there must be a repeating pattern that is then indexed to provide complete coverage of the mandrel. The fibre bands may be impregnated in a wet fashion during the winding process (schematically illustrated in Fig. 3) or the fibre bands may be pre-impregnated by the material supplier with either a thermosetting or thermoplastic resin. The details of the different types of filament winders and their individual components (control of multiple axes, tensioning devices, resin baths and auxiliary functions) are described in a separate publication (Munro, 1988). This paper will focus on various aspects of helical filament winding as depicted in Fig. 3.

CRITICAL MANUFACTURING PARAMETERS

The major manufacturing parameters that affect the mechanical properties and structural integrity of a fibre composite component are the control of the winding path and the control of tension in the band of fibres.

Winding Path Control Using Conventional and Thermoplastic Materials

For wet winding it is crucial that the fibre path be the geodesic path of the surface being wound. The geodesic path is the path which describes the shortest distance between points on the surface such that, if tension is applied, the band of fibres will not move. The path is described by the wind angle which is the acute angle between the fibre band path and the longitudinal axis of the mandrel. For a component, such as a ring or pipe, this angle is constant along the width of the ring or the length of the pipe. For a curved surface this angle changes with radial position. Figure 4 illustrates the geodesic path over the end dome of a helically wound fibre composite pressure vessel mandrel. The constant wind angle of the cylindrical section gradually changes to 90° at the boss (shaft) in preparation for path return. The geodesic path which is determined by geometric parameters (wind angle, vessel radius, etc.) does not generally correspond to the desired principal directions of reinforcement. In order to filament wind off of the geodesic path one must employ winding over pins, rely on friction or compress pre-preg tapes on to the surface. Other techniques include rapid in-situ curing by ultra violet light (if appropriate) or by the winding of thermoplastic pre-preg

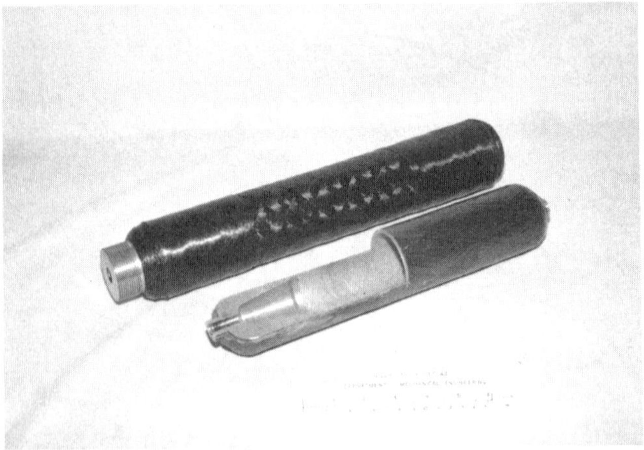

Fig. 1. Proof-of-concept fibre composite pressure vessels for ultra-high internal pressure applications.

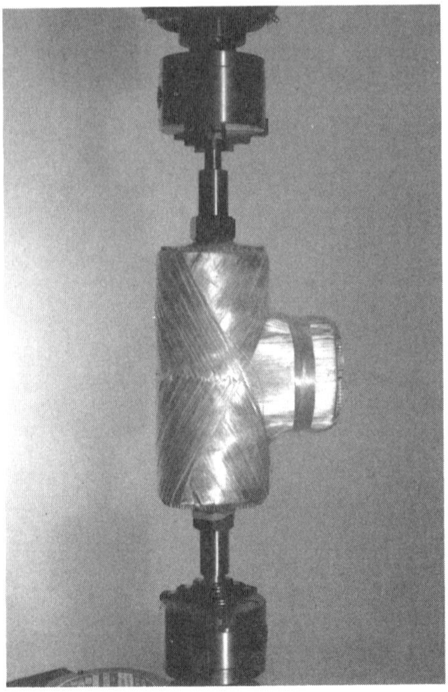

Fig. 2. Proof-of-concept fibreglass pipe tee preform fabricated by a robotic winder.

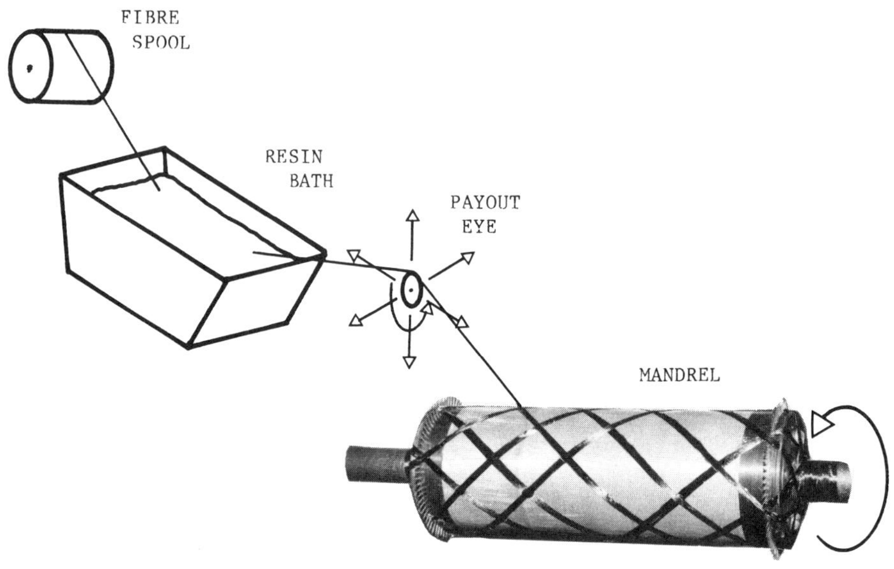

Fig. 3. Basic process of helical filament winding.

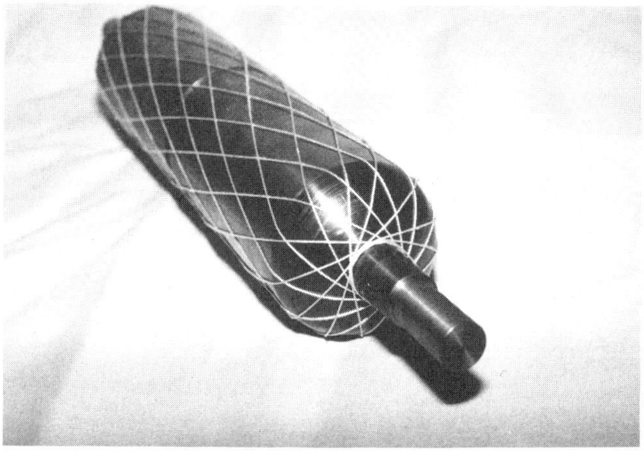

Fig. 4. Typical basic helical winding pattern for a closed end fibre composite pressure vessel illustrated with hard twine on a silicone treated mandrel. This pattern is indexed to provide complete coverage of the mandrel.

tapes where melting and solidification occur at the point of tape laydown on the mandrel before any shifting can occur. Thus this use of thermoplastic tapes, besides allowing for non-geodesic winding, removes the need for a separate cure cycle.

Tension Control

The function and design of tension control devices were discussed elsewhere (Munro, 1988). It is well known that the magnitude of the winding tension affects the volume fraction of fibres and void content of the composite both of which directly affect the mechanical properties. The longitudinal tensile modulus and ultimate tensile strength of a unidirectional fibre reinforced composite material are essentially proportional to the fibre volume fraction. Transverse (to the fibre direction) tensile, compressive and shear properties are functions both of fibre volume fraction and void content as well as the respective properties of the fibre and unreinforced polymer. Variations in winding tension can also affect the distribution of residual stresses (Aleong, 1986).

HYBRID HELICAL FILAMENT WINDING PROCESSES

The process of conventional helical filament winding can also be coupled with other advanced fibre composite manufacturing processes, briefly discussed by Munro (1988). The most promising processes for hybridization are compression moulding, resin transfer moulding, robotic winding, braiding, tape laying and pultrusion (Fig. 5). The hybridization of each of these processes with conventional helical filament winding is due to a number of reasons, however, some of the major driving forces have been indicated in the Figure. The details of the different processes are discussed in the following paragraphs.

One of the most promising processes for fibre composite automotive components is resin transfer moulding. In this process a dry fibre preform is first made. This preform is then placed in a closed mould into which low viscosity polymeric resins are injected. One of the major problems with resin transfer moulding is the preparation of the dry fibre preform. Filament winding can be used where the resin is replaced by a binder solution for the high-speed automated production of the preforms. Wet filament winding can also be used in combination with compression moulding where the filament winding process is used to prepare a wet wound slab-type preform which is then compression moulded to final size and shape.

The maximum speed of filament winding can be limited by the processing speed of the associated computer, by minimum time in the resin bath (for wet winding) and by the physical motions of the mandrel and payout eye. For wet winding the limitation is typically the minimum time in the resin bath rather than a physical limitation of the machine. The limitation for thermoplastic winding might also be similar in that there may be a processing limit to the rate of heating and cooling of the thermoplastic matrix. However, for prepreg tape winding and future advances in wet and thermoplastic winding, the limitation on the process will be due to the maximum speed of the mandrel and, in particular, the maximum speed of the payout eye. Fahim, Munro and Sarrazin (1987) have therefore looked at increasing the winding speed by reducing the mass of the payout eye support structure using graphite fibre composite arms, in this case for an in-house stationary robotic winding facility (Munro, 1988). The graphite fibre composite arms installed on a stationary robot would increase its response substantially although component size would be limited to the working envelope of the robot. For long components it is proposed to use the technology developed during the graphite fibre composite robotic arm programme to develop a conventional lathe-type winder equipped with a lightweight graphite fibre composite carriage, perhaps in the form of a robotic arm.

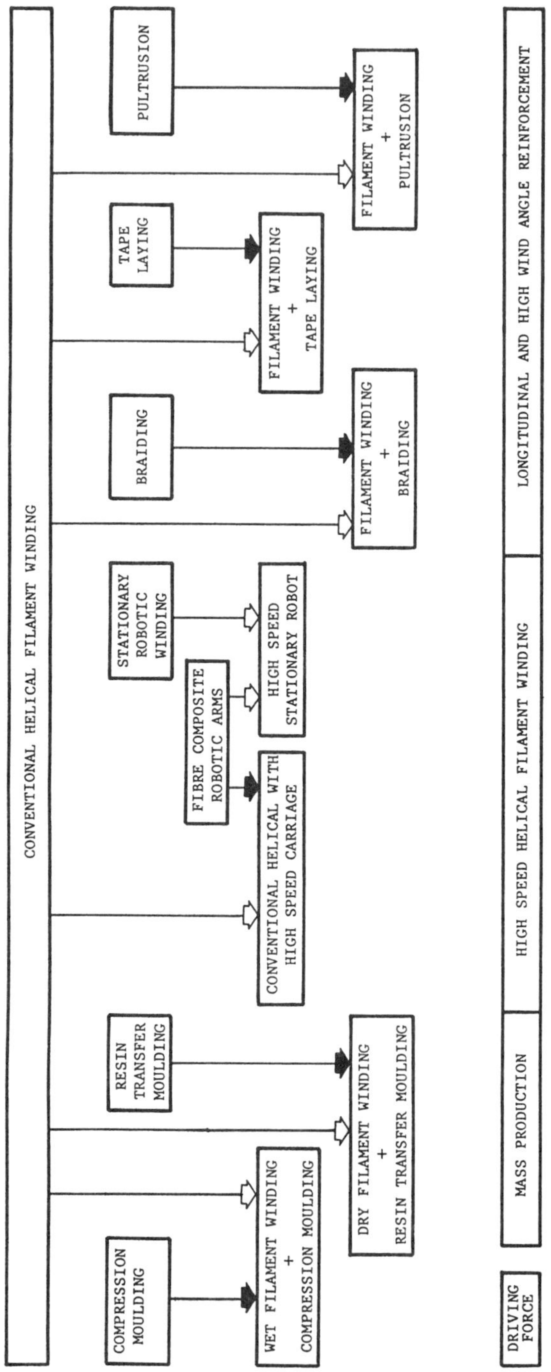

Fig. 5. Hybrid helical filament winding processes for continuous fibre reinforced polymeric composites.

Other future hybrid filament winding processes will be driven by the need for both longitudinal and high wind angle (≥45°) reinforcement in a single component. The braiding process is similar to that used in braiding a rope except that both solid rope-type and very complex thin shell-type structures can be fabricated. Tape laying is a process in which wide flat unidirectional prepreg tapes are laid, using a compaction roller, onto flat or slightly curved surfaces. In the pultrusion process generally unidirectional fibres pass through a resin bath and then into a heated curing die of the shape of the desired profile. Braiding, tape laying and pultrusion are all ideally suited for longitudinal reinforcement. On the other hand, filament winding is not a particularly efficient method for introducing longitudinal reinforcement. Very little material is laid down for full excursions of the payout eye from one end of the component to the other. Typically, a polar or planar filament winder is used for pressure vessels with low wind angles. On the other hand, filament winding is a very efficient process for high wind angle reinforcement including hoop-type winding (ie. wind angles close to 90°). Thus, the hybridization of either braiding, tape laying or pultrusion with filament winding seems appropriate for these type of components.

COMPUTER-AIDED DESIGN AND MANUFACTURING (CAD/CAM) OF FILAMENT WOUND FIBRE COMPOSITE COMPONENTS

When the cost of materials and processing equipment is high there is pressure to use computer-aided design to evaluate sophisticated prototypes "on the board" rather than by testing manufactured prototypes. This is particularly important for filament winding as there are a limited number of geodesic paths that will produce the required repeatable pattern. As mentioned earlier, this pattern is subsequently indexed in order to provide complete coverage of the mandrel. It is therefore desirable, first of all, to mathematically model the surface of the component (Fig. 6). This then allows the simulation of winding to occur in order to ensure complete mandrel coverage and check for any problems areas before actual fabrication. When the component can be manufactured, the loading of the component can then be simulated and resultant stresses and strains can be determined using an in-line finite element package. If no failure criteria are exceeded, then the coordinate data can be downloaded to the filament winder for fabrication. Thus, a computer- aided manufacturing component completes the process. The CAD/CAM approach will prove to be particularly useful for the filament winding of non-symmetric structures where each traverse of the payout eye follows a unique (ie. non-repeating) path.

CONCLUSIONS

Filament winding has been shown for many years to be a very versatile, accurate and cost effective process for the fabrication of fibre composite components. It's advantages will be enhanced in the future with the addition of thermoplastic tape winding, hybrid processes and the CAD/CAM approach.

ACKNOWLEDGEMENTS

The author wishes to acknowledge and thank the students that have contributed to the development of filament winding equipment and techniques in the author's laboratory, in particular, Doug Lindstrom, Etienne Bernard, Chris Aleong, Hugo Sarrazin, François Gosselin, Mike Shea, Richard Morin and Mike Tenace.

REFERENCES

Aleong, C. (1986). Experimental determination of residual stresses in filament wound composite rings. M.A.Sc. thesis, Department of Mechanial Engineering, University of Ottawa.

Fahim, A., M. Munro and H. Sarrazin (1987). High power to weight robots utilizing high performance composite materials. Proceedings of the DREP/RRMC Military Robotic Applications Workshop, Victoria, B.C., 11-13 August, 1987 pp. 67-72.

Munro, M. (1988). Review of manufacturing of fibre composite components by filament winding. To be published in Polymer Composites, Vol. 9, No. 5, October 1988.

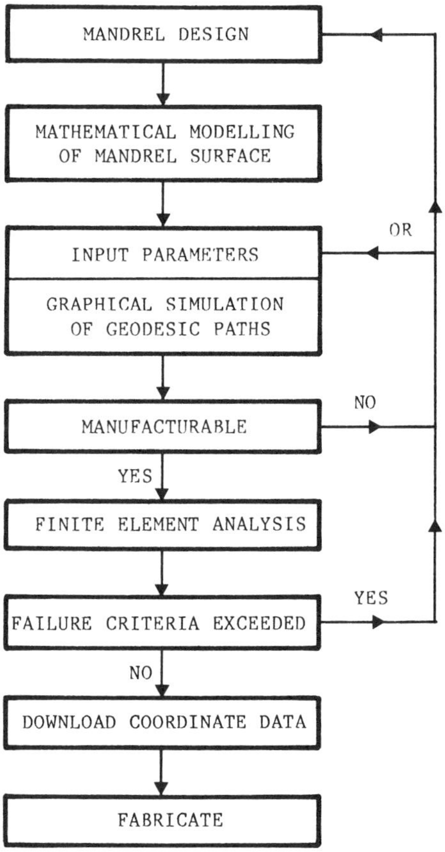

Fig. 6. Flowchart for CAD/CAM by filament winding of continuous fibre reinforced polymeric composites.

Thermoplastic Filament Wound Applications

Malcolm W. K. Rosenow
B.A.Sc., M.A.Sc., P.Eng.
FRE Composites Inc.,
64, rue Wales
St. André-est
Québec, Canada

Abstract

This paper deals with the concept of using a unidirectional fibre reinforced thermoplastic tape, in a filament winding application. The characteristics of the thermoplastic tape and the methods of use, are explained. The example, of an effectively "autofrettaged" fibre reinforced composite pressure vessel without the need to pressurize above the operating pressure of the vessel is described.

Introduction

Filament winding is a manufacturing method in which continuous resin-impregnated rovings or tows are wound over a rotating mandrel, under controlled tension, in a predetermined geometrical pattern. This manufacturing technique has the capability of varying the wind angle, winding tension of the reinforcement, and resin content (wet winding) in each layer until the desired thickness and resin content are obtained in addition to the required directional strengths.

For parts that are simple with repetitive geometries, winding machines are usually controlled by mechanical means such as gears, chains, or cams.

Advances in microprocessor based computer control systems have advanced the filament winding process from winding simple linear geometries or even axisymmetric shapes, to include laying the fibre down on concave surfaces, and winding around openings. The state of the art filament winding machines offer six axis of motion, which has greatly expanded the types of shapes that can be filament wound. Figure 1 illustrates a typical state of the art filament winding machine.

Filament winding is a very efficient means of making fibre reinforced composites, because the fibres can be laid down automatically, rapidly, and be oriented in the direction of principal stress.

A variety of fibre reinforcements are available, depending on the cost and desired level of performance. Table 1 lists a selected variety of fibre reinforcements that are commercially available. Blends of fibres can be used to optimize cost, and performance. In general, glass is the least expensive of the fibres, but has the lowest performance level based on its density. Carbon fibers, in general, are the most expensive but have the highest performance level, except for the intermediate modulus grade of carbon fibre(IM), where the price per pound can be less than half the cost of kevlar.

Epoxy resin systems are the most widely used matrix resins because of their excellent processibility and physical and mechanical properties. Thermal stability and the glass transition temperature limitations restrict the upper continuous use temperature to about 177°C (350°F). Any increase above the glass transition temperature, results in significant changes in the properties of the cured resin system. Other thermoset resin systems, such as polyimides, have an upper temperature limit of 232°C (450°F) for extended time periods.

For applying thermoset resins, wet winding is the most widely used method. In this process the fibres are passed through a heated resin bath to apply the resin. Wet winding is limited to materials with viscosities of 2,000 centipoise, but does offer the most economic means of manufacture.

In prepreg winding, the resin is impregnated into the fibres prior to winding using a hot melt or solvent dip process. Prepreg winding can be used with higher viscosity resins such as novolacs, polyimides, and some epoxies. It has the advantage of a longer working life before curing, and the resin content and fibre quality can be determined before winding instead of after. These materials are generally available through companies that specialize in prepreging.

Thermoplastic materials are receiving more attention due primarily to their unique properties and processing characteristics. These materials offer better mechanical properties at elevated temperatures, have lower moisture retention, and have better chemical and solvent resistance. The manufacturing process allows for significantly faster fabrication rates because no curing is required, and the fabricated part can be reshaped after the initial fabrication by remelting the resin.

The general interest in advanced thermoplastics, is mainly driven by four major advantages, and these are;

(1) Thermoplastics provide service operating temperatures greater than 176 °C (350 °F), and under severe environmental conditions.

(2) The toughness properties, usually measured by interlaminar shear values (G_{IC}), will provide much better damage tolerance over thermoset resins.

(3) The processing technology for thermoplastics provides the potential for low cost manufacturing.

(4) Thermoplastics can be easily repaired due to the remelting characteristics.

Thermoplastic resins range from low melt temperature polymers such as Nylon, to high melt temperature polymers such as PEEK (Polyetheretherketone). A small selection of thermoplastic resin properties is outlined in Table 2.

These thermoplastic materials are now available in several forms including powder impregnated yarns, comingled yarns, melt impregnated yarns or tows, and slit tape. A typical thermoplastic tape consists of several thousand unidirectional filaments bonded together by a thermoplastic resin, and has a tape width and thickness tailored to a specific processing needs. Generally the tape width used in our applications would be 3.2 mm (.125 inches) by .15 mm (.006 inches) thick.

The concept of using thermoplastic impregnated materials in filament winding is simple in principle. The steps required are, guiding the material, preheating the thermoplastic tape, contact point heating of the tape and the bonding surface, consolidating the material, and finally cooling the part. Some of the major obstacles still to be overcome are the higher processing temperatures which can exceed 427°C (800°F), and the higher resin viscosities. A variety of processing parameters must still be optimized before thermoplastic filament winding is as common as thermoset filament winding.

Thermoplastic Filament Winding of High Pressure Containers:

In addition to the general advantage of the fibre reinforced thermoplastic materials, there are a number of specific or component related advantages. One of these, "autofrettage" fibre reinforced composite pressure vessels without the need to pressurize above the operating pressure will be discussed.

Pressure vessels can be categorized according to four types of material classifications. These are all metal containers, hooped wound metal containers utilizing a hooping material that is different than the container, a composite vessel using a thin metal liner such as aluminum, and an all composite container. The subject of this paper is only to deal with a hooped wound metal container that is open ended, similar to a gun barrel.

The process of Frettage is a long standing process of reinforcing a metal container subject to internal pressure with a different material. It consisted of reinforcing the cylindrical part with a material that would absorb a portion of the circumferential stress, therefore sharing the load. It was originally applied to gun barrels, and is centuries old. The parts were made of steel and the reinforcement was obtained either by doubling the thickness of the same material, or by forging, or by wire winding. The longitudinal stresses were resisted by the original component.

The technique of Autofrettage is a process of pre-stressing or over stressing a hollow cylindrical member beyond the elastic range using internal pressure. With a metal cylinder comprising of two different layers of the same material, that is the base material (inner layer) and the reinforcing layer (outer layer), the resultant structure would be pre-stressed, with the inner layer in compression and the outer layer in tension.

This type of system has two limitations. First, it does not offer any weight advantages, and secondly the elastic limit of the reinforcing layer must be much greater than the inner layer to take full advantage of this process.

Another technique is to create an interference fit between the two materials by having the outside diameter of the base material greater than the inside diameter of the reinforcing layer. Again, the main limitation to this process is that, the ultimate tensile strength of the reinforcing material is only marginally different from that of the container.

A particular recurring problem with filament winding over a metal container using an thermoset epoxy is the separation of the composite section from the metal part after curing. Due to the differences in the thermal coefficients of expansion of the composite and the metal, this problem will always exist.

Another problem is the level of stress that is developed in the composite section during the pressurization of the metal component. For a given strain level, the stress level developed in the composite, Sc, is very low, compared to the ultimate stress, Scu, of the material as illustrated in Figure 2. As a consequence, the composite material would be stressed far below its ultimate capacity and is an inefficient use of the composite's high stress carrying capabilities.

We can take advantage of the differences in elastic elongation of each material, by filament winding under very high tension. To take full advantage of the thermoplastic prepreg tape, the fibres can be loaded under very high tension, and melted on the metal part at the same time. By winding under high tension, a desired level of pre-stressing in the metal component can be achieved, with no risk of the fibres migrating close to the metal component. Filament winding with a liquid resin impregnated system under tension merely gathers all the fibres in close contact with the metal part, and forcing the resin to the outer surface. This creates a two layer coating, rather that an absolute composite structure in which the fibers are individually and uniformly embedded in the matrix to achieve structural integrity. Thermoplastic tapes can be stressed in their present form, unlike thermoset prepregs, because they are in a fully consolidated state. That is, the thermoplastic tape is a composite preform, and when stressed, the fibres and the thermoplastic resin share the load.

To utilize the unique characteristics of the thermoplastic tape, the tape is wound under controlled high tension onto the base material. This technique of Frettage as applied to thermoplastics is illustrated in Figure 3. In this case, a pre-stressed liner is obtained through filament winding at very high tension levels. For a given strain level, the stress in the composite section is much higher than would be if it where not wound under very high tension, resulting in a reduced level of stress in the metal section. As a result, the composite section is now contributing significantly to the overall performance to the loading.

There are additional advantages for pressure vessels by applying the frettage process with a thermoplastic tape, and they are;

1) Half the thickness of the pressure vessel can be replaced for the same operating pressure resulting in weight savings.

2) The pressure rating can be doubled by hooping over the existing pressure vessel with only marginal weight gain.

Now with the introduction of advanced thermoplastic materials this century old process of Frettage has now taken on a new dimension. It is possible to limit the stress levels in the hooping material or the metal canister by altering the thickness or winding tension and therefore meet the design requirements. The thickness of the composite section and the winding tension must be chosen so that the pre-stressing combined with the lowest possible operating temperature is always effective, otherwise, the hooping material will separate from the metal body. Also the stress due to the internal pressure loading and the design temperature must remain within the permissible material limits.

Another design tool, allows for optimization of the composite section by choosing the combination of reinforcing fibres and resin to obtain the desired modulus and tensile strength. The modulus of the tape can be varied by using a different reinforcing fibre and/or by varying the thermoplastic resin content.

Summary:

Thermoplastic filament winding techniques can be utilized to perform tasks that are difficult, or impossible to accomplish with thermoset materials or even metals. At FRE Composites, we have responded to this challenge, and have developed techniques that have enabled us to filament wind with thermoplastic materials. The technique of Frettage, using the exceptional properties of the thermoplastic tape, compared to the properties of the metal is one example.

Thermoplastic filament winding is not limited to the over wrapping of metal containers. There are many other applications in which thermoplastics can be used that will not only replace metals, but even the presently used thermosets.

Filament winding with thermoplastic reinforced composites is the up and coming technology for this century.

References:

1 "Modern Plastics Encyclopedia", McGraw-Hill Publications

2 "Engineered Materials Handbook, Composites", Volume 1, ASM International

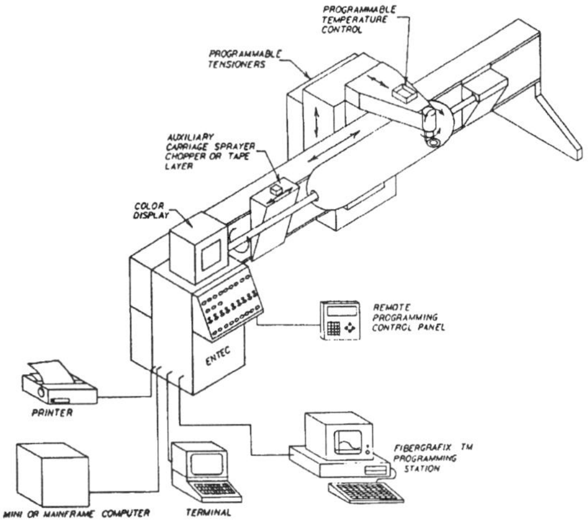

Figure 1. Typical computer controlled filament winding machine configuration (Source: Entec)

Table 1 **Comparison of Reinforcing Fibres and Metals**

Property	Carbon Fibre Ultra High Modulus	High Modulus	Intermediate Modulus	Very High Strength	High Strength	Aramid Kevlar 49	Glass E U.S.A	E-CR Canadian	Metals Aluminum Alloy L65	Titanium Alloy 6A1-4V	Steel 080M46 (BS970:1983)
Tensile Modulus											
GPa.	586	345	295	235	235	124	72	77	72	110	207
psi. x 10^6	85	50	43	34	34	18	10.5	11.2	10.5	16	30
Tensile Strength											
GPa.	1.86	2.2	5.6	4.1	3.5	3.6	1.4	1.5	0.46	0.93	0.62
psi. x 10^6	270	320	817	598	512	525	210	224	67	135	90
Density											
gr/cc	1.96	1.82	1.74	1.8	1.76	1.45	2.62	2.715	2.8	4.5	7.8
lb/in^3	0.071	0.066	0.063	0.065	0.064	0.052	0.09	0.098	0.1	0.162	0.281
Specfic Modulus											
GPa/gr/cc	299	189	169	130	133	85	27	28	26	24	26
10^8in	12	7.6	6.8	5.2	5.3	3.5	1.2	1.1	1	0.99	1.1
Specfic Strength											
GPa/gr/cc	0.95	1.21	3.2	2.3	2	2.5	0.53	0.55	0.165	0.21	0.08
10^6in	3.8	4.8	13	9.2	8	10.1	2.3	2.3	0.67	0.83	0.32

Table 2 Thermoplastic Properties

Properties		Nylon 6	Polycarbonate	Polyether-etherKetone PEEK	Polyphenylene Sulfide PPS
Melting Temperature					
Tm crystalline	C	210-220		334	290
	F	410-428		633	554
Tg amorphous	C		150		88
	F		302		190
Processing Temperature	C	227-288	293	349-382	316-329
	F	440-550	560	660-720	600-625
Tensile Strength	MPa		66	70	66
	psi		9,500	10,200	9,500
Elongation at break	%	300	110	50-110	30,682
Tensile Yield Strength	MPa	51	62	91	
	psi	7,400	9,000	13,200	
Tensile Modulus	MPa	689	2,379		3,309
	psi	100,000	345,000		480,000
Compressive Modulus	MPa	172	2,413		
	psi	250,000	350,000		
Flexural Modulus	MPa	965	2,344	3,000	3,792
	psi	140,000	340,000	435,000	550,000

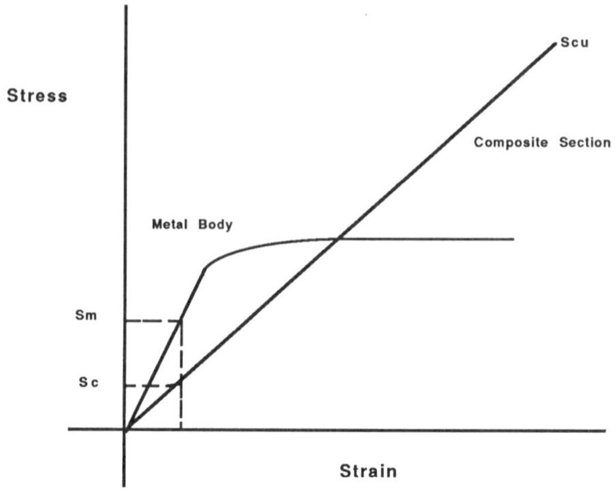

Figure 2. Metal canister filament wound with a composite material under zero tension.

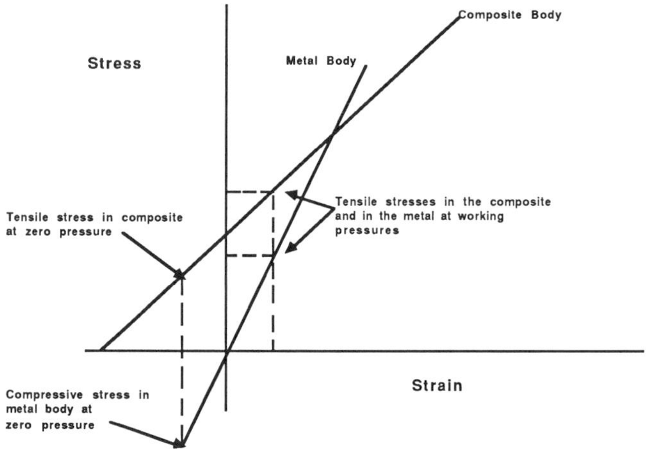

Figure 3. Metal canister filament wound with a thermoplastic tape under high tension.

THE INFLUENCE OF -325 MESH POWDERS ON DC MAGNETIC PROPERTIES

K. Moyer* and J. Ryan**

*Magna-Tech P/M Labs, Cinnaminson, NJ
**Windfall Products, St. Marys, PA

ABSTRACT

Powder metallurgy is widely accepted in today's markets for magnetic devices requiring high quality, stable DC magnetic properties mainly because high compressibility, high purity iron powders are now commercially available. With these iron powders, and additions of fine ferrophosphorus to produce sintered phosphorus irons, it is possible to produce high quality magnets that are virtually non-aging.

Most magnets today are fabricated from relatively coarse iron powders (100-400 mesh). Even though compressibility has been improved to the extent that sintered densities as high as 7.5 g/cm^3 can be realized at practical compacting pressures, some parts require higher magnetic induction than afforded at this density level. Fine -325 mesh powders can be sintered to densities even greater than 7.5 g/cm^3. The DC magnetic properties derived from these finer -325 mesh powders and the possibility of using them for parts requiring higher magnetic induction are discussed in this paper. In addition, the potential for commercial exploitation of the finer powders is considered.

KEYWORDS

Fine powders, magnetic properties, mechanical properties, metal injection molding, sintering, liquid phase, phosphorus iron alloys, powder metallurgy.

INTRODUCTION

With the availability of new, higher compressibility, high purity ferrous powders, powder metallurgy has become of increased interest.(1-2) Not only can parts be fabricated economically, but quality can be engineered as well. Parts with sintered densities exceeding 7.5 g/cm^3 can currently be produced. Since mechanical and magnetic properties are directly related to sintered density, the new powders have opened up markets where quality and economy are required.(3-5)

One such market is the electrical and electronic market, which requires soft magnets with high magnetic induction, fast response, and the ability to be deformed into assemblies. The newer powders satisfy these requirements to the extent that more and more P/M magnets are being specified for emerging devices.

For current markets, however, good is not good enough. There is always a need for higher magnetic induction and faster response to electrical excitation. Traditionally, parts have been produced with an eye on costs and the intention of providing sintered parts to size. Recently, thoughts have centered on producing denser products and controlling shrinkage to achieve higher sintered densities, and improved magnetic induction.(3-5)

Sintered densities can be increased in several ways. With high-temperature vacuum sintering, the melting point of the alloy is the only limit on the sintering temperature. As temperature increases, sintered density increases, which in turn increases magnetic induction. Similarily, Fe_3P powder can be admixed to create phosphorus iron alloys. Alloying with phosphorus improves structure sensitive magnetic properties and promotes liquid phase sintering, which enhances densification and improves magnetic induction.(6-7)

Despite these advantages, more can be done to improve magnetic induction through increased densification. For example, finer powders have higher surface energy. If excess energy is available, sintering can be enhanced, resulting in greater densification, and magnetic induction approaching that of wrought irons. The first part of this paper explores the possibility of employing these finer iron powders to achieve greater densification.

The final part of the study will discuss how improved, fine iron powders can produce improved magnetic parts through metal injection molding. Small, irregularly shaped parts are often specified in designs for magnetic devices. These parts are currently machined from screw machine stock, a labor-intensive process that wastes material through chip formation. Furthermore, stresses introduced by machining degrade magnetic properties. Like P/M technology, metal injection molding, can be employed to produce parts with dimensions close to those specified, without causing mechanical stress or wasting material. The magnetic properties achieved through this process, (that requires fine powders), will be described.

PROCEDURE

The first half of this study centers on conventional P/M technology. For these studies, standard rings, 1.8 inches outside diameter by 1.5 inches inside diameter by ¼ inch thick (5 cm outside diameter by 4.2 cm by 0.7 cm), were used to determine DC magnetic properties. Tests were conducted according to American Society for Testing and Materials procedures.(8) Tensile properties were also derived using standard ASTM procedures.(9) Backup metallography and chemical analyses were conducted when required.

DISCUSSION OF RESULTS

Magnetic Properties

Graphs and predictor equations have been generated for iron and phosphorus irons containing 0.45, 0.8, or 1.0% phosphorus.(5) These graphs are included as Figs. 1 through 4. These curves and equations permit reliable prediction of basic magnetic properties including magnetic induction, residual induction, maximum relative permeability, and coercive force. Of these properties, magnetic induction and residual induction are linearly dependent upon the sintered density. As sintered density increases, magnetic and residual induction increase.(5)

Therefore, anything done to increase sintered density will, in turn, increase magnetic induction or force. The increased surface energy of fine powders (-325 mesh) can provide a means for approaching theoretical density without sacrificing structure sensitive properties or sensitivity of the magnet.

Comparison of Induction of Phosphorus Iron Alloy Compacts

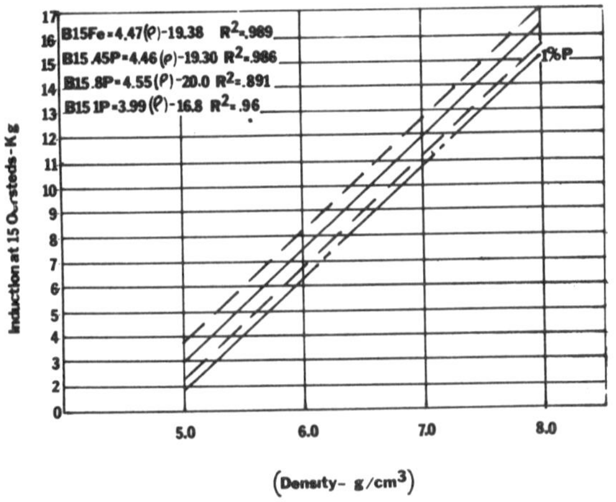

Fig. 1

Remanent Magnetization as a Function of Density for Phosphorus Iron

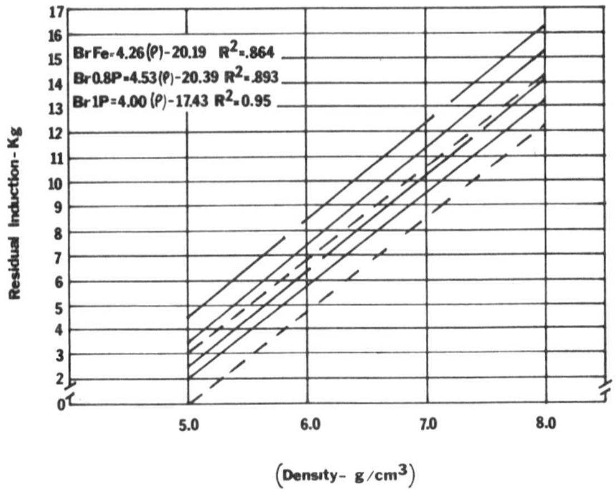

Fig. 2

A Comparison of the Maximum Permeability of Fe+0.45P Compacts With the Predictor Equation for Fe as a Function of Grain Intercept

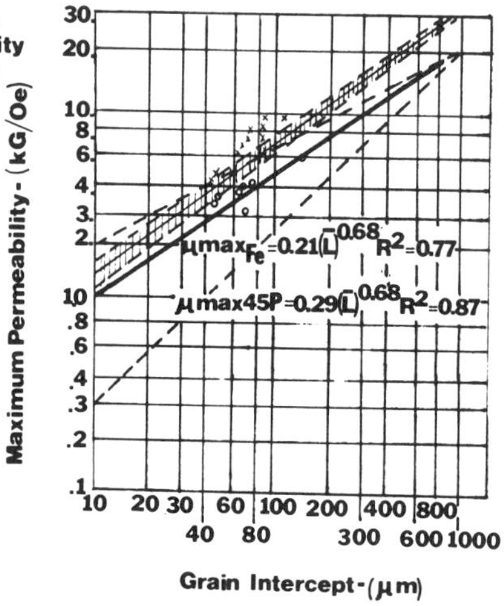

Fig. 3

A Comparison of the Coercive Force of Fe+0.45P With the Predictor Equation for Fe as a Function of Grain Intercept

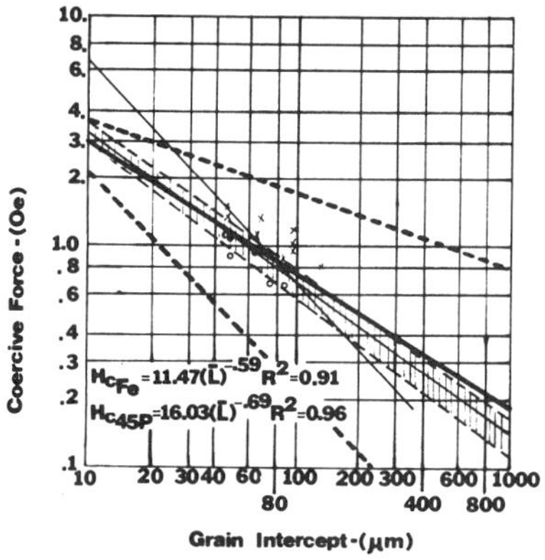

Fig. 4

To demonstrate this phenomenon, individual data is compared in Table I.

Magnetic Properties of -325 Mesh Powders Versus Normal Distribution Powders Compacted at 50 TSI (680MPa) Sintered 1120°C

Property	Iron		0.45P		0.8		1.0	
	100/200	-325	100/200	-325	100/200	-325	100/200	-325
Sintered Density g/cm^3	7.24	7.17	7.40	7.37	7.27	7.30	7.30	7.44
Magnetic Induction at a field of 15 oersteds gausses	12,800	13,000	12,500	13,000	12,900	13,600	12,600	13,100
Residual Induction gausses	11,100	11,700	12,000	12,700	12,500	13,200	12,100	12,700
Maximum Relativity Permeability	3,800	3,200	4,800	4,000	4,500	4,400	4,400	4,000
Coercive Force oersteds	1.2	1.6	1.1	1.6	1.3	1.5	1.3	1.7
Average Grain Size microns	45	33	52	34	55	47	60	40

Magnetic induction and residual induction of specimens sintered at 1120°C in hydrogen are generally higher because the -325 mesh powder sinters to a slightly higher density, owing to greater surface energy. Sintered density is further enhanced through liquid phase sintering as phosphorus content increases. However, structure sensitive properties, relative maximum permeability and coercive force are a function of grain size and purity.(3-5) Since the -325 mesh powder contains a greater number of fine pores, grain growth is impeded and structure sensitive properties are degraded.

With a 1260°C sinter in hydrogen or vacuo, greater densification occurs through increased diffusion, more efficient liquid phase sintering and coelescence of pores. The result is higher sintered density, especially with activated fine -325 mesh powders, increased grain growth, and significantly enhanced magnetic properties. The surface activation of the -325 mesh powder significantly increases magnetic and residual induction, particularly for the iron and lean phosphorus alloy. However, structure sensitive properties are degraded because grain growth is limited owing to the excessive quantity of small pores present, Table II.

In general then, employing a fine -325 mesh powder promotes greater densification which, in turn, increases magnetic and residual induction. Owing to enhanced diffusion high temperature sintering promotes significant densification, particularly the leaner phosphorus alloys. Higher magnetic induction means that greater force is available for magnetically actuated devices. This greater force can be utilized either to reduce the size of the magnet or the amount of copper wire required to achieve actuation.

Magnetic Properties of -325 Mesh Powders Versus Normal Distribution Powders Compacted at 50TSI (680 MPa) Sintered 1260°c

Property	Iron		0.45P		0.8P		1.0P	
	100/200	-325	100/200	-325	100/200	-325	100/200	-325
Sintered Density g/cm³	7.48	7.65	7.24	7.30	7.37	7.47	7.54	7.67
Magnetic Induction at a force of 15 oersted gausses	14,400	15,000	13,100	14,300	13,800	14,200	13,500	13,400
Residual Induction gausses	10,500	10,800	9,900	12,100	13,200	13,800	12,900	12,900
Relative Maximum Permeability	8,000	6,700	5,400	4,700	8,400	6,700	8,200	6,300
Coercive Force oersteds	0.5	0.7	0.8	1.1	0.9	1.0	0.8	1.0
Average Grain Size microns	151	215	69	57	92	94	66	82

Mechanical Properties

Often, in addition to optimum magnetic properties, increased strength or ductility is required. In general, when strength or ductility is required, the alloy containing 0.45% phosphorus is chosen to obtain a compromise between magnetic and mechanical properties because this alloy is not as brittle as those containing 0.8% phosphorus or more. If a -325 mesh powder is used in place of a powder of normal distribution, greater ductility and higher strength will be achieved at a conventional sintering temperature of 1120°C, owing to the greater surface activity of the finer particles. Even if the more brittle 0.8% phosphorus iron is specified, parts made from finer -325 mesh powder will exhibit a greater degree of ductility compared with parts fabricated from a normal distribution, Table III.

Mechanical Properties of -325 Mesh Parts Versus Parts Fabricated From a Normal Distribution

Property	1120°c			
	.45P		.8P	
	100/200	-325	100/200	-325
Ultimate Tensile Strength-psi	52,400	60,100	45,100	56,000
Yield Strength (0.2% offset) psi	34,700	40,600	-----	53,400
Elongation % in 1 inch	13	22	0.3	1.1
Apparent Hardness HRB	30	45	69	75

Metal Injection Molding

Metal Injection Molding (MIM) is beginning to find acceptance in relation to other metal processes. Parts often require complex geometries, or are very small, rendering conventional established processes difficult. In these instances, metal injection molding can be used to advantage because parts can be made close to size and sintered to high densities.

If metal injection molding is to be practical, fine powder such as carbonyl iron is currently selected. These powders are selected because they flow well and, when mixed with appropriate binders, can be readily injected into complex cavities and extracted. These parts sinter to high densities owing to the fineness of the powders, thereby achieving high magnetic and residual induction. In addition, because they are sintered at higher temperatures (in excess of 1260°C), grain growth occurs which results in improved relative maximum permeability and coercive force. Typical DC magnetic properties of a 0.6% phosphorus iron, sintered at 1290°C in vacuo, are shown in Table IV.

DC Magnetic Properties of 0.6P Carbonyl Iron Rings

Property	
Magnetic Induction B_{15} kgausses	12.8
Remanant Magnetization B kgausses	11.4
Relative Maximum Permeability max	5800
Coercive Force H_c oersteds	0.8

The magnetic properties are typical of those of high temperature well sintered parts. The metal injection molding process offers opportunities for fabricating complex parts impossible to fabricate using more traditional processing technologies.

Problems

Although fine -325 mesh powders offer advantages in the production of magnetic parts, particularly higher magnetic and residual induction and greater strength and ductility, for most applications it's impractical to specify the finer powders for the advantage of improved properties. If the powder is water atomized, yields are extremely low as compared to powder of normal distribution. These low yields restrict the quantity of powder commercially available and increase the expense of these powders. The alternatives are powders produced through carbonyl or electrolytic processes. Both are batch operation, which renders these powders even more expensive. Therefore, if -325 mesh powders are to be used to fabricate large volumes of parts, a less expensive powder capable of being produced in large quantities is a prime requirement.

Should such powders become a commercial reality, development of methods to adapt them to processes currently utilizing powders of normal distributions will be necessary. New techniques to improve flow will also be required as well as powders with good compressibility if density levels higher than those realized with currently available high compressibility water atomized powders are to be achieved. Parts fabricators will also have to deal with the greater shrinkage that will result from high temperature sintering these more active powder compacts. Despite these problems, as time progresses and higher quality parts are required, finer powder (-325 mesh) parts should find greater acceptance as a viable resource for providing high magnetic induction, higher strength, and greater ductility.

REFERENCES

1. Herschberger, Ralph and McGeehan, Patrick, "A New Higher Compressibility Iron Powder," Progress in Powder Metallurgy, 1986, Metal Powder Industries Federation, 105 College Road East, Princeton, New Jersey, 08540, pp. 305-320.

2. Gagne, M., Ciloglu, C., Trudel, Y. and Capus J.M., "The Influence of Residual Elements on the Green and Sintered Properties of Water-Atomized Steel Powders," 105 College Road East, Princeton, New Jersey, 08540, pp. 777-788.

3. Moyer, Kenneth H. and Ryan, James, "Extending the Usage of Phosphorus Iron P/M Parts for Magnetic Components," Progress in Powder Metallurgy, 1986, Metal Powder Industries Federation, 105 College Road East, Princeton, New Jersey, 08540, pp. 435-446.

4. Moyer, K. and Ryan, J., "Iron-Phosphorus Alloys for Magnetic Applications-Do We Know Everything About These Alloys?," Progress in Powder Metallurgy, 1987, Metal Powder Industries Federation, 105 College Road East, Princeton, New Jersey, 08540, pp. 323-336.

5. Moyer, K. and Ryan, J., "Extending the Usage of Phosphorus Irons for Magnetic Applications," 1988 International Powder Metallurgy Conference, Orlando, Florida, Metal Powder Industries Federation, 105 College Road East, Princeton, New Jersey, 08540, June 1988.

6. Bozorth, Richard M., Ferromagnetism, D. Van Nostrand Co., Inc., Princeton, New Jersey, 1961, pp. 251-253.

7. Lindskog, P., Tengzelius, J. and Kvist, S.A., "Phosphorus as an Alloying Element in Ferrous P/M," Modern Developments in Powder Metallurgy, Volume 10, Metal Powder Industries Federation, 105 College Road East, Princeton, New Jersey, 08540, 1977, pp. 97-128.

8. Standard Test Method for Direct-Current Magnetic Properties of Materials Using Ring Test Procedures and the Ballistic Method, Designation A-596, American Society for Testing and Materials, 1916 Race Street, Philadelphia, PA, 19103.

9. Standard Material Specification for Ferrous Powder Metallurgy Structural Parts, American Society for Testing and Materials, 1916 Race Street, Philadelphia, PA, 19103.